JUNKYARD JAM BAND

JUNKYARD JAM BAND

DIY Musical Instruments and Noisemakers

DAVID ERIK NELSON

**no starch
press**

First printing

19 18 17 16 15   1 2 3 4 5 6 7 8 9

ISBN-10: 1-59327-611-7
ISBN-13: 978-1-59327-611-9

Text stock is SFI certified

SUSTAINABLE FORESTRY INITIATIVE   Certified Sourcing   www.sfiprogram.org   SFI-00854

Publisher: William Pollock
Production Editor: Riley Hoffman
Cover Design: Beth Middleworth
Interior Design: Octopod Studios
Developmental Editors: Jennifer Griffith-Delgado and Liz Chadwick
Technical Reviewer: Ron Sloat
Copyeditor: Julianne Jigour
Compositors: Riley Hoffman and Susan Glinert Stevens
Proofreader: Paula L. Fleming

For information on distribution, translations, or bulk sales, please contact No Starch Press, Inc. directly:

No Starch Press, Inc.
245 8th Street, San Francisco, CA 94103
phone: 415.863.9900; info@nostarch.com; http://www.nostarch.com/

Library of Congress Cataloging-in-Publication Data

Nelson, David Erik, author.
   Junkyard jam band : DIY musical instruments and noisemakers / by David Erik Nelson. -- 1st edition.
      pages cm
   Summary: "A collection of DIY musical instruments made from everyday materials. Projects include
audio effect generators, a sequencer, an electric ukulele, a PVC slide-whistle, and an acoustic-
electric thumb piano. Features step-by-step, illustrated instructions, modifications, and a soldering
primer. Touches on topics like circuits, harmonics, wind instrument physics, and music theory"--
Provided by publisher.
   ISBN 978-1-59327-611-9 -- ISBN 1-59327-611-7
   1. Musical instruments--Construction.   I. Title.
   ML460.N43 2015
   784.192'3--dc23
                                                            2015018140

About the Author

David Erik Nelson is an essayist, freelance writer, and award-winning science fiction author whose stories have appeared in *Asimov's* and *The Magazine of Fantasy & Science Fiction*, among others. He is the author of *Snip, Burn, Solder, Shred* (No Starch Press). He lives in Ann Arbor, Michigan, with his lovely wife, tolerable children, and aging poodle.

When push comes to shove, these words are my Darkened Violins;
I play them for my Carnal Spider, a Lonesome Lost Nut, and a Zonal Ninja Sneeze.
Otherwise, I wouldn't bother playing at all.

BRIEF CONTENTS

Acknowledgments..xix

Introduction...xxi

Part I: Quick Projects and Tinkering.....................................1

Project 1: The Slinkiphone 3

Project 2: The Plasti-Pickup 15

Project 3: The Elephant Trumpet 27

Project 4: The CPVC Slide Whistle... 37

Project 5: The Scratchbox 57

Project 6: The Droid Voicebox... 71

Project 7: Circuit Bending for Beginners 95

Project 8: Junkshop Percussion 115

Part II: Weekend Projects 129

Project 9: The Playing-Card Pickup 131

Project 10: The Robo-Tiki Steel-Stringed Ukulele.................... 143

Project 11: The Twang & Roar Kalimba.................................. 169

Project 12: The Mud-n-Sizzle Preamp.................................... 197

Project 13: The Universal LFO 219

Project 14: The Twin-T Phaser/Wah....................................... 241

Project 15: The Single-Chip Space Invader Synth 265

Project 16: The Bleepbox 8-Step Analog Sequencer.................. 289

Appendix A: Electronic Components, Tools, and Skills................323

Appendix B: Extra Circuits 357

Appendix C: Music Theory Crash Course.................................369

CONTENTS IN DETAIL

Acknowledgments xix

Introduction xxi
What's in This Book? xxii
A Note on Safety xxiii
Further Reading xxiii
Support and Contact xxv
License xxv

Part I: Quick Projects and Tinkering 1

Project 1: The Slinkiphone 3
Preparation 4
Building the Slinkiphone 6
▶   **Finding Amplifiers** 8
Playing the Slinkiphone 12
▶   **Echoes and Amplifiers** 12
Tips, Tricks, and Mods 13

Project 2: The Plasti-Pickup 15
Preparation 16
Building the Plasti-Pickup 17
▶   **Protect Your Lungs and Brain!** 23
Playing the Plasti-Pickup 23
▶   **Doubling the Vocals** 24
Tips, Tricks, and Mods 25

Project 3: The Elephant Trumpet 27
Preparation 28
Building the Elephant Trumpet 30
Playing the Elephant Trumpet 33
Tips, Tricks, and Mods 35
          Tube Length 35
          Mutes 36
Resources 36

Project 4: The CPVC Slide Whistle 37
Preparation 39
▶   **Buying CPVC** 40
Building the CPVC Slide Whistle 42
          Build the Mouthpiece 42
          Add the Piston and Slide 48
          Finishing Touches 50

Playing the Slide Whistle..51

   ▶   **Pipes and Pitches**...52

Tips, Tricks, and Mods...53

Resources..55

Project 5: The Scratchbox..57

   ▶   **Behold the Magnetic Musicassette!**.......................................58

Preparation...59

Building the Scratchbox...60

   ▶   **Building a Tape Deck Amp**...62

Playing the Scratchbox..67

Tips, Tricks, and Mods...69

Project 6: The Droid Voicebox..71

Preparation...73

   ▶   **On the LM386**...74

Building the Droid Voicebox..76

   ▶   **Connections in Circuit Diagrams**..77

Playing the Droid Voicebox...90

Tips, Tricks, and Mods...90

       Tweak the Body Contacts...91

   ▶   **Resistor Math: Series vs. Parallel**..92

       Use a Different PCB..93

Project 7: Circuit Bending for Beginners...95

Preparation...97

Three Basic Circuit Bends...98

       Modify the Output...100

       Explore Resistor Bends...104

       Add a Power Reset...109

Package Your Project...110

Playing Your Circuit-Bent Instruments..112

Tips, Tricks, and Mods...112

Resources..113

Project 8: Junkshop Percussion...115

   ▶   **Meet Vince: Professional Washboardist**.............................116

Listening to Music..117

Exploring Surfaces and Sounds..118

       Boom...119

       Ting and Ring..119

       Click Sounds...119

       Shake and Rattle..121

       Sticks...122

       Your Hands as Percussion Instruments...............................123

       Multisurface Improvised Instruments.................................123

   ▶   **Washboard Playing Tips**...125

Exploring Techniques 126
    Practice Rudiments with Your Hands 126
    Free Your Fingers and Feet 127

Part II: Weekend Projects 129

Project 9: The Playing-Card Pickup 131
Preparation 132
Building the Playing-Card Pickup 133
▶   **Guitar Pickups Demystified** 136
▶   **The Facts on Wax** 140
Playing the Playing-Card Pickup 140
Tips, Tricks, and Mods 141
Resources 141

Project 10: The Robo-Tiki Steel-Stringed Ukulele 143
Preparation 145
Building the Robo-Tiki Steel-Stringed Ukulele 148
    Build the Pickup 148
    Build the Neck 148
    Build the Body 155
    Add Pegs and Strings 158
    Tune and Test 160
▶   **Tuning by Ear** 161
Playing the Robo-Tiki Steel-Stringed Ukulele 163
▶   **Evolution of the Ukulele** 164
Tips, Tricks, and Mods 164
    Swap Out the Cigar Box 166
    Straight Acoustic or All Electric 166
Resources 168

Project 11: The Twang & Roar Kalimba 169
Preparation 171
▶   **Buying Music Wire** 174
Building the Basic Twang & Roar Kalimba 174
    Prepare Your Tines and Bridge 174
    Drill the Holes 176
    Install the Saddle Blocks and Bridge 179
    Install the Tines 180
Time to Go Electric! 182
    Build the Electronics 182
    Test the Electronics 185
    Install the Electronics 186
Tuning the Twang & Roar Kalimba 189
Playing the Twang & Roar Kalimba 190

Tips, Tricks, and Mods 191
    Troubleshooting.. 191
    Acoustic Mods 192
    Electronic Mods 193
    Expand Your Scale 193
▶   **Jamming in C Major** 196
Resources... 196

Project 12: The Mud-n-Sizzle Preamp 197

Preparation 198
Building the Mud-n-Sizzle Preamp 201
    Prepare the Hardware 202
    Build the Circuit 204
▶   **Bridging Connections the Easy Way** 207
    Complete the Circuit 209
    Install the Circuit in Its Enclosure 212
▶   **On Enclosures** 214
    Test and Install the PCB 214
Using the Mud-n-Sizzle Preamp 216
Tips, Tricks, and Mods 216
    Boost the Fuzz with a Darlington Pair 216
    Experiment with Resistors 217
    Build the Preamp into an Instrument 218

Project 13: The Universal LFO 219

Preparation 221
Building the Universal LFO 223
    Prepare the Hardware 223
    Build the Circuit 225
    Install the Hardware 231
    Troubleshooting and Packaging 233
Using the Universal LFO 234
    Integrate the LFO 234
    Create a Tremolo 236
    Combine the LFO with the Mud-n-Sizzle Preamp 237
Tips, Tricks, and Mods 237
Building on Another Generic PCB 239

Project 14: The Twin-T Phaser/Wah 241

Preparation 243
Building the Twin-T Phaser/Wah 245
    Prepare the Hardware 245
    Build the Circuit 249
    Install the Hardware 255
    Test the Twin-T Circuit 257
    Add the Phaser/Wah 257
    Final Testing and Enclosure 259

Using the Twin-T Phaser/Wah 261
Tips, Tricks, and Mods 262
    Troubleshooting Mods 262
    Tweak Components 263
    Mount the Twin-T Inside a "Broken" Pedal Enclosure 263

Project 15: The Single-Chip Space Invader Synth 265

Preparation 267
Building the Single-Chip Space Invader Synth 269
    Prepare the Hardware 269
    Build the Circuit 272
    Install the Hardware 275
    Troubleshooting 277
    Enclosure 278
Playing the Single-Chip Space Invader Synth 280
    Trigger Mode 280
    Theremin Mode 280
    Experimental Free Style 280
Tips, Tricks, and Mods 281
    Add Speakers 281
    Work with Pedals and Effects 282
    Tone Mods 283
    Pitch Mods 283
    Control Mods 284
    Add a Keyboard 284
Resources 288

Project 16: The Bleepbox 8-Step Analog Sequencer 289

Preparation 290
The Bleepbox Control Breakdown 294
    Output Controls 294
    Modulation Controls 294
    Click-Gate Controls 294
    Step Controls: Pitch and Overdrive Mute 295
    Tempo, Power, and Pause/Reset Controls 295
    Rhythm Control 295
Building the Sequencer 295
    Prepare the Hardware 297
    Build the Circuit 301
    Install the Hardware 303
    Troubleshooting the Sequencer Circuit 307
Building the Modified Single-Chip Space Invader Synth 309
Connecting the Synth 311
Testing the Sequencer and Synth Together 311
Enclosure 313

Playing the Bleepbox 315
　　　Your First Sequence 315
　　　Experiment with the Click-Gate Control and Mute 315
　　　Cut and Loop with Pause and Reset 316
　　　Explore the Modulation and Per-Step Pitch Knobs 316
　　　Change the Rhythm 317
Tips, Tricks, and Mods 317
　　　Strip Away Features 317
　　　Build a 4-Step Bleepbox 317
　　　Add Speakers 319
　　　Add More Sequence Lengths 319
　　　Circuit Bending 319
Resources 321

Appendix A: Electronic Components, Tools, and Skills 323
Guidelines for Sourcing Components 323
Components Primer 324
　　　Resistors: Fixed and Variable 325
▶　　**The Gory Details: Audio Taper vs. Linear Taper** 327
▶　　**The Gory Details: Selecting the Right Photoresistor** 328
▶　　**The Gory Details: Voltage, Current, Resistance, and Ohm's Law** 330
　　　Capacitors 332
　　　Diodes, LEDs, and Transistors 334
　　　Integrated Circuits 335
　　　Wire 336
　　　Quarter-Inch Phone Plugs and Jacks 337
　　　Switches 338
　　　Batteries, Clips, and Holders 339
Tools 340
　　　The Standard Soldering Kit 340
　　　Helpful Additions to the Standard Soldering Kit 343
Skills 346
　　　Soldering 346
　　　Desoldering 349
　　　Using a Multimeter 349
▶　　**Advanced Trick: Using Resistance to Identify**
　　　Mystery Transformers 351
　　　Building a Circuit 352
　　　General Troubleshooting 355

Appendix B: Extra Circuits 357
Super-Basic Continuity Tester 357
Stereo Jack-Power Switch 358
Oscillators and a Metronome 359
　　　LM386 Oscillator Circuit 359
　　　CD4093 Oscillator Circuits 360
　　　555 Timer Oscillator and Metronome Circuits 362

Amps and Preamps 363
    Dirt-Cheap Amp 363
    Basic Transistor-Based Preamp 364
    Two-Transistor Fuzztone 364
Filters, Mixers, Panners, and Splitters 365
    Low-Pass Filter 365
    Extra-Presence Volume Control 366
    Crossfade-Style Mixers 367
    Simplest Passive Multi-Channel Mixer 368

Appendix C: Music Theory Crash Course 369
Beats, Bars, and Time Signatures 370
Notes, Scales, Chords, and Intervals 372
    Keeping Time 372
    Pitches and Octaves 373
    Scales and Chords 373
    Sharps and Flats 374
▶   **The Gory Details: Intervals, Steps, Half Steps, Major, and Minor** 375
The Circle of Fifths 375
Applied Theory: The 12-Bar Blues 377
▶   **The Gory Details: The Music History of Two Hound Dogs** 377

ACKNOWLEDGMENTS

It's a well-worn cliché that any book is a team effort, and even in our magical Information Age of disintermediated self-publishing millionaires, that cliché still has more than a kernel of truth. When I say "This book wouldn't have been possible without the hard work of all those great folks at No Starch Press!", I'm not glad-handing or buttering up No Starch Press to buy another book from me. It's quite literally the truth: I showed up with a jackstraw mash of words and some pretty threadbare photographs. The folks at No Starch Press beat that unwieldy mess into something you can actually use to make some awesomeness. From vague concept through completion, this was a complete collaboration.

First and foremost, I need to thank Jennifer Griffith-Delgado, my editor. Again, I'm not overstating one iota when I say that these were crappy, confusingly written projects prior to her efforts. If you believe in a God Thing, please ask it to bless JGD for her patience and good will. While you're at it, please also ask Your Personal Notion of Grace and Forbearance to rain fortunes down upon Riley Hoffman, who is a great friend and wonderful designer and who worked tirelessly to make this all beautiful. My technical reader, Ron Sloat, brought several welcome improvements to these projects, in addition to generally keeping me honest and figuring out the true build times. Liz Chadwick and Julianne Jigour caught the vast majority of my diverse collection of mistakes, malapropisms, and unnecessarily confusing flights of fancy. Any that remain are my bad, not theirs. The fact that you're looking at these words right now indicates that you somehow heard about this book's existence; I have no clue how that happened, but I know Marlon Rigel did it. He is a peach. Most of all, I need to thank No Starch Press's publisher, Bill Pollock, for his irrational faith in a book that was 26 months late and sort of crazy to begin with.

There are also plenty of folks who've never set foot in the No Starch Press offices who've done me great service: Phil Proefrock lent me his material know-how; Jesse the Guitar Tech at the Southfield, MI, Guitar Center fronted me a geared tuner when I most needed one; Eli Neiburger and the Ann Arbor District Library furnished me with not one but two digital oscilloscopes; Craig W. Van Otteren and Jim Jett shared the intricacies of their DIY tin-can washboards; Joe Neely steadfastly attempted to teach me piano and also introduced me to the Penguin of Fifths (for which I am eternally grateful); Vince Russo, Adam Stein, and John Churchville are all excellent musicians and affable chaps whose insights on practical instrument design and construction, as well as real-world musical techniques, were truly invaluable.

Most importantly, I want to thank my strong, patient, and determined wife—but I have no idea how. If something comes to you, please email me: *dave@davideriknelson.com*.

Finally, I want to reiterate something I brush up against in Appendix C: I need to thank Everyone. The making of music—and, more importantly, the making of musical tools—is a deeply human thing that humans have done and continue to do, often with little compensation beyond the constant reminders that they are "wasting their time" and just "making a bunch of noise" while everyone else is doing something important and earning money. I'm deeply indebted to every anonymous rando who's ever strummed a chord, plucked out a melody, looped a drum cut, posted to a forum about her preferred sine-wave oscillator circuit, or taken a shot at singing that One True Song—which is to say, odds are I'm deeply indebted to you, Gentle Reader and Maker of Things.

Thank you.

INTRODUCTION

Jack White (best known as the front man for The White Stripes) has a custom hollow-body electric guitar with a built-in *bullet mic* on a spring-loaded retractable cable. Bullet mics are traditionally harmonica mics, but White uses his for vocals. Spring-loaded cable reels are traditionally used in canister vacuum cleaners. This same guitar—he calls it the *Triple Green Machine*—also has a built-in phototheremin, which is a simple light-activated synthesizer; you can build one using the schematics and notes found in "Oscillators and a Metronome" on page 359.

In some respects, building a guitar like the Triple Green Machine is patently insane. White paid Randy Parsons—a renowned luthier and master of traditional American hand-building techniques—to take a really nice stock Gretsch hollow body and then cut it up, rout it out, and rewire it in ways that would have likely made Fred Gretsch weep.

You could argue that there was no reason to do this: Jack White made his career playing on a plastic department-store guitar from the 1960s. He could have built his career playing a single guitar string nailed to a two-by-four, something he demonstrates in the opening of the documentary *It Might Get Loud*. He didn't need to hack up a $4,000 guitar and add a bunch of bells and whistles. But it was what he wanted, and one of the great satisfactions in life is getting Exactly What You Want. That's why folks who've made it get custom cars and bespoke suits, handmade furniture and one-off guitars: they get the supreme satisfaction of having it their way.

White and I are both from metro Detroit, but I'm no rock star. I don't have a custom guitar with a rad-roaring bullet mic or built-in synth, or the musicality to justify such a rig. But I do have a Trek mountain bike. I've had it for quite some time, and as my lifestyle has changed—getting married, buying a house, having kids, leaving jobs, starting businesses—I've modded the thing to the point where any Trek engineer would weep to look at it.

I put a hard leather Brooks saddle on it because that's what's comfy for my particular butt. I added a huge full rear fender so the tire would stop tossing broken glass and mud in my toddler's face when I towed him in the trailer, and then I swapped out the off-road nubbies for slick road tires to up my speed and lower my effort while towing said toddler. As a result, the geometry of the bike changed a tad, and the front tire was throwing rocks and mud in *my* face, so I added another big ugly black fender. I've added a goat bell (to ward off texting college kids, who nonetheless still step blindly off the curb and into my path), a computer (I like knowing how fast I'm going), lights (I like living to see tomorrow), mirrors (see previous parenthetical), and an old-school waxed-canvas tool bag that's exactly the right size to hold the lube, rags, and zip-ties I need to keep rolling. I trashed the stock "performance" grips—made of a gooey off-formula silicone that I hated—and replaced them with strips of upcycled tire inner tube, which I can wrap to the just-right diameter. They're an absolutely perfect grip for me, and I love them.

The bike is a totally irrational mess now, but it's Exactly What I Want. I couldn't buy another like it in any store, and I wouldn't trade it for any other bike in the universe. And that's the Big Picture here: one of the essential satisfactions in life is getting Exactly What You Want, and the one way that folks of any income level or renown can be sure to get exactly what they want is by breaking out the multi-pliers and zip-ties and doing it themselves.

What's in This Book?

The 16 projects in this book are broken into two broad groups, with the easier projects toward the front of each section. **Part I: Quick Projects and Tinkering** includes exclusively acoustic projects, all-electric projects, and hybrid electro-acoustic projects. It ends with chapters on circuit bending and improvised percussion; these focus more on exploring the material word and building skills than building a specific instrument. Younger or less experienced makers will be comfortable with any of the projects in Part I. A few, such as the Slinkiphone (Project 1) and Elephant Trumpet (Project 3), make especially good quick-start, rainy-day activities for kids of all ages.

The builds in **Part II: Weekend Projects** are more complex. There's more of a time commitment, and the sawing, snipping, soldering, and measuring take a bit more care and attention. That said, the early projects in this section are still designed for the total beginner. Even the later projects are far from arduous: the Mud-n-Sizzle Preamp (Project 12) and Single-Chip Space Invader Synth (Project 15) are only a touch more complicated than the projects I used to introduce high schoolers to soldering back when I taught at the Hippie School for Troubled Youths.

This book is capped off with three appendices. The first, **Electronic Components, Tools, and Skills**, is a compact primer on electronics, including an introduction to common components and lessons on soldering, desoldering, using a multimeter, prototyping, troubleshooting, and so on. The second appendix, **Extra Circuits**, is just that: about a dozen extra circuits that might be fun or handy for the sort of person who digs *Junkyard Jam Band*. The final appendix, **Music Theory Crash Course**, will bring you up to speed on some musical jargon and concepts that can be a little opaque to beginners.

A Note on Safety

Before you jump in, I want to impress on you a few safety tips that I picked up while teaching at that Hippie School for Troubled Youths:

▶ In order to build real things, you need to use real tools; nothing in this book is dumbed down or babyfied. Show caution, heed the warnings, and wear goggles, masks, and work gloves when advised to do so. Sawdust, metal shavings, and PVC chips can wreck your eyes; soldering irons and fire can burn you; saws and knives can cut you; nails and needles can poke you; electricity can zap you.

▶ Take fumes seriously! Work outside when advised and wear a mask when I suggest doing so.

▶ Just like working out of a cookbook, make a point of reading every step of a project before doing anything. You want to have a full understanding of what you'll be doing and what you'll need at hand before launching into the build.

▶ And lastly, don't let these warnings dissuade you from making these projects! Troubled teens were able to complete them without injury or mishap.

Further Reading

Most "DIY musical instrument" books either limit themselves to a specific pro-quality instrument (as in, build this maple ukulele), focus on a single style of instrument (such as cigar-box guitars, traditional hand drums, and so on), or are a grab bag of grade school activities (involving, say, toilet paper tubes and rubber bands). If you dig what you find in *Junkyard Jam Band*—a wide range of instruments, many idiosyncratic—then here are some other books you might like as well:

▶ *Making Simple Musical Instruments: A Melodious Collection of Strings, Winds, Drums & More* by Bart Hopkin (Lark Books, 1999). This book includes 30 instruments with a wide variety of crazy, innovative designs (see, for example, his insanely awesome "Styrofoam guitar"). There's a mix of kid-suitable projects and those calling for an adult helper, plus lots of great info about the physics of sound.

▶ *How to Make Drums, Tomtoms, and Rattles: Primitive Percussion Instruments for Modern Use* by Bernard Mason. Originally printed in 1938 and still available as a Dover Thrift Edition and ebook, this is a classic when it comes to Native American percussion instruments. It was a school library staple when I was a boy, and it's one of my favorites. Trigger warning: although this book is wonderfully informative and Mason demonstrates a deep and abiding respect and affection for Native cultures and their preservation, he was still a white man writing in the 1930s, and it shows.

▶ *Handmade Electronic Music: The Art of Hardware Hacking* by Nicolas Collins (Routledge, 2009). This is easily my favorite musical electronics book, full of great projects and ideas and jammed to the gills with precious, precious information. Collins covers a wide range of experimental electronic instruments (even simple, high-quality microphones) and how to create sound-art installations. Highly recommended.

▶ Craig Anderton's *Electronic Projects for Musicians* (Music Sales America, 1992) and its companion, *Do-It-Yourself Projects for Guitarists* (Backbeat Books, 1995). Although Anderton's pro-grade designs are a little out-of-date (he relies heavily on a few components that are no longer produced), you can buy complete kits for them online (mostly through PAiA, *http://www.paia.com/*). Both books offer plenty of great design ideas and improvements that can be adapted to a variety of electrified instruments.

▶ *Guerrilla Home Recording: How to Get Great Sound from Any Studio* by Karl Coryat (Backbeat Books, 2008). This is a hardware/software-agnostic primer on audio engineering. If you decide to record, Coryat's book will help you squeeze great results out of any old setup you can muster.

▶ Electronic tinkerers and makers of all levels will also benefit from a handful of books by Forrest M. Mims III. A great place to start is *Getting Started in Electronics* (Master Publishing, 2003), which offers a solid foundation in basic electronic theory and skills as well as scads of great, simple circuits. His books are mostly schematics with little further explanation, but the designs are robust, with lots of opportunities to adapt them to new and novel projects. Circuit benders—folks who make a hobby of torturing new, noise-musical sounds from old electronic toys—love Mims's designs and frequently hack them into their creations.

- Finally, my first book, *Snip, Burn, Solder, Shred: Seriously Geeky Stuff to Make with Your Kids* (No Starch Press, 2011), includes 24 projects; about half the book is dedicated to musical instruments (electric and acoustic, suitable for both kids and adults). Other projects include sock squids and Cthulhus, cardboard boomerangs, water rockets, cheap-and-easy screen printing—if you have busy kids, this book's got ways to keep them busy.

Support and Contact

My blog lives at *http://www.davideriknelson.com/sbsb/* and features videos, tutorials, Q&As, templates, and archived web pages of some DIY instrument resources that have become scarce online. The best way to contact me is through my website. If you drop me links to pictures and videos of what you make, I'll add them to the online gallery.

License

All the designs in this book are mine, but I'd be hard-pressed to claim that they're all that "original." I'm using components in fairly standard ways and riffing on tried-and-true designs from old textbooks, manufacturer datasheets, and the like. The schematics and diagrams themselves, as they appear in this book, are my original work, licensed under a Creative Commons Attribution-ShareAlike 4.0 International (CC BY-SA 4.0) license (*https://creativecommons.org/licenses/by-sa/4.0/*). Share them as you like, as long as you tell folks I made them. Riff on them as you choose, as long as you also allow people to share and riff on your riffs. This license does not extend to the text of the projects themselves. If you want to reprint or distribute the build instructions or collaborate on something extra special, please be in touch; I can always be contacted through my website.

Folks occasionally contact me for permission to build my projects and sell them commercially. If you want to build something using these designs and sell it—or riff off the design and do likewise—then more power to you. Just make sure to give me a shout-out; something as simple as "Inspired by a project in David Erik Nelson's book *Junkyard Jam Band* (*davideriknelson.com/sbsb*)" would be super rad.

PART I

Quick Projects and Tinkering

When you were small and someone showed you a musical instrument, you didn't say, "Rad! How do I play an augmented ninth on this thing?" You touched it with your hands, eager to make the instrument make noise (probably despite brittle warnings of "Be gentle—this is expensive!"). Once you figured out how to get noise out of the thing, your next question wasn't about that augmented ninth, either. In all likelihood, you didn't have a next question. You found a way to make noise, and then you started feeling your way toward making a different noise so that you could make patterns you liked out of those noises. You used your hands and ears and breath to explore the instrument and thus used the instrument to explore yourself. *What am I hearing? What do I want to hear next? Why do I want that? How can I do it?*

Touching objects in a quest to find the Good Noise is the core human relationship to music-making tools. The projects in Part I of this book are designed to lower the barrier between you and the Good Noise: they're quick projects requiring few tools and no previous experience, but they richly reward exploration. Many are also highly kid friendly, even for the youngest. This is a great place to start if you want to build and hone skills—both in construction and in feeling your way toward the Good Noise.

1 THE SLINKIPHONE

The "walking spring" is well-known for its "slinkity" sound, but that metallic ringing is far from the only music it has to offer. The humble Slinky is a veritable wellspring of great sci-fi sound effects. In fact, when properly amplified, it can even serve as a rudimentary vocal echo unit.

This project is quick, and the results are impressive, making it a great one to do with a younger assistant—whose little hands will prove very helpful when working within the confines of a disposable plastic party cup. This is also a good project for a beginning solderer of any age.

If you choose to eschew the electronics, this is a five-minute project suitable for elementary schoolers and scout troops on a budget. Without amplification, you'll lose the zappy undertones and have a much more subtle vocal reverb, but you'll also cut the project's price in half. If you are going to "go electric" with the Slinkiphone, then please heed my warnings (outlined in Step 4) when selecting your piezo.

The finished Slinkiphone is shown in Figure 1-1. Hear samples at *http://www .nostarch.com/jamband/*.

FIGURE 1-1: *The finished Slinkiphone (with pickup installed)*

Preparation

Build Time

▶   5 to 20 minutes, not including drying time for glue

Tools

▶   A standard soldering kit (See page 340.)

▶   A pushpin

▶   A sewing needle (It should have an eye large enough to accommodate your fishing line.)

▶   Needle-nose pliers or very small hands

▶   A hobby knife, utility knife, or pocketknife with a small, sharp blade

Supplies

▶   A disposable 16-ounce plastic cup

▶   A full-sized metal Slinky (This is usually marketed as the "Original" Slinky; a "mini" or plastic Slinky won't work for this purpose.)

- Electrical tape, duct tape, or heat-shrink tubing[1]

- Silicone-based household glue (This is also called *room-temperature vulcanizing rubber*, or *RTV-1*; the 1 means that it's in a single tube, instead of two tubes that you have to mix together. Don't use glues like Super Glue or Krazy Glue here, as some formulations of these cyanoacrylate-based adhesives melt polystyrene plastics. Double-sided foam tape is a suitable replacement.)

- Nylon fishing line (Anything you have handy will do. I used 4 lb. test mono-filament, which is strong but still easy to thread through a normal sewing needle. Even dental floss would likely work in a pinch.)

- A piezo element (I highly recommend using one that is 22 mm or smaller in diameter because it will be easier to install and work with. I used a 20 mm, 6.5 kHz piezo buzzer element, which is Digi-Key part #102-1126-ND.)

- 24-gauge insulated wire (Either stranded or solid core wire is fine, as is 22-gauge speaker wire.)

- A 1/4-inch mono phone jack, also called a *guitar jack*

FIGURE 1-2: *Tools and supplies*

---

1. Bare wire-to-wire and wire-to-component solder joints have a tendency to short against each other when you're rocking out, which sounds just *awful*. You can insulate them with electrical tape or duct tape, but tape tends to either get gummy as it ages or dry up and fall off. Heat-shrink tubing, on the other hand, stays put and doesn't muck up your project's innards. You can buy lengths of heat shrink in a variety of diameters—and even as multicolor precut sampler packs—at many hobby, electronics, or hardware stores.

Building the Slinkiphone

Use your pushpin to pop two holes, roughly 3 mm apart, through the middle of the bottom of your plastic cup, as shown in Figure 1-3. We're using a cheap 16-ounce "party cup" here for two reasons: (1) it's easy to pop a hole through with a pin, and (2) the thin, rigid plastic vibrates readily, making a nice diaphragm. The Slinkiphone ultimately functions as both a rudimentary microphone and an amplifier, so a relatively rigid diaphragm with a large diameter is a plus.

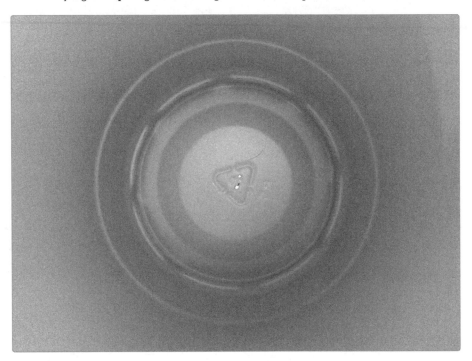

FIGURE 1-3: *The pinholes in the bottom of the disposable plastic cup*

Thread your needle with about 18 inches of fishing line and just let it dangle without tying a knot. (You don't actually need that much line, but you'd go insane trying to work in the cup's confined space with anything much shorter than 10 inches.) Working from the outside of the cup, run the needle through one of your holes from Step 1. Then use pliers (or the tiny hand of a tiny helper) to pull the needle through and send it back out the other hole (as in the left image in Figure 1-4). Finally, while holding one end of your monofilament, pull the needle so that the other end snakes through both holes in the cup and off the needle. The result should be a long piece of monofilament threaded through the party cup with both of its ends hanging outside the cup, as shown on the right in Figure 1-4.

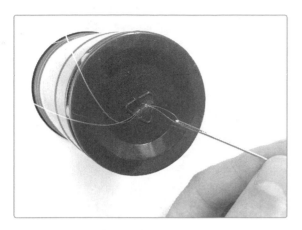

FIGURE 1-4: *Threading the cup*

Step 3   Tie the final coil of the Slinky to the bottom of the cup. Try to make this as tight as possible, but it's perfectly acceptable if there are a few millimeters between the Slinky and the cup when you hold the cup aloft. Triple-knot the fishing line. Thicker fishing line doesn't like to stay knotted, so if you're having trouble, you can secure it with a bead of glue. (Cyanoacrylate-based superglue is okay here—it's perfectly safe on nylon or metal, but keep it off the cup itself.) Once you have a satisfactory knot, snip off the excess fishing line.

Ta-da! You now have the five-minute version of this project, guaranteed to entertain and delight children of all ages along the inverse bell curve shown in Figure 1-5: a kindergartner will laser-blast this acoustic Slinkiphone all afternoon, middle-schoolers will shout into it for a half hour, teens will disdain it and then warm up once they see their kid sister belting "Party Rockin'" into it, and grown men will laser-blast it all afternoon. But if you want to push this into rock 'n' roll territory, then it's time to break out the soldering iron and add a pickup to your Slinkiphone.

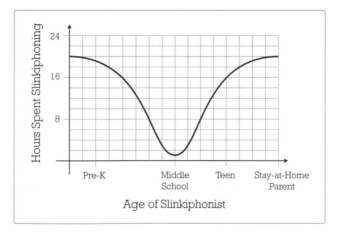

FIGURE 1-5: *Slinkiphone appeal correlated to age of player (confidence interval: 97%)*

Finding Amplifiers

The majority of the projects in this book require amplification, and most of them should be safe to use with any run-of-the-mill guitar amp or PA system. My testing amps include an old Gorilla GG-10 bass practice amp, a Roland KC-100 keyboard amp, a Behringer Eurorack UB1204 mixer, and a slew of little DIY amps. However, a modified or flubbed circuit might result in a noise toy that could damage your amp (which is why I tend to test everything on my home-brew amps and the beat-up old Gorilla first).

You might not want to chance your '58 Fender Tweed Champ on the first little synth you've ever built. Fortunately, there are plenty of low-budget options for amplification:

▶ **Old computer speakers**: Most of us have a few sets of these kicking around in a milk crate in the basement. If the speakers have a wall-wart power supply, then they have a built-in amplifier. Many of these offer sur-prisingly good bass and high-end response. Computer speakers are a good item to watch for at garage sales and resale shops, as they go cheap.

▶ **Cheap practice amps at garage sales**: Lots of kids pick up guitars, and 90 percent of them lose interest. There are plenty of dinged-up—but perfectly serviceable—*practice amps* (smaller, usually less expensive, amplifiers, around 10 watts) sitting around garages and attics. You could probably snag one for $30 or less while eating an ice cream bar in some empty nester's driveway this summer.

▶ **RadioShack Mini Audio Amplifier**: This is part #277-1008, which RadioShack has made for decades. It's ugly and sort of tinny, but it's cheap (they currently sell for $15) and easy to get, and it doesn't burn through batteries. It weathers plenty of abuse and can output to external speakers, making it a fairly serviceable *preamplifier* (a subject we'll hit again in Project 5, the Scratchbox).

▶ **Old tape decks**: Many old dual-cassette boomboxes (the ones designed for copying tapes) will pump whatever is plugged into the mic input through the speakers. If you want to get a little crafty, you can always convert any old tape deck into a grimy little amplifier (see "Building a Tape Deck Amp" on page 62).

▶ **DIY**: You can build a perfectly serviceable test amp using $5 in parts. Plus, it's great practice for soldering and reading schematics. You'll find a circuit diagram for my "Dirt-Cheap Amp" on page 363, and full step-by-step build instructions in my first book, *Snip, Burn, Solder, Shred*. This is a great amp for beginners to build, and it makes a wonderful test amplifier for dicey projects. There are also *tons* of simple amp designs floating around online.

Step 4   If this is your first time soldering, then plug in your iron and, while it's warming up, peruse the soldering primer in Appendix A ("Soldering" on page 346). When you're ready, consider the piezo element shown in Figure 1-6.

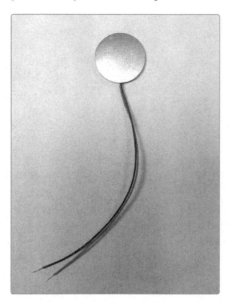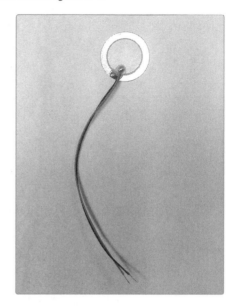

FIGURE 1-6: *The 22 mm piezo element, front (left) and back (right)*

For our low-fi, high-abuse purposes, pretty much every piezo element is the same. Lots of folks who don't like waiting for a mail-order delivery run out to the hobby store, purchase a "piezo buzzer," crack it open, and pull out the element. I don't approve of this (it's a little pricey and a big pain), but grant that it's an acceptable way to source a piezo if you're in a hurry.

That said, for this project you really want to order something a touch smaller. Our piezo pickup needs to rest mostly on the cup's flat, slightly raised bottom without blocking the pinholes or touching the edge of the cup. The piezo elements in most prefab buzzers tend to run close to 30 mm, which is too big. You'll cover the pinholes, which will prove bothersome down the road when your kid finally breaks off the Slinky and you want to reattach it. Also, harvesting a piezo from the buzzer's plastic housing risks breaking the already overpriced component (annoying!).

Step 5   To build the piezo pickup (also called a *contact pickup* or *contact mic*), start by cutting two 5-inch lengths of insulated hook-up wire. Strip 1/4 to 1/2 inch of insulation off each end and tin the ends of the wire. (All of this is covered—and illustrated—in the soldering primer in Appendix A.)

Step 6   You'll note that your 1/4-inch "guitar" jack has two lugs. If you like, you can tin these lugs, too (although with thicker-gauge wire like this, it's less necessary). When you're ready, solder one wire to each lug. Finish by slipping a 3/4- to 1-inch snip of heat-shrink tubing over each wire (see Figure 1-7).

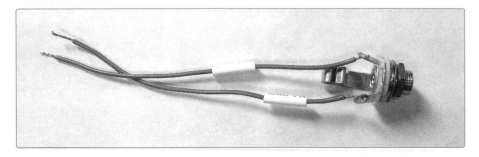

FIGURE 1-7: *The prepared jack*

Step 7   Strip another 1/4 inch of insulation off each lead on the piezo element and then solder one lead to each of the wires you connected to the jack in Step 6. Because this piezo element isn't polarized, it doesn't matter which lead goes to which lug (although technically the black lead is the ground and should go to the sleeve lug on the jack—phone jack anatomy is covered in "Quarter-Inch Phone Plugs and Jacks" on page 337). After the solder joints cool, slide the shrink tube up over the joint and snug it down to size by running the barrel of your soldering iron over it a few times, as described in the soldering primer. See the results in Figure 1-8.

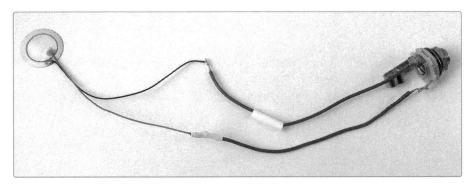

FIGURE 1-8: *Our pickup. Note that only one piece of shrink tube is in place over its soldered joint.*

Step 8   Now, let's install the pickup. Squeeze a pea-sized dollop of silicone-based glue onto the brass front of the piezo element, and press it into the bottom of the cup. The edge of the piezo should be close to the pinholes from Step 1 without covering them, as shown in Figure 1-9. Set this aside. Most brands of silicone-based glue will set up in 30 minutes, although you want to allow a full 24 hours for it to dry completely. If even 30 minutes is too long to wait, then it's fine to use double-stick tape instead.

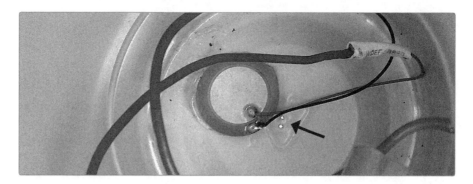

FIGURE 1-9: *Gluing down the pickup's piezo element. Note that the two pinholes have been left clear.*

Step 9   The final step is installing the jack itself. Remove all the nuts and washers from your jack (most will have a single mounting nut and one or two washers). Use your hobby knife to cut a pickup hole in the cup, roughly 1 1/2 inches from the rim. Start small, with a hole not bigger than a 1/2 inch in diameter, and carefully enlarge it, periodically checking to see whether your jack will fit yet. Going a *little* big is okay because the cup will split if you force the jack through, but don't make the hole too large. It's easier to trim than backfill. Once you have a suitable hole, mount the jack from the inside. If it came with two washers, put one on the inside of the cup wall and the other on the outside. If you have only a single washer, put it on the outside. Finish by locking that nut down nice and tight (this will prevent the cup from splitting). The guts of the finished Slinkiphone are shown in Figure 1-10.

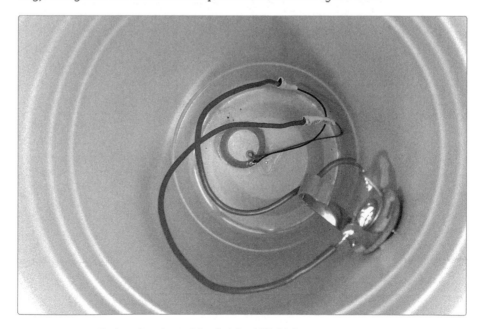

FIGURE 1-10: *An interior view of the finished Slinkiphone*

Playing the Slinkiphone

Once you tied that Slinky to the cup, this noise toy probably became pretty self-explanatory: shake the Slinky, and you'll hear a reverberant "laser blast" rattle of booms and zaps. The cheap plastic party cup vastly magnifies the sound of the metal coils bumping against each other (working much like the belled horn on an old-timey gramophone). Plug the Slinkiphone into an amplifier (taking care to support the jack as you do so—remember, this is a disposable plastic cup), and you'll get a louder version of the same thing, with more undertones and stray zaps and twangs.

Pressing the Slinky's connecting coil firmly against the cup's bottom will muffle some of the booms, emphasizing the higher pitches (a trick that works in both acoustic and electric Slinkiphones). Your best laser blasts will come from holding the cup in one hand, resting the Slinky in the other, and then dropping the Slinky. When the free end hits the ground, it will send a compression wave traveling up the spring, which will rebound between floor and cup several times, making a terrific series of echoes of the initial metallic blast.

You can also use your Slinkiphone as a vocal echo effect, though the sound will be much more pronounced (and more impressive) if you've elected to build the full electric version. Plug in your Slinkiphone, press your mouth firmly into the cup, and hold the other end of the spring at arm's length, letting it dangle slackly from the cup to your hand. Give a hoot. Shout some threats at He-Man. Cackle like a witch. Experiment with changing the tension in the spring: bring your arm in closer to add slack, or increase tension by gathering some coils into your hand. You could also loop the end of the spring to a doorknob and take a few steps back to give the Slinky some tension, or let it dangle directly from the cup with the end hanging in mid-air or resting on the floor (shorter Slinkiphonists may need to stand on a chair to keep the free end aloft).

Echoes and Amplifiers

A naturally occurring echo is the result of a sound (like your voice) reflecting off a distant hard surface and coming back to your ear, somewhat delayed by its travels. Something similar happens in the Slinkiphone: your voice vibrates the bottom of the cup, and this vibrates the spring, setting it all a-jitter. These jitters travel up and down the Slinky, resulting in an artificial echo. For years, a contraption very similar to this was the only way to get an echo effect without actually recording or performing in a grand hall or abandoned cave. Even now, many guitar amps—especially those aimed at classic rock, blues, surf, and rockabilly traditionalists—include a *spring reverb* effect that actually uses springs to create a booming, sproingy ersatz echo.

Tips, Tricks, and Mods

If you have a buddy to work with, you can use the Slinkiphone as a funky outer-space reverb pickup on an acoustic instrument like a guitar or hand drum. While one person plays, the other presses the Slinkiphone against various points on the instrument in either a "cup up" or "cup down" position, as shown in Figure 1-11. The "cup up" position—which puts the piezo pickup most directly in contact with the instrument—is especially effective where vibrations are most intense, like along the bridge of a guitar. The Slinkiphone handler should be encouraged to experiment with different spring tensions as well as cup orientations and placements.

FIGURE 1-11: *Using the Slinkiphone as a reverb pickup in both "cup up" (top) and "cup down" (bottom) orientations*

2 THE PLASTI-PICKUP

The world is full of wonderful aural nuggets. If you have a good tool to capture them, you can mine a lot of little sounds out of everyday life. The Slinkiphone is just one specific case of this untapped wealth, and the piezo-based pickup we built for that project is commendable in many ways:

- It picks up vibrations only in objects it's touching, so avoiding background noise is easy.

- It's super sensitive, bringing out subtleties that would be lost to a normal microphone.

- It's super cheap at under $5 in parts, even if you're stuck buying pricier off-the-rack components instead of ordering them by mail.

However, the Slinkiphone's basic piezo pickup is also *highly* prone to feedback and *extremely* delicate. The thin leads on the piezo element feel like they'll snap off if you just look at them funny. Fortunately, these problems are easily addressed, which is what we'll focus on for this project.

If we rubber-coat our little piezo pickup, we can eliminate casual feedback without noticeably deadening the response. This makes for a really hearty little contact mic: I've soaked my Plasti-Pickups in water overnight, stepped on them, lightly chewed them, clamped them to dog bowls, squashed them between boards, taped them to drums and guitars, dropped them, kicked them, neglected them, and forgotten their anniversaries and birthdays—all with no ill effects.

The finished Plasti-Pickup is shown in Figure 2-1. Hear samples at *http://www .nostarch.com/jamband/*.

FIGURE 2-1: *The finished Plasti-Pickup*

Preparation

Build Time

▶   About 10 minutes (plus 4 to 24 hours of drying time)

Tools

▶   A standard soldering kit (See page 340.)

▶   A hobby knife, utility knife, or pocketknife with a small, sharp blade

▶   Needle-nose pliers

▶   A respirator mask suitable for filtering organic solvent fumes (If you're working outside, this is optional. Use a mask if you're working inside!)

▶   (Optional) A 1/4-inch mono phone "guitar" cable (You can get by with a piece of string and a bent paper clip.)

Supplies

- A piezo element, such as Digi-Key part #102-1126-ND (Any piezo element will do, but getting one with pre-soldered leads will save your sanity.)
- An enclosed 1/4-inch mono phone jack, also called a *guitar jack*; specifically Digi-Key part #SC1316-ND
- Plasti Dip Multi-Purpose Rubber Coating
- Duct tape

FIGURE 2-2: *Tools and supplies (not shown: mono phone "guitar" cable)*

Building the Plasti-Pickup

Step 1 Let's start with the jack. You'll note that this has more lugs than we have wires; that's because it's technically a *switch jack*, designed for situations where, for example, you have a built-in speaker and you want that to cut off when you plug in an external amplifier. This project doesn't call for any switching, so let's take care of that spare lug. It's the middle lug in Figure 2-3, but keep in mind that if you buy a different jack, you'll likely have a different number of lugs in different orientations. Take hold of the spare lug with your needle-nose pliers, twist it back, and then smoosh it flat against the plastic enclosure.

FIGURE 2-3: *The lugs of our 1/4-inch jack, before and after*

Your jack almost certainly came with some washers and a nut for securing it inside a case. You won't need those for this project. Set them aside, but keep track of them: almost all instrument, effect, and amplifier jacks use the same size nuts and washers, which tend to work their way loose over time and drop off at inopportune moments. A plastic baggie of spares is often a blessing.

Step 2   Now we need to finish sealing the enclosure, lest our jack flood with Plasti Dip in Step 10, making a useless mess. Tear off a piece of duct tape roughly 1 1/2 inches wide and a bit more than 3 inches long. Push the two remaining lugs through the middle of this piece of tape. (This will be easiest if you pierce the tape with the tip of a hobby knife first. Don't worry about making an airtight seal; Plasti Dip is very viscous, and we aren't going to give it a chance to seep into a little gap like this.) Bring the two free ends of the tape up along the sides of the enclosure, making a sort of taco with a guitar jack filing (see Figure 2-4a).

Step 3   Fold the tape snugly up along the jack, covering the bent-back switch lug (see Figure 2-4b). This process should look a lot like the second step in making a loop-the-loop paper airplane.

Step 4   Next, fold in one wing, a little like wrapping a tiny present (see Figure 2-4c).

Step 5   Finally, fold in the other wing, making your jack into a bashful vampire bat (see Figure 2-4d). Then, flip it over and do the same steps with the other side. When you're done, the tape will peek up over the top of the plastic enclosure by a few millimeters. This excess tape is fine: it'll help keep the Plasti Dip out of the jack in Step 10, and it can be trimmed away once the coating dries.

a) Making the guitar-jack taco

b) The taco becomes a flightless, paperless paper airplane.

c) Wrapping the tiny present

d) The bashful vampire bat

FIGURE 2-4: *Sealing the enclosure with duct tape*

Step 6   Use your soldering iron to tin both lugs. (If you need a soldering refresher, check out "Soldering" on page 346.) This process can be a pain because the lugs are big and suck away a lot of heat. However, when you're trying to solder together something beefy, like a lug, and something shrimpy, like pre-soldered piezo leads, it's a good idea to tin the lugs first. The amount of heat it takes to bring the untinned lugs to temperature for a good solder joint can totally melt and mangle thin-gauge insulated wire.

Step 7   Tin both of the leads connected to the piezo, but be careful and work quickly; those little wires get hot *fast*. Now solder one lead to each lug, as shown in Figure 2-5.

Officially, the black wire is the ground lead and should be connected to the sleeve of the jack.[1] If you're using the SC1316-ND jack from Digi-Key, then the sleeve lug is the one mounted diagonally on the jack's enclosure.

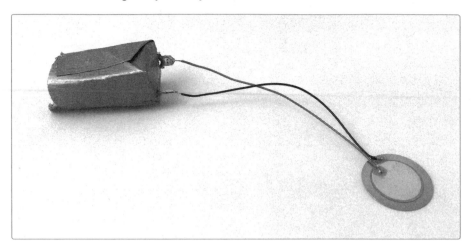

FIGURE 2-5: *Your piezo pickup is now operational.*

Step 8   At this stage, your contact mic is functionally complete! Let's take a second to test it before we coat the whole thing in rubber. Power up your amp and turn the volume just a notch or three above zero (a piezo contact mic can be *loud* and, prior to dipping, very prone to feedback). *Carefully* plug in your pickup. You'll find that the tip of the guitar cable presses against the duct tape from the inside. The duct tape shouldn't split at this point, but take care nonetheless. If you do manage to bust a seam, just cover the breach with a little square of tape and move on. It'll all be covered in rubber soon enough.

Give the piezo a gentle tap. You should hear a surprisingly loud rasp from the amp. Bring up the volume on your amp if you like, but the fact that the pickup is functioning should be pretty unambiguous. If the pickup is silent, take a second to be sure that all of the wires are still connected (those leads on the piezo are pretty delicate). Also, make sure there isn't a stray snip of wire or glob of solder forming a bridge between the lugs inside the jack. If the pickup is working, then carefully twist the piezo leads together, taking care not to break any connections. Retest once you've twisted the wires, as shown in Figure 2-6, and then move on to plasti-dipping.

---

1. The *sleeve* connection of the jack is the one that comes into contact with the barrel or body of the guitar-cable plug, as opposed to the tip. This barrel connects to the shielding inside the cable and then to the ground connection in the amplifier. If any of this is confusing, check "Soldering" on page 346 for illustrated clarification.

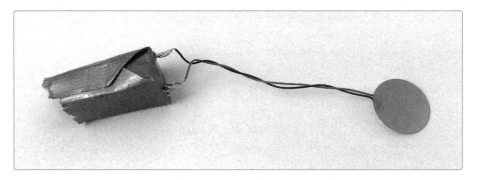

FIGURE 2-6: *Twisting the two leads makes it easier to coat them in Plasti Dip.*

Step 9   Once your pickup checks out, it's time to rubberize it. You'll want to hang your pickup up to dry, so if you have a spare guitar cable, now is the time to *carefully* plug it into the jack. Otherwise, you can fashion a workable dipping line from some string and a paper clip bent into a hook. Both options are shown in Figure 2-7.

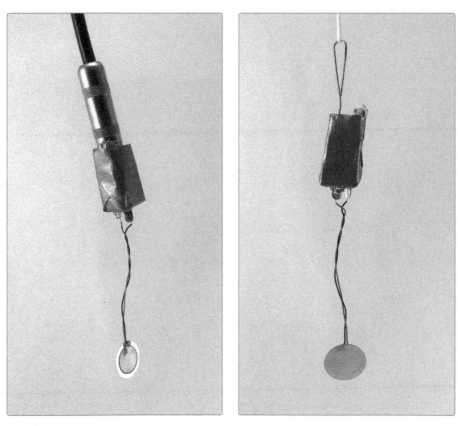

FIGURE 2-7: *Ready to dip, using a guitar cable (left) or a paper clip and string (right)*

Step 10   We'll use Plasti Dip mostly according to the manufacturer's instructions, so give it a few hard shakes, don your respirator, and pop the can. If you're using an old can that's been sitting around the workroom, the Plasti Dip may have separated, leaving a clear, oily layer resting on top. Stir it a few times with an old pencil or chopstick to recombine the layers. (If you do use a chopstick, *throw it out*. Plasti Dip is *not* food safe.)

You are going to slowly lower the pickup into the Plasti Dip, stopping when the liquid is just below the top edge of the jack's plastic enclosure. Try not to overtop it, as there's a small risk the Plasti Dip could run into the jack and foul things up. The manufacturer advises lowering whatever you're coating at a rate of 1 inch per 5 seconds, but that might drive you nuts. I've found that lowering the pickup over the course of a 10-Mississippi count (and raising it at the same rate) is sufficient for a good first coat.

Once you've done the first coat, let the pickup hang over the can for a second or two and drip. (Plasti Dip is pretty viscous, so it shouldn't drip much.) If the wires and jack enclosure are fully coated at this point, with no bare spots or gaps between wires or lugs, then the first coat is perfect. The brass edge of the piezo element itself *might* peek through, but that's okay because we're going to double-coat it anyway. If your wires and jack are coated and gapless, then give just the piezo element itself another leisurely dunk, and hang it up to dry. If you have gaps and thin areas along your wires and jack body, then dunk the whole thing again. This one-two dunk should be sufficient, but if you feel paranoid, you can dunk the whole thing several times. I've gone up to five complete coats without seeing any loss in performance—although the pickup does start to look like a goofy rubber lollipop. You can always wait until the Plasti-Pickup is completely dry and add layers later if you have second thoughts.

The Plasti Dip coating should be tack-free in about four hours, although it's best to let it dry overnight. Once the rubber is totally dry, you can trim away any stray duct tape poking up over the edge of the coating.

FIGURE 2-8: *Ready for the second dip!*

Playing the Plasti-Pickup

The Plasti-Pickup is meant to amplify things you wouldn't normally think to amplify. You can obviously clamp one to a Slinky for an ad hoc Slinkiphone, but don't neglect the percussive possibilities of wooden tabletops, plate-glass windows, and metal handrails and fences. Metal mixing bowls and dog dishes can also be lots of fun: they sing beautifully and are easier to record using a contact pickup than a conventional microphone. Experiment with clipping or taping your Plasti-Pickup to various points on the bowl (along the rim, underneath, and so on), or try filling the bowl with different levels of water. Blow into that water with a length of tube to make some whale noises.

The Plasti-Pickup is great for capturing some conventional musical sounds that are hard to record, like the sound of stomping out a kick-drum rhythm on a porch. You can emulate this by laying your Plasti-Pickup on the ground and covering it with a length of 1×6 or 2×4 board. Experiment with doing this over both hard surfaces (wood floors, a piece of scrap wood, in the kitchen) and soft ones (on the lawn, over the rug) to get a variety of easily recorded rollicking hoedown beats. You can also pinch the Plasti-Pickup between two boards (the pickup's thickness will keep one end separated a bit) and then tap on the free ends for a "wooden tuning fork" sound or a simple "plank xylophone" marimba.

Yelling at the Plasti-Pickup won't get you anywhere because the vibrations in the air are too subtle for it to pick up. However, you can press it to one side or the other of your larynx—that is, your Adam's apple—and it'll amplify your voice accurately without risk of pops and hisses, emphasizing the low end and losing most of the sibilant "mouth business." Using this in addition to a conventional vocal mic is a quick-and-dirty way to get a richer vocal sound, akin to what you might produce by doubling the vocals in a studio recording.

Doubling the Vocals

Doubling the vocals (also called *doubletracking*) is an incredibly common trick for beefing up vocal tracks on the cheap: the singer records several very similar takes, which are then mixed together to create a very rich choral stereo effect. Classically, the best take is centered in the stereo field, with the top two runners-up panned hard left and right. This allows a singer to be a little inconsistent in each take and still sound good because, at any given instant, the listener's ear naturally focuses on the most on-key version of the three tracks. Doubling also provides a "fatter" sound because the takes are not identical and thus don't perfectly mesh. If you record vocals with both a conventional vocal mic and a Plasti-Pickup on the throat, you'll get a similar effect: each mic gets a slightly different version of the take, and the two signals fill and enrich each other rather than just overlapping and blending. Not too shabby for a few bucks in supplies and some soldering.

Finally, the Plasti-Pickup is a perfectly serviceable general-use acoustic pickup to amplify cellos, violins, guitars (steel or nylon string), or even percussion instruments. In essence, it isn't at all different from the vintage Dean Markley acoustical transducer pickup I paid $40 for back in the early 2000s (when I was a foolish and callow youth). See Figure 2-9 for a side-by-side comparison.

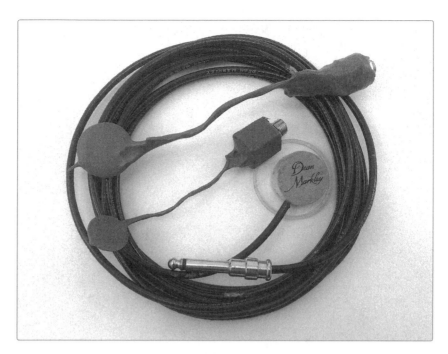

FIGURE 2-9: *My pricey Dean Markley acoustical transducer and a couple of homebrew Plasti-Pickups*

Tips, Tricks, and Mods

You have lots of options when it comes to components for this project. Basically, any piezo element and audio jack should work. When it comes to piezos, I advise buying those with pre-soldered leads because soldering to a piezo element can be very frustrating. You should also buy cheap ones rather than tearing them out of prefab piezo buzzers; piezos harvested from buzzers will work just fine, but it's a wasteful way to get supplies. In general, the larger your piezo, the more low end you'll pick up, so keep that in mind if you're planning to build an acoustic pickup for your double-bass, viola, or big fat djembe drum.

As for jacks, I favor 1/4-inch audio phone jacks because that's what most amps, effects pedals, and mixers use. This project can just as easily be built around a 1/4-inch plug instead, as is the case with my factory-made Dean Markley transducer, or it can be built using an 1/8-inch mini-jack or plug (that's the size of the standard headphone jack on your laptop, tablet, and MP3 player and on most cell phones). Heck, you can use any esoteric jack you've got that plugs into the things you want to plug into.

If your jack doesn't have any sort of enclosure, then you can always fake it by wrapping a strip of duct tape around the jack and soldering points just prior to dipping. That's what I did with the lower Plasti-Pickup in Figure 2-10, and it works great, despite being ugly as sin.

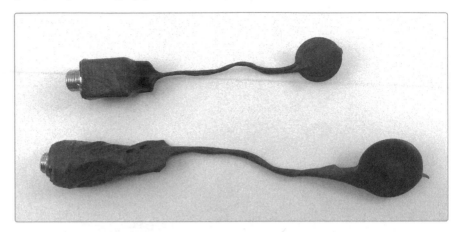

FIGURE 2-10: *The top pickup was made using the piezo and partially enclosed jack from the parts list. The bottom was made using a larger piezo element and a regular guitar jack, like the common Switchcraft jack listed for most projects in this book. Both work great!*

3 THE ELEPHANT TRUMPET

Conventional wind instruments fall into three broad groups: reed woodwinds, such as clarinets and saxophones; reedless woodwinds, such as flutes; and brass instruments, such as trumpets, tubas, bugles, and so on. This last category is a bit of a misnomer because many "brass" instruments are not made of brass: didgeridoos, for example, are considered brass instruments, as are obscure instruments like the serpent (a twisty hybrid brass-woodwind distantly related to the tuba), the Russian rozhok (an archaic wooden trumpet), the damnable plastic vuvuzela, and our own Elephant Trumpet (see Figure 3-1).

Whatever the materials, all of these "brass" instruments are *labrosones*: they make musical tones by harnessing the power of farty lip noises to excite a column of air.

Folks often call horns like the one we are about to build "bugles" because these ad hoc kitchen-supply instruments are valveless, and the bugle is the most common valveless brass instrument. But the Elephant Trumpet is technically a natural horn[1] and closer to a trumpet or trombone than a bugle. Like a trumpet or trombone, our Elephant Trumpet has a largely *cylindrical* body, meaning it has a uniform diameter down most of its length. Subsequently, it has the bright, brassy, burring voice of a plus-sized trombone. In contrast, bugles, tubas, cornets, flugelhorns, French horns, shofars, and so on are fundamentally *conical*, with their diameter steadily increasing along their entire length, from mouthpiece to bell. This shape gives them their full, round, mellow tone.

The finished Elephant Trumpet is shown in Figure 3-1. Hear samples at *http://www.nostarch.com/jamband/*.

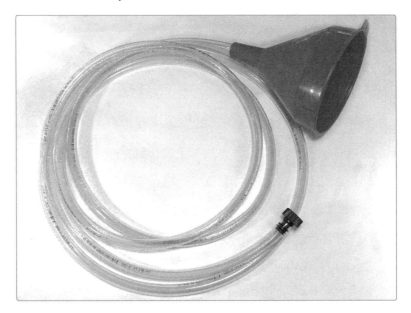

FIGURE 3-1: *The finished Elephant Trumpet*

Preparation

Build Time

▶ About 5 minutes, plus 30 minutes to 24 hours of drying time for any glue

---

1. The *natural horn* is the valveless predecessor of the modern French horn, to which it bears a passing resemblance. You see natural horns in Renaissance paintings now and again.

Tools

▸ A hobby knife, utility knife, or pocketknife

▸ A ruler or tape measure

▸ (Optional) Needle-nose pliers

Supplies

▸ Two garden-hose quick-connect couplings (Both should have male quick-connect fittings, but one should have a male garden-hose thread and the other a female garden-hose thread. See Figure 3-3 for clarification.)

▸ A 2-quart funnel with an opening around 7 inches in diameter

▸ 16 1/2 feet of 3/4-inch *OD (outer diameter)* × 5/8-inch *ID (inner diameter)* flexible vinyl tubing, or something similar that accommodates your funnel and hose coupling

▸ (Optional) Duct tape or silicone-based household glue (This is also called *room-temperature vulcanizing rubber*, or *RTV-1*; the 1 means that it's in a single tube, instead of two tubes that you have to mix together.)

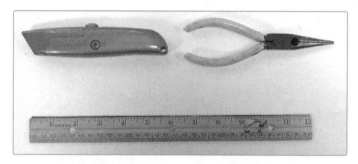

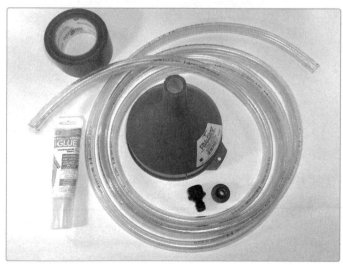

FIGURE 3-2: *Tools (top) and supplies (bottom)*

Building the Elephant Trumpet

Step 1

Constructing a good Elephant Trumpet is an exercise in proper parts sourcing, so the first few steps will focus on finding the three major components of your trumpet. Let's start with the mouthpiece, which you'll make from a garden-hose quick-connect coupling. You'll find the couplings shown in Figure 3-3 in the garden supplies aisle of the hardware store. While both have male push-connector couplings (the little end), one has a male garden-hose thread (the threads are exposed), and the other has a female garden-hose thread (the threads are concealed). I specify two couplings because different sized mouths have an easier time with different sized mouthpieces: most adults seem to favor the female-to-male coupling, while many kids have an easier time getting a good trumpet blast from the male-to-male coupling. Conveniently, these two couplings are often sold in a set together. They're usually available in both brass and plastic, but plastic will be much more comfortable to play.

Your female-to-male connector will probably have a rubber grommet in it, and now is a good time to pull the grommet out. Needle-nose pliers will make this easier. You won't need the grommet for this project, but it might come in handy when the washer on your garden hose goes bad, causing your outside spigot to spray water everywhere when you water the tomato plants.

FIGURE 3-3: *A pair of garden-hose quick-connect couplings: male-to-male (lower left) and female-to-male (upper right)*

Step 2   Next up is the bell, which is the amplifier for your trumpet. There are two places in the hardware store you're likely to find a 2-quart funnel: the automotive aisle and the canning or kitchen supplies aisle. Two-quart kitchen funnels tend to have long, narrow throats that will fit snuggly inside the vinyl tubing that forms the body of your trumpet. They also tend to be made of bright tin and are thus a bit pricier. You can find a cheaper plastic 2-quart funnel in the automotive aisle. Both work great, and in either case, the funnel mouth should be about 7 inches in diameter. A touch smaller or larger is fine (ranging between, say, 6 and 8 inches). Figure 3-4 shows examples of both funnel types.

If you go with a plastic funnel, it's okay to get one with a built-in screen because the screens are pretty easy to tear out (as discussed in Step 4). But avoid metal funnels with screens, as these are usually welded in place.

FIGURE 3-4: *A 2-quart kitchen funnel (left) and an all-purpose plastic funnel (right)*

Step 3   Finally, pick up about 16 1/2 feet of vinyl tubing with a 3/4-inch outer diameter and 5/8-inch inner diameter. With those dimensions, the tube should fit easily into the narrow end of a heavy-duty plastic funnel and snugly hold the smaller male end of the garden-hose quick connector, and it will also accommodate narrower-throated funnels, like the kitchen funnel. You can go with any other tubing you find that strikes your fancy and will fit both your mouthpiece and funnel, but still be sure to get at least 16 feet of it. You can always cut it down, but you can't make it longer.

Step 4   Building your first Elephant Trumpet will take less than a minute. If the funnel you've chosen has a built-in screen, use needle-nose pliers to tear it out now (see Figure 3-5). Once the throat of the funnel is clear, slide the final inch of the vinyl tubing into the funnel. The tube may be snug enough to stay put on its own, but you'll probably want to secure it with a strip of duct tape or a smear of silicone-based household glue.

FIGURE 3-5: *Descreening your funnel*

Step 5 Finish off your trumpet by adding the mouthpiece to the other end—*but don't glue the mouthpiece in place!* Being able to easily swap out mouthpieces is a feature, not a bug. Once the glue attaching the tubing to the funnel has dried, hop to "Playing the Elephant Trumpet" on page 33 to give it a test toot. You might choose to finish your Elephant Trumpet by securing its coiled body with a few windings of duct tape, a little twine-wrapping, or even a couple of hose clamps, making it look more like the Renaissance-era natural horns the design riffs on (see Figure 3-6). I prefer to keep the tube loose, as you saw in Figure 3-1, because that makes improvising mutes and other novel effects—such as dangling the bell down a well while tooting—easier. It's also funny to tangle lil' trumpeteers in lengths of tubing while they blow themselves blue in the face.

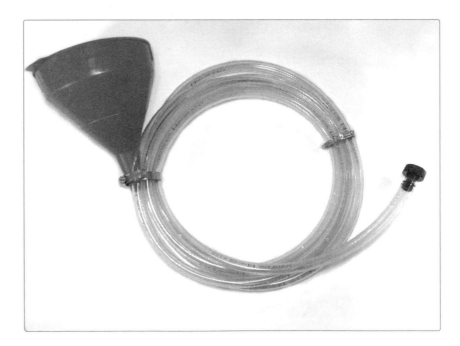

FIGURE 3-6: *The finished Elephant Trumpet, coiled "natural horn" style*

Playing the Elephant Trumpet

If you've never played a horn before, the Elephant Trumpet is a shockingly good place to start. The secret is to find the *embouchure*, or mouth shape, that works for you. You need to buzz your lips in the mouthpiece, making a sort of controlled angry-hornet-fart noise. Every human has a unique combination of mouthparts, so this can involve some fiddly experimentation. A few pointers:

▶ Keep your lips moist.

▶ Keep the center of your lips loose but the corners tight. Try frowning slightly.

▶ Place the mouthpiece so that it's outside your lips. That is, the mouthpiece should make contact with the skin above and below your lips, not with the fleshy part you use for kissing.

▶ Press your lips firmly to the mouthpiece, but don't clamp down.

▶ Keep your cheeks firm and flat, not ballooned out.

▶ Search *embouchure* on YouTube. You'll find a lot of conflicting advice, but that's because there are a lot of different kinds of faces.

You'll almost certainly get the best results if you stand tall with your shoulders back, make a point of pulling your breath all the way down to your belly, and push your air out using your abs instead of your chest. To begin, pinch off the airflow through your lips by placing your tongue behind your teeth, a move horn players call *stopping*. Pressurize the air inside your lungs by tightening your stomach, and then release the air in a lip-buzzing burst by saying "ta." This is basic *tonguing*. Practice double-tonguing (say "ta-ka" while blowing) and triple-tonguing (say "ta-ta-ka"), too.

Try both couplings, and stick with it for a few minutes. You should get a good toot sooner rather than later. Once you've got that, a good first exercise is to focus on sustaining a single, unwavering pitch for as long as possible. Visualize blowing straight through the horn, rather than into it.

A natural horn like this, with its fixed length, is basically limited to playing notes that are in the *harmonic series* (a set of notes that are mathematically related according to their wavelength—something we'll discuss in "Pipes and Pitches" on page 52, and touch on from a different angle in Appendix C). For example, on a bugle—which is operationally identical to our Elephant Trumpet—you can play the notes described in Table 3-1.

TABLE 3-1: Bugle Notes

Tone	1st	2nd	3rd	4th	5th	6th	7th
Bugle note	C_4 (middle C)	G_4	C_5 (one octave up from middle C)	E_5	G_5	Bb_6	C_6 (two octaves up from middle C)

Given the length and width of your Elephant Trumpet, you'll tend to get a harmonic series somewhere in the neighborhood of the one in Table 3-2.

TABLE 3-2: Elephant Trumpet Notes

Tone	1st	2nd	3rd	4th	5th	6th	7th
Approximate Elephant Trumpet note	F_2	C_3	F_3	A_3	C_4	$E\sharp_4$	F_4

Odds are that as you experiment, the first note you'll find is the *second tone*, which is middle G on the bugle and usually something like a low C on an Elephant Trumpet. You can go up and down in pitch by pinching and relaxing your lips. If you relax your lips a touch, you should get that lower *first tone* note. If you pinch them a little, you'll pop up to the *third tone*. All pitch changes can be accomplished with just lip power; avoid the temptation to modulate the pitch by squeezing the mouthpiece tighter against your lips. It's a tricky balancing act. Focus on increasing your lip tension by pinching your lips together while still keeping them loose enough to buzz.

With a half-hour's monkeying around, you'll likely be able to find the fourth tone. With no prior horn experience, I can rarely hit the higher tones, so they may evade you for quite a while, too—and that's okay: if you can hit tones 2–4, you've got most of what it takes to play the classic "Taps" (see Figure 3-7). Tight lips and a hearty blow will help you fake your way to that high fifth tone in the sixth bar.[2]

```
Tones:
22 3   23 4   23 4   23 4   23 4   34 5          4 3 2 22 3

Bugle notes:
GG C   GC E   GC E   GC E   GC E   CE G+          E C G GG C

Elephant Trumpet notes:
CC F   CF A   CF A   CF A   CF A   FA C          A F C CC F
```

FIGURE 3-7: *Sheet music (from the* US Navy Manual for Buglers NAVPERS 10137-B) *and several different tablatures for "Taps"*

---

2. No clue what I mean by "sixth bar"? Flip to Appendix C for a crash course in music theory.

Tips, Tricks, and Mods

Based on woodcuts from the early 1500s, the lowest-pitched natural horns of that period appeared to be just shy of 20 feet long. As it turns out, this is a fairly fun column of air to excite; it's in the realm of the modern tuba, which is about 16 feet long when uncoiled. This is why I chose such an extremely long length of vinyl tubing for the Elephant Trumpet.

Tube Length

For a brighter sound, you might consider cutting your tube down to 12 feet, like a French horn—bearing in mind that the French horn is a conical instrument, so even if you match its length, your horn's voice will likely still be fairly different. Both the trombone and trumpet, on the other hand, are cylindrical-bore horns. When unwound, the average trombone is around 9 feet long and has a 6 1/2-inch to 8 1/2-inch bell opening. A trumpet body tube is around 6 feet long, with a 4-inch to 5-inch bell. A trumpet-like Elephant Trumpet configuration can have a body tube as short as 4 1/2 feet and still have a viable brass sound. Figure 3-8 shows a few different Elephant Trumpets with different bells and tube lengths.

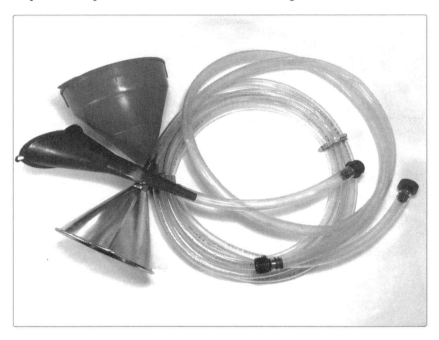

FIGURE 3-8: *A variety of rubber horns*

In addition to playing lower-pitched notes, a longer tube is going to give you a fuller, brassier, more sonorous tone. Shorter tubes are easier to play but sound "fartier." A smaller bell is less loud but more defined, which is to say, it's less rich and more like a trumpet.

Mutes

There are many types of mutes for brass instruments, but the only one that really interests us is the *plunger mute*, which lets you actively articulate the horn's voice by closing off and opening up the bell as you play. The plunger mute is vital to, among other effects, the infamous "sad trombone" sound effect. As its name implies, the plunger mute is conventionally made by pulling the cupped rubber end off of a *brand-new* toilet or sink plunger (see Figure 3-9, far left). Get a plunger with a diameter *smaller* than that of your horn's bell. A similar muting technique can be accomplished with other items small enough to fit in the horn's bell. Try disposable paper and Styrofoam coffee cups, mason jars, your closed fist or bunched fingers, and so on.

New Orleans Jazzers of the '20s would get a more subtle effect using a Derby or bowler hat large enough to cover the opening of the bell of their horns. Try using various small cardboard boxes, kiddie sand buckets, and food storage containers in this manner. If you've left the coils of your Elephant Trumpet loose, you can go one further than King Oliver and Louis Armstrong by using your own belly or a big bowl of water as a mute.

FIGURE 3-9: *A variety of improvised mutes*

Resources

In "Playing the Elephant Trumpet" on page 33, I gave a *very* brief overview of how to coax a noise out of this horn. For more details on playing, the *US Navy Manual for Buglers NAVPERS 10137-B*—drafted just after World War I and last revised in 1953—is a surprisingly informative source. You can find a copy online by searching for the title.

4 THE CPVC SLIDE WHISTLE

Woodwinds are generally broken into two groups: those that produce sound when a physical reed vibrates a column of air (as is the case for harmonicas, saxophones, clarinets, party horns, the Chinese sheng, and so on) and those that accomplish the same task without the physical reed (think flutes, pipe organs, beer bottles, referee whistles, and so on). In this second group, the instrument splits a stream of air, and that split stream acts as a reed. Such reedless woodwinds are among the oldest known human-crafted instruments, dating back at least 30,000 years, and were once crafted from cave bear femurs or swan bones and mammoth ivory.

Reedless end-blown flutes—like the recorder, penny whistle, Native American flutes, lots of Cracker Jack novelty whistles, and our own humble CPVC Slide Whistle—are all *fipple flutes. Fipple* refers to a type of mouthpiece, illustrated in Figure 4-1. The player's breath is forced down a narrow passage—the windway—before crashing into the sharpened windcutter at the far end of the open voicing window. The windcutter splits the air stream, creating turbulence that then causes the stream to oscillate between flowing over and under the windcutter. This air reed vibrates the column of air within the body of the whistle, making a single clear pitch. The length of the air column determines that pitch: a longer column of air makes for a lower pitch, while a shorter column of air gives a higher pitch. The slide whistle has a moveable piston[1] so that the musician can temporarily change the length of this column of air—and thus the pipe's pitch—without getting out a hacksaw.

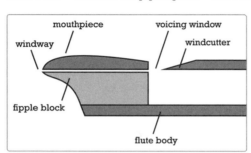
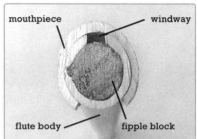

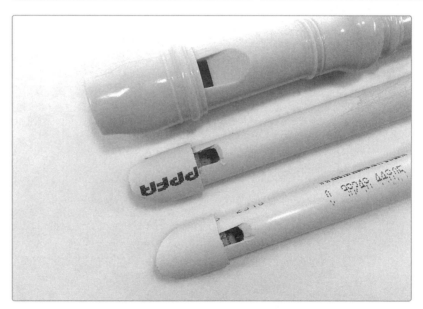

FIGURE 4-1: *Fipple flute anatomy (top left), the finished mouthpiece of our slide whistle (top right), and a comparison of homemade and commercial fipple flute mouthpieces (bottom)*

---

1. Hence the slide whistle's alternate name, the *piston flute*, which invites double entendre almost as much as *jazz flute*.

The finished CPVC Slide Whistle is shown in Figure 4-2. Hear samples at *http://www.nostarch.com/jamband/*.

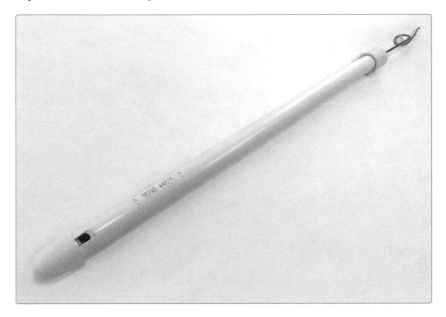

FIGURE 4-2: *The finished CPVC Slide Whistle*

Preparation

Build Time

▶ Under an hour

Tools

▶ A hacksaw

▶ An electric drill with small bits (Probably 3/32 inch is about right, although it depends on the gauge of your wire coat hanger.)

▶ Needle-nose pliers

▶ A hobby knife, utility knife, or pocketknife with a small, sharp blade

▶ A ruler or tape measure that shows 1/16-inch increments

▶ Medium-grit sandpaper (Even something as fine as 100-grit will work.)

▶ A miter box

▶ (Optional) A tapered half-round file or other metal file with a flat face

▶ (Optional) A small flathead screwdriver

▶ (Optional) A table vise or several clamps

- ▶ (Optional) A pair of safety goggles (These are useful for protecting your eyes while sawing wood and plastic.)
- ▶ (Optional) Metal snips (Protip: many sets of pliers have decent integrated snips.)

Supplies

- ▶ About 18 inches of 1/2-inch diameter CPVC (Ideally, you want *SDR 11* CPVC. Note that 1/2-inch *CPVC* is different—and significantly narrower—than 1/2-inch *PVC*. See "Buying CPVC" below for details.)
- ▶ A 1/2-inch CPVC end cap
- ▶ A few inches of 7/16-inch dowel
- ▶ Duct tape
- ▶ A wire coat hanger
- ▶ A small screw eye (The hardware store is full of workable options. I used a 13/16-inch-long screw with a 3/16-inch eye capable of supporting 20 pounds of hanging weight.)
- ▶ (Optional) Petroleum jelly

Buying CPVC

You need only about 18 inches of pipe for this project, but narrower gauges of PVC and CPVC are almost always sold in 10-foot lengths. Fortunately, that will only set you back about $4, and you'll have plenty of spare pipe for mistakes and experiments. In some hardware stores, you might have several different *schedules* of CPVC to choose from. You *can* make a fipple flute from any of them by following these instructions, but the specific dimensions listed in this project will work best if your flute body has the same ratio of internal diameter and wall thickness as my model. That's why PVC—which is measured differently—won't work with these measurements. If you have a choice, pick SDR 11 CPVC. If you don't have a choice, you're almost certainly looking at SDR 11. It could also be SDR 13.5, whose walls are about 0.2 mm thinner. That should be well within design tolerance—this ain't rocket surgery.

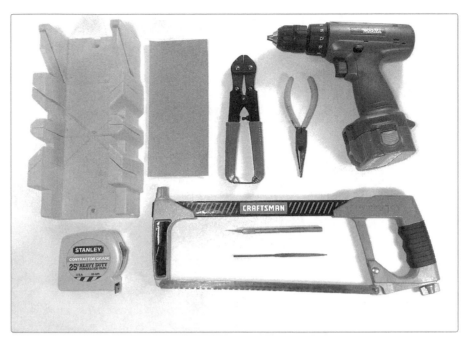

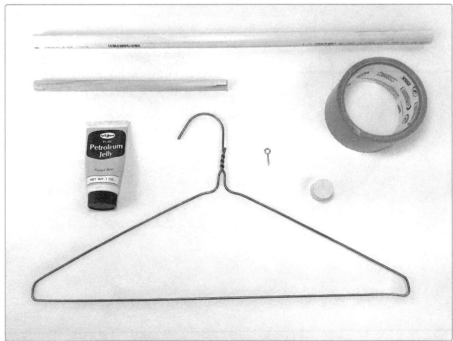

FIGURE 4-3: *Tools and supplies (not shown: vise and safety goggles)*

Building the CPVC Slide Whistle

This project can be broadly broken into three parts. The first—and most time-consuming and fiddly—is concerned with building the instrument's mouthpiece. After that, you'll add the piston and slide and then do a little finish work.

Build the Mouthpiece

Step 1 Start by cutting the CPVC. This single piece will supply us with material for the body of the whistle and the mouthpiece (shown in Figure 4-4). Your first cut should be 1 inch from the end of the pipe and angled 45 degrees, such that the longest edge of the smaller piece of CPVC is 1 inch long. You could freehand this cut, as it's not vital that it be exactly 45 degrees, and it will automatically match the angle of the flute body because a single angled cut forms both. But you also want this angle to match the angle of the fipple block you'll cut in Step 2, and that might prove a pain without a miter box.

If you don't have a miter box, you can still approximate a 45-degree angle. Just mark one side of the pipe 1 inch from the end and the other side 5/8 inch from that same end, and then cut an angle that hits both marks. That said, consider dropping a few bucks on a miter box, especially if you plan on building the Robo-Tiki Steel-Stringed Ukulele (Project 10) or the Twang & Roar Kalimba (Project 11).

⁎ **NOTE:** *This project doesn't involve a ton of sawing, but you'll produce a fair number of sharp plastic flecks when cutting the CPVC. If you're making a bunch of slide whistles (and sawing like a maniac), goggles are advised.*

Once you've cut the angled mouthpiece, trim down the whistle body so that it measures 14 inches at its longest (that is, from the tip of the pointy bit down its length to the straight-cut end; see Figure 4-4).

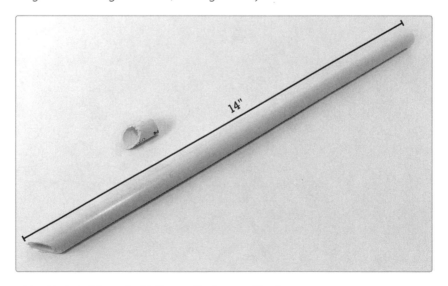

FIGURE 4-4: *The cut whistle mouthpiece and body*

Step 2 Cut your fipple block and piston. The piston is just a straight-cut 1/2-inch-long piece of the 7/16-inch dowel; it's the lower piece shown in Figure 4-5. The fipple block is a 1-inch-long piece of dowel with the same angle as your mouthpiece and flute body. (Protip: if you freehanded the cut in Step 1, you can get a pretty good matching angle by using the mouthpiece as a guide—the point, after all, is for the two to match.)

FIGURE 4-5: *Your piston (lower) and fipple block (upper). Note that the fipple block's longest (i.e., lower) edge is 1 inch long.*

Cut the Windway

Step 3 Prepare to cut the walls of the windway. This is the trickiest part of building the slide whistle, but it's still not all that tricky. You want the windway to be about 1 1/4-inch long, a bit over 1/4-inch wide, and centered on the long side of the whistle body. The diagram in the first panel of Figure 4-6 shows the desired result. To accomplish this, start by drawing some guidelines. First, draw a line down the center of the long edge of the flute body (a clear plastic ruler will help). Then add a crossbar 1 1/4 inches from the sharpened tip of the CPVC, as shown in the middle panel of Figure 4-6. The crossbar corresponds to the farthest edge of the voicing window, which we'll ultimately sharpen into a windcutter. Now add parallel lines 1/8 inches from either side of your centerline, as in the final panel of Figure 4-6; these will be the sides of your windway.

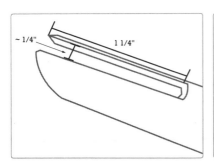  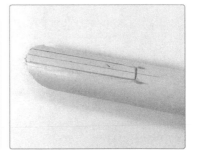

FIGURE 4-6: *A diagram of a finished windway slot, including dimensions (left) and guidelines for your windway cuts (middle and right)*

Step 4   We're going to cut the windway in several steps. First, use your hacksaw to cut a shallow slot in the pipe at the end of the voicing window, as shown in Figure 4-7. You'll need to make only three or four strokes; don't cut completely through yet.

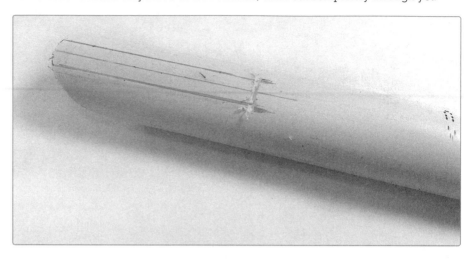

FIGURE 4-7: *Slotting the end of the voicing window*

Step 5   Next, cut the long edges of the windway. If you have a table vise, locking the pipe into it will make your life a little easier for this step. If not, don't fret: CPVC cuts easily, so you should be able to do this without a vise. Either way, *be careful*! It's really easy to slip when you're cutting along the length of a cylinder.

Cut along both long edges of the windway, following the parallel lines you drew in Step 3, without touching the bottom of the whistle body (see Figure 4-8). Do your best to make your cuts parallel to the centerline and each other. This should keep the walls of the windway parallel to each other, just like the hallway walls of a well-built house.

FIGURE 4-8: *Cutting the windway*

Step 6   Once you finish these cuts, you should have a tongue of CPVC in your windway, just barely connected to the rest of the pipe (see the leftmost image in Figure 4-9). Grab hold of this, bend it back (see the middle image in Figure 4-9), and break it off (see the rightmost image in Figure 4-9). This method is a touch sloppy, but it gives us a decent foundation for the proper windcutter angle.

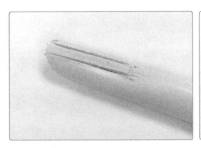  

FIGURE 4-9: *Your windway is cut (and sloppy)!*

Step 7   It's time to split the mouthpiece, which is the final sawing step. (Again, a table vise will make this easier.) You want to split this piece along its shortest edge, as shown in Figure 4-10. This doesn't have to be precise, so you don't need to draw guidelines.

FIGURE 4-10: *The prepared mouthpiece*

Step 8   Clean up your cuts. The pieces you just made are probably a mess of splinters and burrs, so take a second to sand down the rough edges.

  ✳ **WARNING:** *Don't be super aggressive as you sand! If you're sanding hard enough to smell burning plastic, you're (1) probably about to take too much off and (2) risking your health. The friction from sawing or sanding can heat PVC or CPVC enough for it to release phosgene, a chemical cousin of chlorine gas, best known as a WWI-era chemical weapon. Very little gas will be released in this case (you aren't heating the CPVC much), but if you make a lot of flutes or work in a small, closed environment, then slip on a respirator mask like the one shown in Figure 2-2 on page 17. It's better to be safe than sorry! If you're making only one flute, you should be safe if you go slowly and work outdoors or in a well-ventilated space.*

Once you've sanded those edges, you need to clean up the end of the windway that will ultimately form the windcutter (see the before and after pictures in Figure 4-11). Ideally, the end of the windway should have square corners, and the underside of the windcutter lip should be curved (matching the interior of the pipe). The exterior of the windcutter needs a smooth ramp and sharp leading edge.[2] If you're confused, refer to the whistle anatomy diagram in Figure 4-1.

For this more detailed sanding, you should either cut a smaller rectangle of sandpaper and fold it around a stiff, narrow tool or use a tapered half-round file. The flat side of the file is good for squaring off those corners and smoothing the windcutter's ramp, while the curved side is nice for restoring the smooth underside of the windcutter lip.

When you're done shaping the windcutter, smooth the vertical walls of the windway.

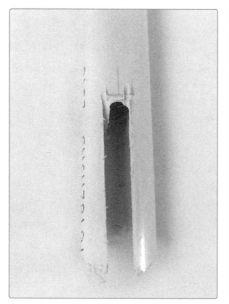 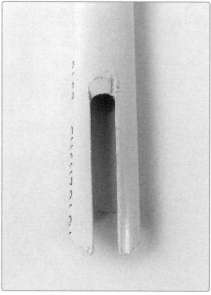

FIGURE 4-11: *The windway before (left) and after (right)*

---

2. Recorders and penny whistles usually have a 30-degree angle on the windcutter, but if you make the ramp less than 30 degrees, you'll generally have an easier time bending pitches and hitting overtones (see "Playing the Slide Whistle" on page 51 for more on these playing techniques).

Install the Fipple Block

Step 9   It's time to place the fipple block inside the whistle body. You want a snug fit, so wrap the fipple block in a double layer of duct tape. Don't use more than two layers, or the fit will be *too* snug. Then, push the block into place so that the window is about 1/4 inch deep, as shown in Figure 4-12. At this stage, the angled end of the fipple block should be more or less flush with the angled end of the whistle body.

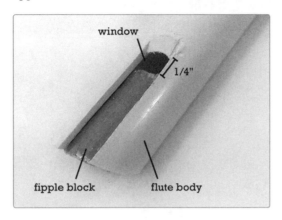

FIGURE 4-12: *The fipple block is in place (note that the window is about 1/4 inch deep).*

Step 10   Add the mouthpiece and test the whistle. Don't pinch your fingertips as you snap the mouthpiece into place! This part will call for a little brute force, but don't be afraid of breaking anything; CPVC is pretty tough stuff. Once you've got the mouthpiece on the whistle body, align the angled ends. Give the whistle an exploratory toot, blowing gently and steadily into the windway. You'll probably get a somewhat breathy, perhaps weak, whistle sound. Aspiring fipple flute players tend to start out blowing *way* too hard, which won't give you a good tone. If the whistle is breathy or squeaky, try blowing more gently. Playing a slide whistle is a lot like coaxing a hoot from a bottle.

Once you have a sound, experiment with clarifying that tone by backing the fipple block out a touch (thus making the window a little larger). You'll find moving the fipple block much more challenging now that the mouthpiece is clicked into place, so try using a small-bladed screwdriver (or your metal half-round file) as a prying tool. Insert that tool into the voicing window and lever it against the fipple block until you find a window size that gives a full, solid, low tone, somewhat like that of a recorder. Fine-tune that sound with minor adjustments in the placement of the mouthpiece; slide it down or back slightly to shorten or lengthen the window. A good final alignment will probably have the window measuring around 5/16 inch from the end of the fipple block to the sharp edge of the windcutter, and the mouthpiece should cover all but about 1/16 to 1/8 inch of the fipple block (see Figure 4-13 for clarification).

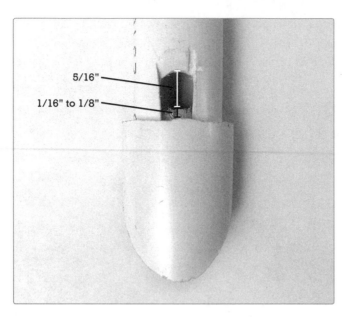

FIGURE 4-13: *Final mouthpiece to fipple block to windcutter alignment (with approximate lengths)*

Add the Piston and Slide

Step 11   Prepare the piston. If you didn't sand it earlier, do so now, just to smooth off the rough edges. Then, drill a small hole in the middle of one face of the piston. This hole should have a slightly smaller diameter than the threaded body of your screw eye. Unless your dowel is a *very* hard wood, you can probably brute-force the screw eye into the wood without drilling, but the hole helps. Either way, screw in the eye. Finally, cut a 3/4-inch-square piece of duct tape, stick it to the flat face of the piston (the one with no screw eye in it; see Figure 4-14), and smooth it down over the edges. The goal is for the face of the piston to have a nice, smooth surface and the piston itself to fit firmly, but not too tightly, into the whistle body. A sloppy piston face will make the whistle really breathy over large portions of the piston's travel, while a loose piston will be wheezy and may unpredictably *flip between registers*—that is, jump up to a higher note.

FIGURE 4-14: *The finished piston*

Step 12   Now install the slide (to control the piston) and an end cap (to hold the whole mess in place). Your slide is a 16-inch length of wire coat hanger or a similar rod. Cut the longest straight section you can from your wire hanger, and you should be all set. Bend it into a hook, as shown in Figure 4-15, and connect the piston. Then, drill a hole through the center of the 1/2-inch CPVC end cap and thread the slide through. This hole will probably be around 3/32 or 1/8 inch, but hangers come in various gauges; just use the smallest drill bit that's still larger than the wire itself.

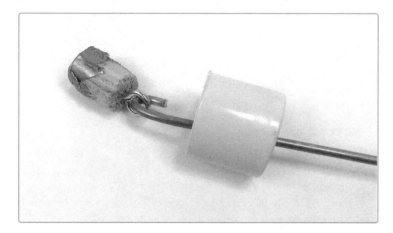

FIGURE 4-15: *The slide and piston, with end cap, ready to install*

Step 13   Put it all together. Thread the piston into the whistle body and snug the cap into place. Once again, give a few exploratory toots, this time with the slide out as far as it will go. You'll almost certainly get a recognizable tone, about an octave lower than what you had when the pipe was open-ended. The sound may be somewhat breathy, and it's not unusual at this stage to need to adjust the position of the mouthpiece by 1/16 inch or so. You might even need to slightly reposition the fipple block, backing it out a touch farther. In any event, it should take no more than a few seconds to get a tone you like. This timbre should be different from the one you heard with the open-ended whistle. The slide whistle's tone is a dulcet version of what you'd get by blowing over a bottle.

Once you're content with the tone, you may wish to sand down the face of the mouthpiece because the end of the fipple block will likely protrude by up to 1/16 inch, as shown in Figure 4-16. This part is optional, though: leaving the fipple block alone doesn't hurt the playability of the instrument.

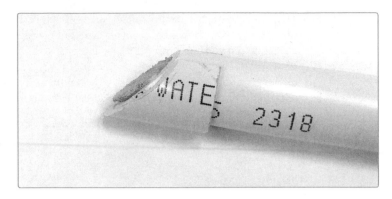

FIGURE 4-16: *The slightly protruding fipple block*

Finishing Touches

Step 14   First off, a 16-inch slide is too long. A nice way to finish off this project is to run the piston up until it's just clear of the window (the whistle can't make any sound once the piston crosses the windcutter). Then, mark this position on the coat hanger, run the slide back out, and curl the slide around one full twist, as shown in Figure 4-17. This makes a nice fingerhold for playing.

The slide action may feel pretty stiff. To counter that, pull off the end cap and fully remove the slide and piston from the flute body. Smear a thin layer of petroleum jelly (or similar nontoxic lubricant) around the sides of the piston and then put the instrument back together. If the slide is too loose and you're getting a breathy tone that disappears or flips to a higher register over certain lengths of the whistle, try adding another layer of duct tape to the piston face and then greasing the piston with petroleum jelly again.

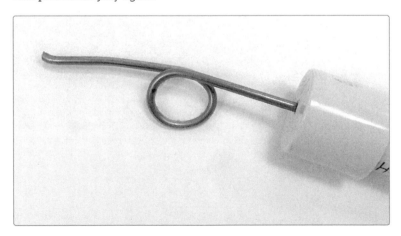

FIGURE 4-17: *The finished slide handle*

Playing the Slide Whistle

To get the best sound out of any fipple flute—including store-bought recorders and penny whistles—start by blowing a steady, gentle stream of air directly into the windway. Pretend you're trying to cool off a spoonful of hot soup without spattering it all over your guests, and you'll have the right *embouchure* (mouth shape), pressure, and speed. Although slide whistles are generally thought of as novelty instruments, toys, and sound effects, folks with a decent ear will find it possible to pick out familiar tunes.

Air reeds, like the one at the heart of our little slide fipple flute, are often characterized as *flow-controlled valves*. With an air reed vibrating a given column of air, it's the velocity of the air stream, not its pressure, that determines the pitch. If you want to coax out a higher pitch at any position on the whistle, blow *faster* (called *overblowing*), not harder. With the particular whistle we've built here, these overblown notes tend to be about a fifth higher than the octave of the fundamental note.[3] That's pretty handy in American folk music, where the root and its fifth are your go-to notes. (See Appendix C for a crash course in music theory if these terms are unfamiliar.)

Once you've got a decent "cooling-the-soup" tone, you can experiment with overblowing by removing the slide altogether and blowing straight into the open pipe, which is easier to overblow than the closed slide whistle. As you practice blowing faster without blowing a lot harder, you can eventually coax up to four distinct notes from this open pipe: the fundamental (for a 14-inch pipe like this, that will probably be the B just below middle C), the same note one octave higher (in this case, the B above middle C), the fifth up from that (F#), and then the next B up from that. (Again, if this is all Greek to you, check out Appendix C.)

Getting the whistle to jump registers like this with the slide in place is a little more challenging—in part because it's a closed rather than an open tube. Our narrow windway and relatively steep windcutter don't help; a recorder, which has a wide and shallow voicing mouth, is much easier to overblow. If you want to learn to overblow the slide whistle, start with the slide all the way out so that the whistle is playing its lowest note. Once you can overblow that note, begin experimenting over the rest of the whistle's range. You should be able to run the slide all the way in—to the point where there is no note to be had by just using your "soup-cooling" blow. Then, you should be able to coax out a terrifically piercing note by overblowing.

One final note: if you're recording, keep in mind that the sound of your whistle is actually being made by the air reed vibrating at the window, so that's the point on the instrument you want to mic. This applies to fipple flutes of any kind, including most whistles and even open-ended flutes, like recorders.

---

3. Fun fact: An open-ended flute of these dimensions will overblow a full octave.

Pipes and Pitches

Our slide whistle is a *closed pipe*, a tube open at just one end—like an organ pipe, a flute, or a soda bottle. By blowing across the top of that tube (or into a specially-shaped end piece, as is the case with our slide whistle), you set up vibrations at the open end of the tube and cause the air trapped inside to vibrate at a certain frequency, as shown in the far left of Figure 4-18. The *wavelength* of that vibration, or the distance between peaks in the wave (denoted by the funky lil' Greek letter *lambda*, λ) determines which note you hear. The longer the wavelength, the lower the note will be. If you keep blowing at the same speed but make the vessel smaller (by, for example, pushing in the slide of your slide whistle, as illustrated in the middle diagram of Figure 4-18), then that wavelength gets squished shorter, and the pitch goes higher.

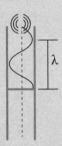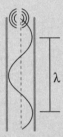

FIGURE 4-18: *The sound waves inside a closed pipe (left), the same pipe shortened (middle), and an equal-length open pipe (right)* *

Remember how your whistle had a higher pitch before you added the cap and slide? That's because without the slide, the whistle is open at both ends and is thus an *open pipe*, like the one in the final panel of Figure 4-18. This open pipe is the same length as its closed cousin shown at the far left, and is being blown at the same speed, but as you can see, it has an extra half wave crammed in there. The result is that its wavelength is a bit squished. The squished wavelength is shorter, and shorter wavelengths mean higher pitches—which is why the open pipe has a higher pitch than the closed pipe.

---

* Note to physicists, music PhDs, and sticklers: at first glance, this image may seem to contradict something mentioned in Step 13, that a closed tube is reliably about an octave lower than an open tube of the same length. You might expect the illustration to show the open pipe playing a note with a wavelength half that of the closed pipe, which would correspond to a note one octave higher in pitch. Instead, I'm showing the open pipe playing a note with a wavelength five-sixths that of the closed pipe. That's because the one-octave difference between closed and open pipes only holds at the pipes' *first resonance,* the note you get when you use that gentle "cooling-the-soup" blow. For clarity, the diagrams here show the pipes' *third resonance,* which is the third overblown tone mentioned in "Playing the Slide Whistle" on page 51. This allowed us to show complete wavelengths in each diagram, which are a lot easier to understand if this is your first run-in with the physics of hooting bottles.

But why does an open pipe need that extra half wave? In a closed pipe, the vibrating air bangs against the sealed end, making that an area of high pressure (which, in the diagram, corresponds to a peak or trough in the wave). Meanwhile, in an open pipe, there's nothing for the air to push against, so it has to return to baseline atmospheric pressure (represented by the *zero crossing*, where the wave crosses the dashed line corresponding to baseline atmospheric pressure). This need to return to baseline invariably means cramming in one more half wave, which means making the wavelength shorter, which results in a higher pitch. This is why an open pipe will always be higher pitched than a closed pipe of the same length, even though the tubes are otherwise exactly the same and the columns of air they contain are equal.

Tips, Tricks, and Mods

One place to experiment is in the overall length of the whistle body, because that determines the lowest note you can play. If you go much over 17 inches, however, you're going to need to find different wire stock for your slide, as coat hanger wire is both too short and too easily bent. Another place to tweak, as mentioned in the build instructions, is the angle of the windcutter. A shallower ramp is significantly easier to overblow, for better or worse. Professional flautists get great effects by overblowing, while many beginners struggle with unintentional register jumping spoiling their melodies.

You can also use this same fipple mouthpiece to make fun little whistles out of shorter scraps of CPVC and nubs of dowel, with no slide required. Recall that the closed tube is reliably about an octave lower than the open tube. Also, the first overblow on the open tube is one octave above its fundamental, while the first overblow of the closed tube is a fifth above its octave. These can be fun building blocks, as you can close off the end of the whistle with your hand to trill between a note and its big brother and then overblow to get a note an octave higher and a fifth over that. That's one simple, fun way to accent a raucous little jam session, and it's easy to teach to total musical novices. If you have a mess of kids to entertain, you can make a terrific racket this way. Such simple two-tone whistles—which are similar to the eagle- and elk-bone war whistles used by many Plains tribes to direct battlefield maneuvers—seem to work best with a shorter flute body, maybe around 6 to 8 inches long. For an example, check out Figure 4-19; the two-tone whistle is at the bottom.

The best conversion for our CPVC Slide Whistle, by far, is to make it into a water whistle by pulling off the slide assembly and replacing it with a water-filled pipe (see the middle whistle in Figure 4-19). It's a little messy to work with as an instrument, but it sounds great and is a huge hit during bath time or out in the kiddie pool.

A smooth piston face and tight seal are instrumental to a good-sounding slide whistle. A liquid will naturally form a smooth face under the even pressure you create when blowing the whistle, and it likewise maintains a perfect seal. The result is a terrific, rich, even tone.

To build a water whistle, just grab a foot-long piece of 1/2-inch PVC, slap an end cap on it, and fill it with water. Now remove the end cap and slide from your CPVC Slide Whistle and slide the whistle body into the water-filled tube.[4] Play it just as you would the standard CPVC Slide Whistle.

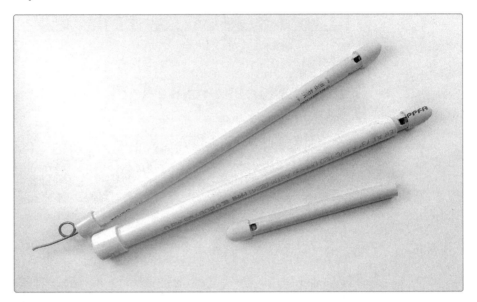

FIGURE 4-19: *A finished CPVC Slide Whistle (top), water whistle (middle), and two-tone whistle (bottom)*

4. If you're wondering how a 1/2-inch piece of CPVC can fit into a 1/2-inch piece of PVC, recall that CPVC is labeled according to its nominal *outer* diameter, while PVC is sorted by its nominal *interior* diameter. It's a match made in the plumbing aisle!

Resources

There are lots of online resources for making fipple flutes and many competing designs using cheap, convenient PVC and CPVC. I was first drawn to this project by Tim Escobedo's "Home De(s)pot Flute" (from his now defunct, and much mourned, FolkUrban DIY musical instrument site). Escobedo's design is an easy-to-play pentatonic[5] PVC fipple flute that's built like a penny whistle—that is, it has actual finger holes. His fipple is fundamentally different from what I've used here and is now popularly referred to as an *exo-fipple* (although I don't know that Escobedo ever called it that himself). Escobedo's "Home De(s)pot Flute" design is archived here: *http://www.oocities.org/tpe123/folkurban/flute/homedespot.html*.

My CPVC Slide Whistle's fipple owes a huge debt to Guido Gonzato's work. Gonzato makes and sells really nice PVC penny whistles and has made his core design, including his hard-earned whistle wisdom, freely available as *The "Low-Tech" Whistle: How to Make a PVC Whistle* (*http://www.ggwhistles.com/howto/*). This is an excellent resource if you want to bring your whistle up a notch, build a flute with a broader and lower range, or use larger-bore PVC.

Finally, if this project starts you down the fipple flute path, then the message boards at the Chiff & Fipple website will be invaluable: *http://chiffandfipple.com/*.

---

5. It plays five notes that would sound good in basically any combination; dip into Appendix C for a little more on scales.

5 | THE SCRATCHBOX

Hip-hop turntabalism arose from folks abusing their LPs and record players in an age when most casual music listening was done via cassette. Now that we're two iterations of musical media down the road, it's high time to bust up your old tape deck and find some new sounds.

A tape deck is actually a relatively simple device built around an amplifier hooked to a fairly sensitive *electromagnetic transducer*, a component that translates variations in a magnetic field into sound. With a little soldering, you'll be ready to coax a whole host of old-skool skritches and scratches from expired credit, gift, and loyalty cards.

Nearly any cassette player will work for this project: you could use a Walk-man or a boom box, with or without a recording function—even an old car stereo would do the trick. It's totally fine, even preferable, for the tape deck to be "broken" because the part we want—the transducer itself, called a *tape head*—is most likely intact. If you happen to choose a working tape deck with a built-in speaker, and you don't wreck it too badly while extracting the tape head, you can convert it to a quick-and-dirty amp for projects in this book or even "real" guitars and other instruments you have kicking around.

Behold the Magnetic Musicassette!

The cassette tape—christened the *Compact Cassette* by Philips when they released it in the United States in 1964—is essentially a miniaturized reel-to-reel setup, and it works along the same principles. A strip of plastic tape coated in *ferric oxide* (laboratory-grade rust, which is magnetic) runs at a constant speed across a *tape head*. During the recording process, an audio signal is pumped into the head, which imparts a magnetic field of varying intensities to the tape, analogous to the varying intensities of the sound waves. During playback, the moving magnetic tape induces its fluctuating magnetic field in the play head. An amplifier built into the cassette player boosts this signal and sends it to a speaker, which reproduces the original audio.

Although their fidelity was relatively poor, these so-called Musicassettes were the dominant medium for prerecorded music for about two decades. Their portability, convenience, and relative indestructibility—especially when compared to vinyl records or early compact discs—trumped audio quality. The format itself was designed from the ground up to make both playback and recording cheap, easy, and highly portable, which inevitably led to the folk art of self-expression via mixtape. The longevity of the mixtape highlights the utter absurdity of modern patent trolls claiming they invented "playlists" a good decade *after* teens across America agonized over whether Bonnie Tyler's "Total Eclipse of the Heart" should be followed by Whitney Houston's "How Will I Know" or Kenny Loggins's "Highway to the Danger Zone."

The finished Scratchbox is shown in Figure 5-1. Hear samples at *http://www.nostarch.com/jamband/*.

FIGURE 5-1: *The finished Scratchbox*

Preparation

Build Time

▶   About 30 minutes

Tools

▶   A standard soldering kit (See page 340.)

▶   An electric drill with bits (You'll probably need a 1/2-inch or 3/8-inch bit, a 1/4-inch or 3/16-inch bit, and a 1/16-inch bit.)

▶   Needle-nose pliers

▶   A set of small Phillips-head screwdrivers for disassembling the cassette player (Often called *jeweler's screwdrivers*, these little guys are required equipment for voiding warranties on modern electronics.)

▶   A small flathead screwdriver for prying open plastic cases

Supplies

▶   Any old audio cassette player

▶   A stack of old credit cards, gift cards, loyalty cards—anything with a magnetic stripe

▶   A 1/4-inch mono phone jack, also called a *guitar jack*

▶   A small enclosure (Metal is fine here; I used a standard Altoids tin. See "On Enclosures" on page 214 for more options.)

- 24-gauge insulated hook-up wire (22-gauge speaker wire is fine, too. It's stranded like hook-up wire, only slightly thicker.)[1]

- (Optional) Any *normally on* momentary pushbutton switch, also called a *normally closed* SPST switch (For more on switches, see "Switches" on page 338.)

FIGURE 5-2: *Tools and supplies*

Building the Scratchbox

Step 1   Explore your cassette player. Every tape deck is different, and there's really no map to this country, so start with some preplanning. We want to pull out the playback head, or *play head* (circled in Figure 5-3), which is usually secured to the cassette player's carriage from the front by two screws. In many cases, you can remove the player's door and have access to these screws, but you might have to at least partially disassemble the cassette player. Most players will have a set of Phillips-head screws on the back or bottom; remove every screw you can find. If you're dealing with a boom box or tabletop model, pull off any little plastic feet; there are

---

1. These two wire types are basically interchangeable if you are connecting hardware to hardware or other off-board components. The only place it really makes a difference is when a connection is going to a printed circuit board because the predrilled holes usually won't accommodate the thicker 22-gauge wire. For a full discussion of wire gauges and types, check out "Wire" on page 336.

often screws hiding underneath. Once you've pulled out the obvious screws, try to force the case open. You'll probably need to slide a flathead screwdriver into the seam and twist to break any adhesive used to seal the case in the factory.

* **WARNING:** *Be careful not to cut yourself! A flathead screwdriver seems innocuous, but with a little muscle behind it—like when you're trying to force apart a recalcitrant cassette deck case—it can easily slash and pierce you. Don't become a statistic!*

FIGURE 5-3: *The elusive play head is circled. On the right, notice the same tape deck from a different angle (including a period-appropriate cassette, for reference).*

Step 2   Extract the play head. Once you get inside the cassette player, you'll find one or two tape heads. The one you want for this project is the *play head*, which will be all metal and centered. (The other, which is probably mostly plastic, is the *erase head*; in Figure 5-4, the play head is circled and the erase head is squared.) You'll find the play head held in place by a pair of small Phillips-head screws and soldered to two or four wires. Loosen the screws, saving them for later, and clip the wires.

FIGURE 5-4: *Play head (circled) and erase head (boxed)*

Building a Tape Deck Amp

If your tape deck has a built-in speaker and still basically works after its surgery, you might as well make it into a spare amp. You just need to grab another 1/4-inch jack and solder it to those wires you just clipped from the tape head, one wire to each lug (as shown in Figure 5-5).

FIGURE 5-5: *Making a tape deck amp: before (left) and after (right). I reinforced the connections to those dinky little play-head leads with shrink tube. This particular Walkmanesque recorder has a built-in speaker on the back.*

If your tape head has four leads, like the one in Figure 5-11, try wiring both the left wires to one lug and both the right wires to the other. It shouldn't matter which pair goes to which lug on the jack, although it's hard to be certain. There are *lots* of half-broken tape decks in the world, and I've mucked around with only a very small percentage. If your first wiring doesn't work, swap 'em, or try just using a single wire from each pair. The worst-case scenario is that something mostly broken becomes completely broken. At that point, you can just tear out the speaker and use it in a new amp, such as the "Dirt-Cheap Amp" on page 363.

To use your tape deck amp, power up the tape deck, plug in your instrument, and press play. FYI, these amps can be loud; the signal coming from a cassette tape is pretty low, so tape decks normally have beefy amps. The upside is that if you crank it up, you can often get rad, crunchy British Invasion distortion.

Gutting a Walkman or other speakerless tape deck? Don't fret; you can use the same trick to transform it into a cheap-and-dirty preamplifier. (There's more on preamps in "Tips, Tricks, and Mods" on page 69.) When building a Walkman preamplifier, the new 1/4-inch jack is your *input*, and the original headphone jack built into the unit is your *output*.

Step 3   Prepare your hook-up wire. Cut one 6-inch length of wire and two 3-inch lengths. Strip and tin both ends of each wire (refer to "Soldering" on page 346 if you need a refresher). If you're leaving out the switch, then just prepare two 6-inch lengths of wire because you'll be wiring the play head and jack directly to each other.

Step 4   Wire up the jack. Solder a short wire to the jack's tip lug, which is connected to the long, hooked metal tongue that makes contact with the plug's tip, and solder the long wire to its sleeve lug, which is also its ground. Jack anatomy is illustrated in "Quarter-Inch Phone Plugs and Jacks" on page 337, and the wiring of this jack is shown in Figure 5-6. If you're skipping the switch, then just solder one wire to each lug of the jack.

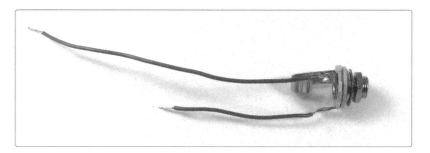

FIGURE 5-6: *The wired audio jack*

Step 5   If you're adding the cutoff switch, you'll do that now. Switches from the hardware store often have screw terminals (as mine does). If your switch has solder lugs, that's also fine. What matters is that the switch is a *normally closed* one, meaning it lets electricity through under normal conditions and is momentarily turned *off* by pressing it. (Many everyday pushbutton switches, like doorbells, are *normally open*, which means pushing the button momentarily turns the switch *on*, not off.) Whichever style lugs you have, connect the remaining 3-inch wire to one terminal and the 3-inch wire connected to the jack to the other, as shown in Figure 5-7. It doesn't matter which wire goes to which terminal.

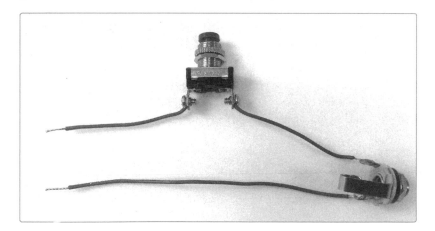

FIGURE 5-7: *The wired cutoff switch*

Step 6   Set all of your wiring aside and prepare the enclosure. The jack will require a 3/8-inch or 1/2-inch hole. Drill this hole on the short side of the tin. In Figure 5-8, this hole is the one farthest to the left. If you're using an Altoids tin–style enclosure, place the hole as far from the rolled upper lip as possible, otherwise you won't have enough clearance for the lid to close once the hardware is inside.

Step 7   Next, drill a 1/4-inch hole in the bottom of the tin, opposite the jack hole. The 1/4-inch hole allows your leads to connect to the play head. As Figure 5-8 shows, this hole is about an inch from the nearest edge of the case and centered. Add a pair of 1/16-inch screw holes near the edge of the tin to secure the play head. Play heads are fairly standard, but just in case, check first to be sure that the screws you retained in Step 2 are slightly *larger* than 1/16 inch. Also, measure the spacing of the mounting holes; they tend to be around 3/4 inch apart.

Step 8   Finally, add a hole for the switch (if you're using one—otherwise, skip to the next step). Most large switches, like the one I used here, will fit in a 1/2-inch hole, while smaller "mini pushbuttons" usually need a 1/4-inch hole. Either way, you should drill the switch hole on one of the long sides of the tin, again leaving as much room as possible for the lid to fully close.

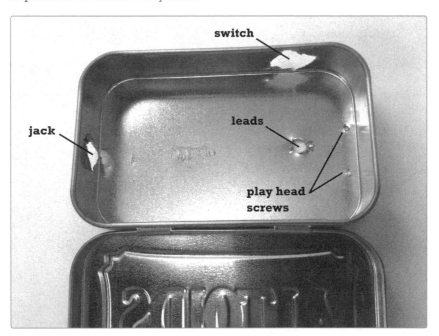

FIGURE 5-8: *The distribution of holes in the enclosure, for the jack, the switch, the leads, and the play head screws. (Raggedy holes should work just fine.)*

Step 9   Mount the hardware in the tin and string the two loose wires through the 1/4-inch hole in the back of the enclosure, as shown in Figure 5-9.

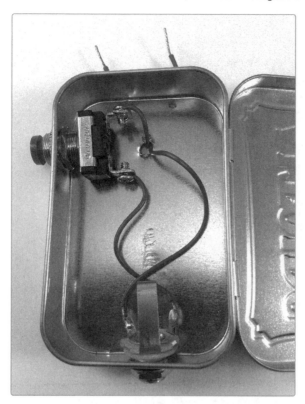

FIGURE 5-9: *The hardware in place*

Step 10   Time to prepare the play head. Almost all play heads have a set of metal prongs sticking up past their top surface. These functioned as tape guides in the head's old life playing "Living on a Prayer" and "Power of Love" singles. We don't need them, so grab them with a pair of needle-nose pliers and bend them back out of the way, as shown on the right in Figure 5-10.

FIGURE 5-10: *Before (left) and after (right) shots of the modified play head*

Step 11 Next, we need to solder the play head to the wires sticking out of the case. If your play head has only two terminals, then solder one wire to each and be done with it. If it has four, solder one of the two wires to *both* the terminals on one edge of the tape head (see the left image in Figure 5-11). Repeat this step for the other pair (see the right image in Figure 5-11).

FIGURE 5-11: *Wiring the four-terminal play head*

Step 12 Finally, mount the play head firmly to the outside of the enclosure using the two screws you salvaged in Step 2, as shown in Figure 5-12.

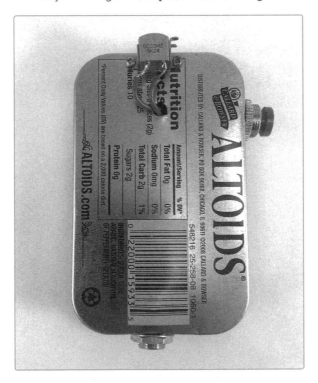

FIGURE 5-12: *The back of the finished Scratchbox, showing the mounting of the tape head*

Step 13   The completely wired guts of the Scratchbox are shown in Figure 5-13. Now that the wiring is done, plug it into an amp, power the amp up, and rub a credit card magstripe along the play head. You shouldn't have to search long to find some zips and growls. If you don't hear anything, then flip to "General Troubleshooting" on page 355 and start checking connections. This is the first full-fledged circuit we've built, but it's still a very simple one. If it's not working, you probably have a simple short that you can clear up with a little electrical tape and maybe some clipping and resoldering.

FIGURE 5-13: *The guts of your completed Scratchbox. (The play head, which is mounted to the back of the tin, isn't visible. Flip back to Figure 5-12 to see how that fits onto the case.)*

Playing the Scratchbox

Much like the Plasti-Pickup project, the Scratchbox is best thought of as an exploratory instrument. Start by just rubbing a credit card's magnetic stripe back and forth along the tape head. You'll likely get some neat zippy sounds. That's because there's *a lot* of info crammed onto those stripes, and it's arranged in three very narrow horizontal bands that run along the length of the stripe.

There's a lot to sift through, so work slowly over short lengths of stripe to really get a feel for the aural possibilities. You'll get very different dynamics working fast or slow. These are influenced by acceleration, pressure, and how well you track a given band. I find that most credit cards have a good whoop at one end of the stripe or the other, which can become nice sharp chirps if you flick the card aggressively. There's also often a long, low purr through the middle section of the stripe that you can bring out with a slow, firm push. Try swiping the card in different ways to find the sounds and movements you like.

In terms of musically working a groove, I find it easiest to hold a credit card cupped in my left hand with the stripe facing out and to wield the Scratchbox with my right hand, controlling the cutoff switch with my thumb. (I'm right-handed, so if you're left-handed, you might try reversing this.) Create rhythms by building up patterns of long and short, or fast and slow, swipes. In turntable lingo, these unarticulated scratches are *baby scratches*. You can spice these patterns up with a *scribble scratch*, created by keeping the play head on one especially noisy spot on the stripe and jiggling or vibrating your hand as quickly as possible, allowing only the tiniest movement along the stripe.

The switch we've installed is a simple mute. Music is fundamentally about repeating patterns of sound and silence; the mute is handy for dropping in those *rests* without disengaging from the card's stripe and risking losing your track. This cutoff switch also allows for the rudimentary emulation of a few classic turntable building blocks created by working the *crossfader*, a control that brings the volume of one turntable down and the other up simultaneously in a single movement. For example, by cutting a sound off just after it begins, we get a *chirp*, which is that classic, sharp *wsht-wsht-wsht*.

Remember that long whoop that usually occupies the middle of the stripe? When you get adept at steadily tracking that sound, you can glide back and forth along it while rhythmically jabbing the cutoff switch, creating an effect known as a *transform* (so-called because it's reminiscent of the sound cue that accompanied the Autobots' and Decepticons' forward-flip-vehicular-shapeshifting in their 1980s cartoon). The trick to a transform is coordinating the glide and the muting: aim for about four cutoffs per card glide while working back and forth at a steady pace. Musically speaking, the cutoff jabs create little eighth-note bursts against the steady half-note swipes. Search YouTube for *turntable chirp*, *turntable transform*, or *scribble scratch* to familiarize yourself with the aural palette offered by hand-modulating, near-random data streams.

Tips, Tricks, and Mods

If you want even more control, you can replace the mute button with a volume knob—try using a 10k ohm potentiometer with an audio taper (flip to "Resistors: Fixed and Variable" on page 325 if this sounds like technobabble). Connect it by removing the switch and soldering one of the wires to the potentiometer's center lug and the other to either end lug.

Another good mod is to add a *preamplifier*. Preamplifiers, often just called *preamps*, increase signal voltage, while *power amplifiers* increase the current so that the signal can drive a speaker. In practical terms, preamps tend to bring a low-level signal, like the one coming out of a standard electric guitar, up to line level, which is what something like a mixing board expects. Placing a preamp close to a sensor bumps up the signal-to-noise ratio by amplifying the signal as close to its source as possible, before random electromagnetic waves have much chance to seep into the chain. The result is a fuller, more aggressive sound with more details for you to pick out and explore. Try using the Scratchbox with the Mud-n-Sizzle Preamp (Project 12). Crank the volume up, and push the tone to the "sizzle" end of the spectrum. The "Two-Transistor Fuzztone" on page 364 is also an effective pre-amp for most Scratchboxes. Or you can build a basic 9-volt powered preamp into your Scratchbox; the "Basic Transistor-Based Preamp" on page 364 is a perfect fit.

<table>
<tr><td>

6

</td><td>

THE DROID VOICEBOX

</td></tr>
</table>

The Droid Voicebox is a descendant of the venerable talk box guitar effect. This is an effect made famous by Peter Frampton, who used an early model to make his guitar "talk" in his signature hits "Do You Feel Like We Do?" and "Show Me the Way." (If you've tragically avoided these seminal Frampton hits, please take a moment to swing by YouTube and educate yourself.[1])

In a commercial talk box effect, a specialized speaker called a *compression driver* pumps the sound from the guitar into a tube that the musician holds in his or her mouth (I know this sounds nuts; please stick with me).

---

1. Peter Frampton is the most egregious user of the guitar talk box, so the talk box has become synonymous with his 1970s good-time stoner brand of pop. But the effect is fairly widely used—often to more subtle effect. My favorite subtle applications include the lead guitar riff in Steely Dan's "Haitian Divorce" and the opening of Bon Jovi's "Livin' on a Prayer."

The guitarist is then able to use his or her mouth to sculpt the guitar riffs into flutters, wows, trills, and even fairly intelligible speech, which is picked up by a plain old vocal mic and then amplified.

The guitarist's talk box uses guitar licks as its sound source. We'll build a small square-wave oscillator to generate those sound waves and a homebrew compression driver to get them to our mouths. The result is a very inexpensive, quick-build "robot voice" effect. You'll sound sorta like Leia masquerading as the Ubese bounty hunter Boushh at the beginning of *Return of the Jedi*. Many folks initially mistake this effect for a much more expensive, trickier-to-build vocoder device.

In terms of build complexity, this project is designed with beginners in mind—especially those who've never worked with an *integrated circuit (IC)* before. The instructions assume no prior knowledge, and the components themselves are fairly robust, likely to weather the fumbles of folks new to soldering. These components are also all extremely common: any hobby shop that stocks electronic components will have the ones listed here, and the funnel, tubing, glue, and so on can be found at any hardware store. Expect the total cost to be less than $20. If you order your electronic components online, or scrounge items like speakers and switches from broken toys and home electronics, you can substantially reduce that price tag. The Voicebox is also great for getting kids pumped about electronics because it (1) makes you sound like a robot and (2) demands you place foreign objects in your mouth and act goofy.

The finished Voicebox is shown in Figure 6-1. Wanna hear it before you build it? Check out samples at *http://www.nostarch.com/jamband/*.

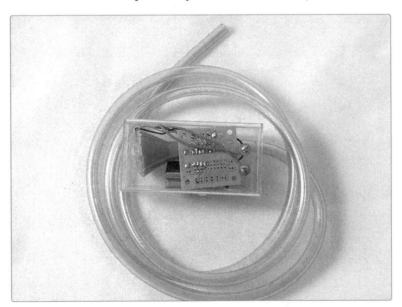

FIGURE 6-1: *The finished Droid Voicebox*

Preparation

Build Time

▶ About an hour, plus drying time

Tools

▶ A standard soldering kit (See page 340.)

▶ A saw (You'll be cutting a plastic funnel, so basically any old hacksaw, wood saw, keyhole saw, or jigsaw will do.)

▶ An electric drill with bits (You'll need 3/8-inch, 1/4-inch, and 1/8-inch bits.)

▶ A fine-point Sharpie or other permanent marker

▶ A small screwdriver, either flathead or Phillips—whichever matches your 4-40 machine screws

▶ Needle-nose pliers

Supplies

▶ An LM386 op-amp IC, such as Digi-Key part #LM386N-1/NOPB-ND (See "On the LM386" on page 74 for more information about selecting an op-amp IC.)

▶ An 8-pin IC socket

▶ A red LED

▶ A 470 ohm resistor (yellow-violet-brown stripes)[2]

▶ A 10k ohm resistor (brown-black-orange stripes)

▶ A 1k ohm resistor (brown-black-red stripes)

▶ A 100 μF electrolytic capacitor

▶ A 0.1 μF capacitor (marked *104*)

▶ A general-purpose IC PC Board (These instructions show RadioShack part #276-159. Some alternatives are listed in "Use a Different PCB" on page 93; if you choose the Adafruit 1/4-Sized Perma-Proto PCB, read "Building on Another Generic PCB" on page 239 before getting started.)

▶ A small 8 ohm speaker (I used RadioShack part #273-0092, which is roughly 1 1/8 inches in diameter, but you can also scrounge perfectly suitable speakers of this type from many toys.)

▶ An SPST pushbutton switch, such as Mouser part #103-1012-EVX or a non-illuminated doorbell button

▶ A 9-volt battery

▶ A 9-volt battery clip

▶ 24-gauge insulated wire (Either stranded or solid core wire is fine.)

---

2. Resistor codes are discussed more fully in "Resistors: Fixed and Variable" on page 325.

- Two 1/4-inch-long 4-40 machine screws (Be sure they're long enough to go all the way through your case's wall; you might need 1/2-inch screws for a beefier enclosure.)

- Two 4-40 nuts

- Two 4-40 washers

- A plastic funnel (I used a 4-ounce funnel with a 3 1/8-inch inside diameter mouth and a 3/8-inch inside diameter throat, but there's room to improvise. Just make sure the mouth of your funnel can accommodate your little speaker and the throat will snugly hold your plastic tubing.)

- Several feet of flexible plastic tubing (I used about 4 feet of tubing with an outer diameter of 3/8 inch.)

- A small, sturdy, nonmetallic enclosure (You can buy a suitable project box at many hobby stores or use something you find around the house or garage. See "On Enclosures" on page 214 for more information.)

- Silicone-based household glue (This is also called *room-temperature vulcanizing rubber*, or *RTV-1*; the 1 means that it's in a single tube, instead of two tubes that you have to mix together.)

- A clean sheet of paper

- (Optional) 22- or 24-gauge bare bus wire (This is uninsulated solid core wire. Because you need only three little snips of bus wire for this project, you can get away with using a scrap of wire, a leftover bit snipped from a component lead, or even a piece of a paper clip.)

On the LM386

There are several versions of the LM386 with varying output power, differentiated by a suffix like *N-1* or *N-3* after the *LM386*. Brick-and-mortar hobby electronics stores tend not to differentiate these well, with the several variants all floating around together in the same bin for the same price. You'll typically find the LM386N-1 (Digi-Key part #LM386N-1/NOPB-ND), which drives an 8 ohm speaker at around 0.325 watts. But it's not unusual to find an LM386N-3 (roughly twice the output power; 0.7 watts at 8 ohms) or LM386N-4 (which puts out 1 watt at 32 ohms—that can amount to 4 watts at 8 ohms!). I've tested the N-1 and N-3, which operate within the same supply voltage range, and both work fine here. Droid Voiceboxes built around the LM386N-3 tend to have a slightly lower base pitch but are otherwise indistinguishable. I haven't tested this design with an LM386N-4, but it should work fine. The LM386N-4 also tolerates a slightly wider range of voltages, which can be handy in amplifier projects. Given the option, I'll grab an LM386N-3 or LM386N-4 when I can.

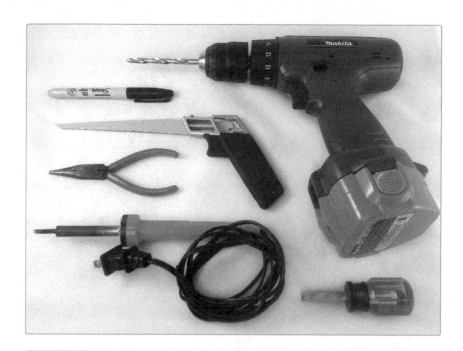

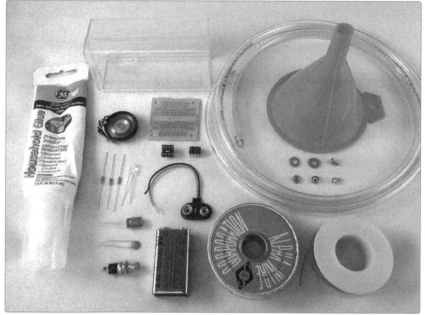

FIGURE 6-2: *Tools and supplies (not shown: sheet of paper)*

Building the Droid Voicebox

Step 1   First, familiarize yourself with the LM386 op-amp IC. The number one stumbling block in your first IC project is accidentally orienting the IC incorrectly in Step 1. You'll get to about Step 13 before you realize what you've done, and then the cursing starts. So, at the risk of being tedious, let's start slow.

Take out your LM386 op-amp IC (some folks generically call these *computer chips*, which is mostly inaccurate in this case) and place it on your clean sheet of paper. One end of the chip will be marked with a gray stripe, a little half-circle divot, a little circle near one corner, or all three. This can vary by brand, fabrication facility, or batch. Position the chip so that this marking is to the left; you should also be able to read any writing on the chip.

Next, number the IC's legs counterclockwise, beginning with the lower-left leg, as shown in Figure 6-3 (left). Note that in this orientation, the IC corresponds to the IC illustrated in the circuit diagram (see Figure 6-4). This is also how you view the IC if you're looking at a finished circuit but not how you look at the IC during many parts of the build, especially when soldering. Flip over the IC so that the notch that was on the left is now on the right and all the pins are sticking up like legs (i.e., dead-bug style). Now renumber the legs, going clockwise from the lower right, as shown in Figure 6-3 (right). This dead-bug view is how you see the IC while soldering and what you have to keep in mind throughout the build. You might want to label the second set of numbers *dead bug orientation* and keep it as a reminder on your workbench. Remember: you only solder to the dead bug.

FIGURE 6-3: *An LM386 with its legs numbered, in live-bug position (left) and dead-bug position (right).*

Step 2   It's time to look at our schematic (shown in Figure 6-4). If this is your first time seeing a circuit diagram, don't fret. The thing to keep in mind is that a circuit diagram is a lot like a subway or bus route map: it doesn't show you how things are actually laid out in the physical world but instead shows you how the parts functionally connect to each other; in other words, it highlights the *scheme* that makes the circuit work. Figure A-28 on page 353 shows which component each symbol denotes.

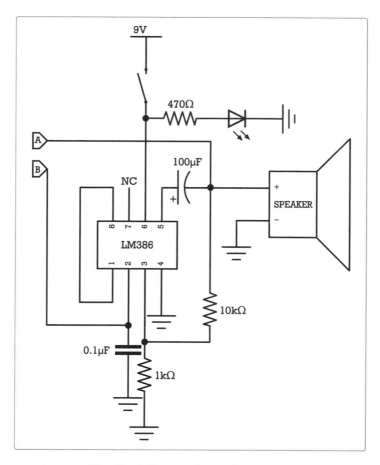

FIGURE 6-4: *The circuit diagram for the Droid Voicebox*

Connections in Circuit Diagrams

Throughout this book, I've done my best to make the circuit diagrams as clear as possible, which means, among other things, that I've avoided having lines cross unnecessarily. For the sake of clarity, I've followed the convention that crossed lines with a dot at the junction represent actual electrical connections. For example, see the right edge of Figure 6-4, where the speaker, a capacitor, a resistor, and a lead all come together. There's a dot at the junction because those elements are all wired together. Lines that simply touch without a dot— like the line connecting the body contact labeled *A* to the speaker, which crosses the line going from pin 6 to the power supply and switch—don't represent a connection.

Step 3   Let's start by mounting the IC socket on the *printed circuit board (PCB)*. The top of the PCB is plain, and the bottom has lots of shiny copper pads and traces. *Pads* are copper ovals and squares with one or more holes through their middles where you can solder the legs of components, and *traces* are the lines connecting various pads. Viewing the board from the top—that is, the side with *no* shiny copper pads or traces—slide the socket into the set of holes all the way to the left (see the left-hand image in Figure 6-5).

These sockets are technically symmetrical, but most models will have a little notch that corresponds to the stripe/notch/divot end of the IC. If your IC socket has a notch, then put that notch to the left. Holding the IC socket in place with one finger, flip the PCB over. You can bend one or more of the legs out to hold the PCB in place if it doesn't want to stay on its own. Set the PCB down and solder each leg to its solder pad.

The LM386 is a pretty hardy little guy, and it's highly likely that even a soldering newbie can join components directly to the IC without frying it. Nonetheless, it's good form to always socket your ICs for two reasons. First, plenty of ICs are easy to damage but costly to replace. Second, freakish things, like static electricity zapping an IC, do happen. The IC is basically the only component in this circuit that's liable to get damaged, so being able to easily swap it out might be quite a blessing someday.

FIGURE 6-5: *The socket mounted on the PCB and soldered down, as viewed from the top (left) and bottom (right)*

Step 4   Next, we'll solder the jumper wires, which are little snips of bus wire connecting various points of the PCB. We'll start by connecting pins 1 and 8 on the IC socket. These are the top and bottom pins farthest to the left when viewed from the top. You'll be soldering from the bottom, however, so this first jumper is actually connecting the solder points farthest to the *right* in the right-hand image in Figure 6-6.

FIGURE 6-6: *Running the jumpers, as seen from the top and bottom. The solder joints added in this step are circled in the right-hand image. The PCB will quickly become crowded with solder points, so we'll use a clean diagram to show the solder points henceforth.*

You'll immediately see why I love this particular PCB: it makes it *really* easy for even the fattest fingers to connect lots of little components together without creating a snarl of short circuits. To connect pin 1 to pin 8, you'll need to snip a 1-inch length of bus wire (or a little chunk of a metal paper clip). Bend the wire into a flat-bottomed *U* and run it vertically between the two middle holes closest to the left edge of the PCB (as viewed from the top). This wire is the single vertical jumper shown in Figure 6-6, along the left edge of the PCB.

At this stage, we're also going to run jumpers for our *common ground*, which is the reference point and return path for electricity in the circuit. (For the purposes of all the circuits in this book, *ground* means the negative battery terminal.) Snip two more lengths of bare bus wire, each 1/2 inch long, and bend them into *U* shapes. Slip one into the open PCB holes at the tops of the two columns below and just to the right of the socket—that is, the fifth and sixth columns of holes on the lower half of the PCB. Run the other between the first open hole below pin 4 of the IC and the hole immediately to its right. This should be very clear once you see it; check out the overhead view of the PCB on the left side of Figure 6-6.

Flip the PCB over and solder these jumpers into place. (Bending the legs of the jumpers out prior to flipping will help hold them in place.)

Step 5   Now we'll prepare the two leads that will run to your Droid Voicebox's body contacts. Cut two 6-inch lengths of insulated wire, strip about 1/4 inch of insulation from one end of each, and tin these two bare ends. (If you're foggy on how to do any of this, flip to "Skills" on page 346.) Set aside one lead and then slide the tinned end of the other into the first open PCB hole below pin 2 on the IC (see Figure 6-7). Solder this wire in place.

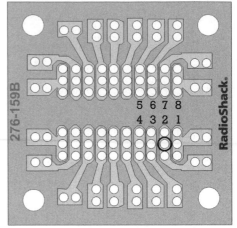

FIGURE 6-7: *The first body-contact lead connects to pin 2. The image on the right shows the underside of the board, indicating the hole and solder pad for this connection; the numbers correspond to the pins on the IC.*

Step 6   Our first capacitor is a little 0.1 µF ceramic disc capacitor, labeled *104*. This sets the overall pitch range for the Droid Voicebox. Run one leg to the next open PCB hole below pin 2 on the IC and the other leg to the first open hole on the ground we designated in Step 4 (see Figure 6-8). Just as with the socket and jumpers, bending out the capacitor's leads after you mount it on the PCB will make it easier to keep it in place when you flip the board[3]—which you'll do now. Solder this cap at pin 2 (we'll solder all the grounds at once in Step 13).

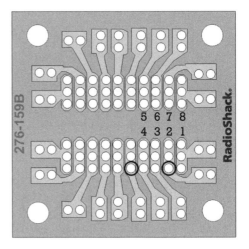

FIGURE 6-8: *The 0.1 µF capacitor is ready to solder. The circles on the diagram (right) show the location of the capacitor legs; for now, you'll solder only the capacitor leg that lines up with pin 2.*

---

3. This practice applies for pretty much any electronic component.

Step 7   Next, we'll install the other capacitor (that little blue barrel), which is a 100 µF electrolytic capacitor. Unlike the ceramic disk, this cap is polarized—the negative leg is marked with a black stripe. Run the positive (nonstriped) leg to pin 5 of the socket. Pin 5 is the IC's output, which is the upper-right pin when you view the chip from the top.

The capacitor's negative leg (the one with the racing stripe) goes to the top hole of the seventh column of PCB holes—that's the currently unused column to the right of the two ground columns we established in Step 4 (check out Figure 6-9 for clarification). Solder the capacitor leg connected to pin 5 of the IC but leave the other alone for now.

You'll note on the circuit diagram (Figure 6-4) that things get a little complicated between the IC's pin 5 output and the actual speaker. That's why we're running the negative leg of the cap over to its own column of holes—so we have plenty of room to wire up those connections.

 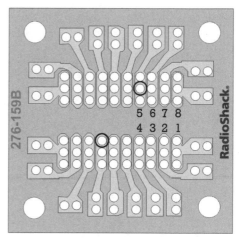

FIGURE 6-9: *Placement of the 100 µF electrolytic capacitor*

Step 8   Now we'll wire the IC's two resistors, which both connect to pin 3. The first is a 10k ohm resistor, coded with brown-black-orange stripes. (If you need details on reading resistor codes, flip to "Resistors: Fixed and Variable" on page 325.) This resistor runs from pin 3 of the IC to the negative leg of the 100 µF capacitor, but we're going to connect it in a tricky way to avoid short circuits and troubleshooting headaches.

See the parallel double row of holes along the bottom edge of the PCB? Run this 10k ohm resistor from the first hole to the last hole along the top row. The other resistor is 1k ohm (coded brown-black-red) and runs along the lower row, from the first hole to the second-to-last hole—that is, it runs from pin 3 to the ground we established in Step 4. Check Figure 6-10 for clarification if any of this is confusing (the 10k ohm resistor is above the 1k ohm resistor). Once the resistors are positioned, solder the two leads that connect to IC pin 3 but leave the other two alone; we'll solder the rest in later steps.

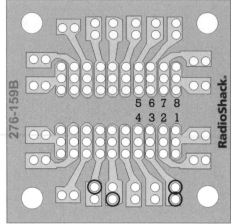

FIGURE 6-10: *Placement for pin 3's two resistors; solder the two leads that share a copper pad now (that's the two circles all the way to the right in the underside-view illustration).*

Step 9   Now it's time to finish off wiring up the output at column 7 on the PCB. Insert the other tinned body contact lead (prepared in Step 5) into the first open hole along column 7. (There should be a space right below the negative leg of the 100 μF electrolytic cap.) Then, take the positive lead from your speaker,[4] strip away about 1/4 inch of insulation, tin the bare wire, and insert that into the next open hole. You can see all of this in position in Figure 6-11.

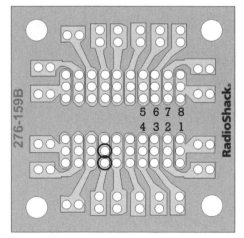

FIGURE 6-11: *The output components are all in place. The underside diagram (right) highlights the placement of the second body contact and the positive speaker wire.*

---

4. On many speakers, the two leads are basically interchangeable. If yours are differentiated (by either being different colors or being marked by a + or − sign), then you might as well honor that. The red lead is positive, as is the + sign.

Carefully flip the PCB over and solder all four wires at column 7: the negative electrolytic capacitor lead, the body contact, the positive speaker lead, and the other leg of the 10k ohm resistor.

Step 10   The LM386 op-amp takes power on pin 6, which we'll wire now. Technically, all you really have to do is insert the red lead from your battery clip into any hole in the column connected to pin 6 and then solder it down. However, if you do that, you'll have to pull the battery out when you aren't playing with the Voicebox, and you won't be able to get some transform effects.[5] That's why we're getting a little more complicated and adding both a switch and an indicator light.

Start by shoving the PCB out of the way and digging out the 470 ohm resistor (coded yellow-purple-brown) and the LED. Solder the resistor to the positive leg of the LED (that's the one lead that's either a little longer, positioned opposite the flat-edge side of the LED, or—most likely—both). Figure 6-12 shows the prepared LED indicator light.

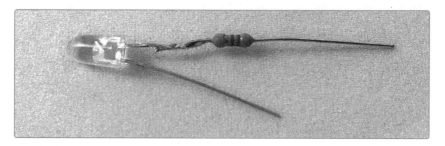

FIGURE 6-12: *Our indicator LED assembly*

* **NOTE:** *It's important to use a red or yellow LED here because we've chosen a resistor appropriate to the limitations of that color LED. An improperly buffered LED powered by a 9-volt battery can get really frickin' hot and even melt down. If you really want a different color here, that's totally doable, but flip to "The Gory Details: Voltage, Current, Resistance, and Ohm's Law" on page 330 and "Finding the Right Resistance" on page 331 because you might need a different resistor.*

Step 11   Because I've chosen a transparent case, I'm going to mount my indicator light right onto the PCB. This should also work fine for many other enclosures; you'll just end up mounting the PCB with the LED peeking out through a 1/4- or 3/16-inch hole in the case. Look at your case now and decide whether you need to add a couple lengths of insulated wire in order to place your LED where you want it. Either way, the resistor side of the indicator light assembly will go to the upper-left PCB hole, as shown in the right-hand image of Figure 6-13. The other, which is the LED's negative lead, goes to the first open hole in the ground connection. (For example, I had room at the second hole in column 6.) Solder the lead that's connected to pin 6—the one in that upper-left hole—and leave the ground lead to be soldered later.

---

5. For more on transforms and other neat little gating tricks, see "Playing the Scratchbox" on page 67.

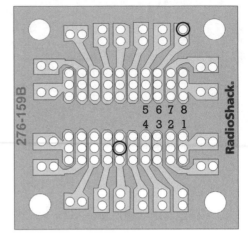

FIGURE 6-13: *Placement for the indicator LED assembly*

Step 12   Now for power! Take out your pushbutton SPST switch and the 9-volt battery clip. If you haven't already stripped and tinned the ends of the battery clip leads, do so now, removing about 1/4-inch of insulation. Then solder the red lead to one of the two terminals on your SPST switch. Now, cut a length of wire (probably less than 4 inches long, depending on your case), strip 1/4-inch of insulation from either end, tin both ends, and solder one end to the other switch terminal. The remaining end of this wire runs to the first open hole in the column connecting to pin 6 of the IC socket, as shown in Figure 6-14. Remember that, viewed from above, IC pins are numbered counterclockwise starting from the lower left (check Figure 6-3 for clarification)—so pin 6 is the third pin from the left on the IC's top row as viewed from the top.

Now, looking down at the circuit from the component side of the board, this wiring looks kinda crazy: we know from the circuit diagram that the LED's resistor-buffered lead goes straight to the power supply, but it's pretty obvious in Figure 6-14 that the LED's resistor and the power lead don't line up. But flip the PCB over, and you'll see how the copper trace for pin 6 snakes around and connects with that top corner hole in the PCB. Again, this is why I love this PCB for simple one-IC projects: it gives you plenty of space to work.

Now that we've got that clarified, feel free to solder that power connection to the PCB.

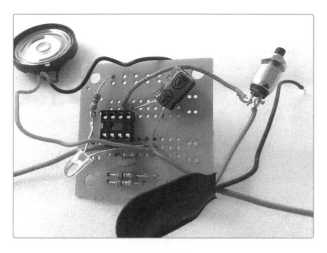
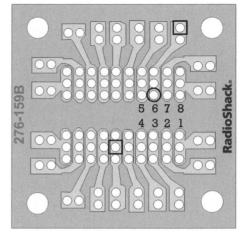

FIGURE 6-14: *We have power! The circle on the underside diagram (right) shows the power connection; the squares indicate the LED's connections.*

Step 13   Now we'll finish the circuit by soldering the common ground from Step 4, which occupies columns 5 and 6 of the PCB. There should be three loose leads there right now: one from the ceramic disk capacitor connected to pin 2, one from the 1k ohm resistor connected to pin 3, and one for the negative leg of the LED. Add the negative (black) lead from the battery clip and the loose lead from the speaker, which will share a hole with the 1k ohm resistor. Solder them all down. The completed circuit is shown in Figure 6-15.

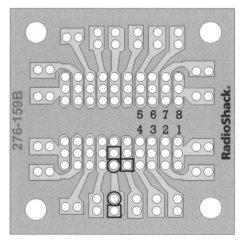

FIGURE 6-15: *The completed circuit (note that the IC hasn't yet been inserted). In the underside diagram (right), the squares indicate the three leads that have been waiting since prior steps to get soldered, while the circles show the two leads you just added.*

Step 14   Let's take this circuit for a test drive. Insert the LM386 op-amp IC into its socket, making sure that the stripe/divot/dot is to the left. Connect a fresh 9-volt battery to the battery clip and then strip about 1/2-inch of insulation from the loose ends of the two body contact leads. Next, hold those leads between your thumb and forefinger so that they *aren't touching!* Squeeze tight and push the pushbutton switch. The LED should come on, and the speaker should buzz or groan fairly loudly (it'll probably rattle around on the table like a jumping bean). Try adjusting how tight you're squeezing; this should alter the pitch of the tone.

　　　　　If the LED lights but there's no sound, check to be sure the wires you're pinching aren't touching; your body acts as the final resistor the circuit needs to determine the pitch, and if you short the wires together, you'll likely get a pitch too high to hear. If there's no noise and no light, ask yourself, "Did I ignore the warning from Step 1 and build this whole stupid thing upside down?" If you're still having trouble, flip to "General Troubleshooting" on page 355.

Step 15   Once you have the noise-making circuit sorted out, it's time to add the funnel so that you'll actually be able to harness the funk of your Droid Voicebox. Start by slipping your speaker into your funnel so that the speaker's cone points straight into the throat of the funnel (in Figure 6-16, my left index finger is pressing the speaker into the funnel). Mark the location of the back of the speaker on the funnel (feel free to add 1/4 to 1/2 inch here, giving you a little extra funnel behind the speaker). For my 4-ounce funnel, this meant cutting off about 2 1/2 inches of the funnel, as measured from the rim of its wide mouth.

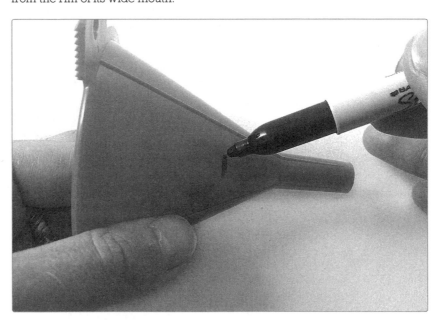

FIGURE 6-16: *Marking the funnel*

Step 16   Use your saw to cut the funnel at the line you drew in Step 15. *Be careful and go slow!* You're attempting to saw a straight slice across a cone; conic sections are advanced geometry, and this is an excellent opportunity to slip and cut yourself. If you're bothered by the rough edge left by the saw, you can smooth it out with medium-grit sandpaper (such as 100), a metal file, or a few stiff swipes along a concrete slab.

Step 17   For the Droid Voicebox to inject its racket into your mouth, you need a nice airtight seal around the speaker and funnel. Silicone-based household glue is basically the same as the clear caulk you might use around your bathtub or a leaky window. It makes a great seal and, as an added bonus, stays soft and rubbery, thus absorbing sound. This will help to minimize stray, unsculpted tones leaking from your Voicebox and distracting folks or mucking up recordings.

Place the speaker facedown into the funnel and give it a nice coating of glue all around the edge (an old pencil, chopstick, matchstick, or popsicle stick is great for applying this glue). Let the glue set up for a few minutes and then coat the entire back of the speaker with an even layer of glue for added sound insulation, as shown in Figure 6-17. Be aware that as you add glue, you may muffle the speaker, thereby decreasing overall volume even as you increase clarity. So you might want to pause, finish the enclosure, and then decide later whether you want to layer on more glue to tighten the seal.

This glue is usually tack-free (no longer sticky) in about half an hour, but it really needs to set for 24 hours to fully dry. Let it set undisturbed for at least half an hour before mounting everything in the case.

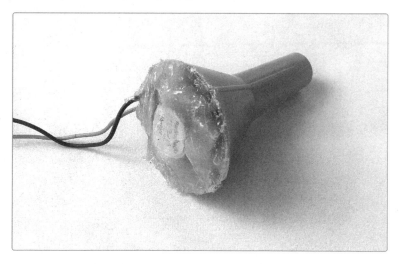

FIGURE 6-17: *The glued speaker with a pretty thick layer of insulating silicone*

Step 18   While you're waiting for the glue to dry, you can drill the holes in your case. Take out your enclosure and decide where you want to mount the two screws for the body contact leads, the pushbutton switch, and the talk tube. Mounting the body contacts and switch opposite each other on the front and back of the case makes for the easiest operation in my opinion, but you're welcome to mount these however you choose (see Figure 6-18, which shows a completed Droid Voicebox sans talk tube).

Drill two 1/8-inch holes about 1/2 inch apart, one for each body contact, and a single 1/4-inch hole for the pushbutton switch. You'll almost certainly want the talk tube to come out of the top of the unit because it needs room to connect to the funnel. If you've used the 3/8-inch tubing listed in "Supplies" on page 73, then drill a 3/8-inch hole in the top of the case for the tube (you can see the tube and how it connects to the funnel in Figure 6-20).

FIGURE 6-18: *Note the opposing orientations of the body contacts (the two screws on the left face of the case) and the power switch (on the right face). You can also see the larger hole at the end of the case, which is for the talk tube.*

Step 19   Now it's time to mount everything in the case. Slip the 4-40 machine screws through the 1/8-inch holes from the outside, place a washer on each screw, wrap a lead around each screw, and then tighten the nut down, as shown in Figure 6-19. The easiest way to tighten these screws is to hold the nut with your needle-nose pliers and use a screwdriver to turn the screw from the outside.

Install the pushbutton switch in the 1/4-inch hole. Carefully feed the talk tube through the big hole (for best operation this should be a fairly tight fit; if you've got lots of wiggle room, you'll probably want to seal around the edge of the tube using more silicone-based glue). If your speaker-funnel assembly is tack-free, then it's safe to carefully slide the tube into the funnel. This should be a really, really tight fit; if it isn't, then you *definitely* want to use some glue to caulk around the edges.

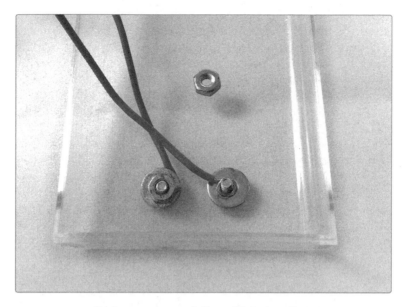

FIGURE 6-19: *Body contacts partially and fully installed, as viewed from the inside of the case. Note that the washer is on the inside of the case.*

Step 20   Once all of the glue is dry, slide the electronics into the case and close her up (see Figure 6-20). You are ready to funk out!

FIGURE 6-20: *The finished Droid Voicebox from a different angle, showing off our hacked-together compression driver*

Playing the Droid Voicebox

As you probably already noticed while testing the circuit, pushing the button activates the Droid Voicebox's square-wave oscillator, but you can only hear a pitch if you've also bridged the two body contacts with your fingers. This is because your body serves as the final resistor in the circuit. A light touch conducts poorly between the contacts, which translates to higher resistance and thus a lower pitch (perhaps so low a frequency that it sounds more like individual clicks than an actual continuous buzz). Pressing harder makes a better connection, which means lower resistance and thus a higher pitch.

To play the Droid Voicebox, place the tube in your mouth, lay a finger across those body contacts, and push the button to get a pitch. Start moving your lips, tongue, and jaw in order to "sculpt" that sound (very similar to the mouthings used when playing a harmonica, trumpet, or Jew's harp). Don't neglect the tube orientation, as this can have a profound impact on the sound.

In my experience, the most intelligible talking effect is achieved by laying the tube along the roof of the mouth (so it doesn't get in the way of the tongue) and closing off the larynx and nasal passage (in other words, hold your breath). Now get a buzz going and mouth some words. Start out with something that's largely defined by lip and jaw movements—such as "Bow-wow-wow yippie-yo yippie-yay!"[6]—and then work your way up to phrases that require more complicated tongue maneuvers. Words and phonemes that rely on back-of-mouth intonations will be the most challenging to achieve.

Using a microphone? Just sing into the mic as you normally would, but with the tube stuck in your mouth—which, yes, will feel awkward at first. By design, talk boxes (and this Voicebox) work excellently with all sorts of vocal mics, both for recording and in live performances. The long tube allows you to get great separation between the mic and the synth, massively reducing bleed-over. This unleashes your inner robot voice without subjecting your listeners to an annoying perpetual buzz.

Tips, Tricks, and Mods

The 0.1 μF ceramic disk capacitor from Step 6 sets the overall pitch range for the Droid Voicebox, with the pitch going down as the value of that capacitor goes up. In my experience, a lower pitch gives a more intelligible talking effect, although a higher pitch range means more energetic sound waves and thus a higher apparent volume. But swapping the capacitor isn't the only great mod you can try.

----

6. With apologies and appropriate credit to George Clinton, the Parliament-Funkadelic Collective, Bridgeport Music Inc., Snoop Dogg, and any of the hundreds of other artists and corporate entities that might make substantive claims to the use of this phrase in robo-funk contexts of any sort.

Tweak the Body Contacts

Body contacts can be a little tricky because skin resistance varies a great deal by individual. It can be further complicated if you live in sweaty climes, have lotion on your hands, and so on. As a rule, a finger will give you between 3M ohms and a few hundred ohms of resistance, with younger skin putting up less resistance and older (and especially callused) hands a good deal more. If you're having trouble with responsiveness from the body contacts, consider bridging them with two or three 1M ohm resistors wired in *series* (that is, in a row, as in Figure 6-21; FYI, 1M ohm resistors have brown-black-green stripes). This puts the circuit right at (or just over) the threshold for producing an audible tone.

FIGURE 6-21: *A set of bridging resistors connect the body contacts, making it easier for my callused old hands to play the Droid Voicebox (the resistors are mounted inside the transparent case, with their leads sandwiched between the washers and nuts).*

Resistor Math: Series vs. Parallel

When resistors are connected in series—as in Figure 6-22—the same current passes through all of them. In order to calculate the *total equivalent resistance* for a set of resistors wired in series, you just add up the resistance of each individual resistor:

$$R_T = R_1 + R_2 + R_3 + R_4 + \ldots$$

So, for example, by connecting two 1M ohm resistors, like the ones in Figure 6-22, we end up with a total of 2M ohms of resistance bridging the body contacts.

FIGURE 6-22: *Series resistors (and the circuit diagram representing the same)*

Resistors connected in *parallel*—like those in Figure 6-23—split the current. Their total equivalent resistance is calculated by adding together the *reciprocals* of each individual component's resistance:

$$\frac{1}{R_T} = \frac{1}{R_1} + \frac{1}{R_2} + \frac{1}{R_3} + \frac{1}{R_4} + \ldots$$

If my thumb offers 1M ohms of resistance, then when I set it across the modded body contacts, it's in parallel with the 2M ohms' worth of resistors I have bridging the body contacts shown in Figure 6-21. Subsequently, you can find the total equivalent resistance from this expression:[*]

$$\frac{1}{R_T} = \frac{1}{R_1} + \frac{1}{R_2} = \frac{1}{1} + \frac{1}{2} = \frac{2}{2} + \frac{1}{2} = \frac{3}{2}$$

That result is the *reciprocal* of the total resistance, so flip it over and you have 2/3 of 1M ohm, which is something like 666k ohm—not a standard resistor you can just run out and by at your local hobby store. This highlights the value of polishing up on your resistor math: it makes it possible to fake together resistances that either you don't have handy or can't actually get from a single store-bought resistor.

FIGURE 6-23: *Parallel resistors*

---

[*] If you're just running two resistors in parallel, there's a shortcut equation: $R_T = \dfrac{R_1 \times R_2}{R_1 + R_2}$.

Now when you lay your finger across the contacts, your body and the bridging resistors form a *parallel resistor network*. As a rule, the equivalent total resistance of such a network is always *less* than the resistance of the *smallest* resistor in the network, so even a light contact from a high-resistance thumb will be able to nudge the total resistance low enough for the circuit to start producing audible tones.

Alternatively, you can replace the body contacts altogether and instead use a 1M ohm (or higher) potentiometer (such as Mouser part #313-1000F-1M). Using a variable resistor in place of the body contact allows sure-fire pitches that are easy to dial in.

Use a Different PCB

Although I love the RadioShack General-Purpose IC PC Board (good ole part #276-159) for single-chip circuits like this one, it might not be the best choice, for one reason: RadioShack has significantly shifted its business model, and no longer has local stores in many smaller communities—which makes this little board less convenient to source. As noted at the beginning of this project, you have several alternatives to this RadioShack-specific PCB. First, you could purchase a clone: an increasing number of small suppliers have begun offering identical workalikes. Search online for *IC PCB Breadboard*, *experimenters breadboard 21-4595*, or *prototyping board 21-4595* for examples. As of this writing, MCM Electronics (*http://www.mcmelectronics.com/*) and SMC Electronics (*http://www.smcelectronics.com/*) consistently stock boards very similar to RadioShack's part #276-159.

Alternatively, if you're a slightly more experienced builder, you can use the schematic in Figure 6-4 as a guide and lay this circuit out on any generic pad-per-hole PCB. (Skim Project 11, the Mud-n-Sizzle Preamp, if you'd like to get a feel for working with a pad-per-hole circuit board.) Any of these generic grid PCBs will work fine: SparkFun PRT-08811, RadioShack's #276-148, Jameco part #105100, or Digi-Key part #V2025-ND. You could also get something larger, like Jameco part #2191445 or Mouser part #574-8015-1, and cut out a 2-inch-square piece to use for your Voicebox circuit.

Finally, I've prepared an alternative layout that will work on any breadboard-style PCB. These are designed to mimic the layout of a prototyping solderless breadboard (described in "Building a Circuit" on page 352), which makes it *very* easy to transfer an experimental test circuit to a permanent PCB.

I've used the Adafruit 1/4 Sized Perma-Proto PCB shown in Figure 6-24. It's a great board: reasonably priced, very well made, and very widely available online (you'll find a full list of sellers in "Guidelines for Sourcing Components" on page 323). For years, a slew of manufacturers have made comparable, if somewhat larger and clunkier, permanent breadboard-style PCBs (even RadioShack has one, part #276-170), so even if Adafruit stops making its 1/4 Sized Perma-Proto PCB, you'll still be able to use the layout shown in Figure 6-24.

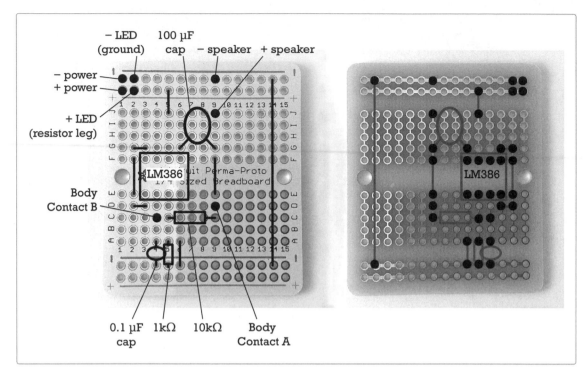

FIGURE 6-24: *Top (left) and bottom (right) view of an alternative perma-proto breadboard layout for the Droid Voicebox. Unlabelled straight lines between solder pads are jumpers. I've highlighted the solder points on the right-hand image.*

To build the circuit using this alternative layout, just go through the numbered steps listed in "Building the Droid Voicebox" starting on page 76, but use Figure 6-24 as your guide for placing the components on the PCB.

7 CIRCUIT BENDING FOR BEGINNERS

The Slinkiphone (Project 1) and Scratchbox (Project 5) hint at a fundamental truth of the modern age: there are lots of really neat sounds hiding inside your toys and consumer electronics. Cracking open cheap electronics and feasting on the sonic goo within is called *circuit bending*. This sort of exploration was pioneered by artist Reed Ghazala, the "Father of Circuit Bending," who stumbled across his first bends as a teen in the mid-1960s. Ghazala has tirelessly promoted the creative potential of musically upcycling consumer electronics ever since. In this project, you'll get your feet wet with three entry-level bends.

Thrift stores, garage sales, and remainder bins are full of electronic toys ripe for exploration. When choosing a toy to tweak, use only battery-powered toys, and keep an eye peeled for older toys. Older toys are usually built with largely generic components and offer more circuit-bending options than newer toys, which tend to offload almost all of their functions to purpose-built microprocessors. That said, bending new toys can still be quite fruitful. For example, most pip-squeak electronic toys—like those shown in Figure 7-1—can generate big, rich sounds with impressive bass, but their puny built-in speakers just aren't up to the task. A little bending can unleash all that good noise.

FIGURE 7-1: *A selection of circuit-bent instruments and toys. Note the black box on the keyboard and the new bits on the other three: the knobs, switches, and bare metal contacts allow us to tinker with the toys' pitch and tone on the fly, while the jacks allow us to amplify the output (and, in the case of the megaphone, pump our own signal into the input).*

✳ **WARNING:** *Only bend battery-powered electronics! Household batteries are generally safe to work with: they're small and supply relatively low voltages and currents. For example, a 9-volt battery will reliably supply under 100 mA and shouldn't hurt you (unless you try to swallow it, I suppose). In contrast, the AC electricity available from your wall outlets is dangerous, offering dozens of times as much voltage and over 100 times the current—120 volts and 10 to 15 amps. Wall current can easily burn your house down and will certainly kill you given the opportunity. Nothing in this book is intended to ever encourage you to work on any electrical device that plugs into the wall!*

Hear my circuit-bent keyboard in action in the samples at *http://www .nostarch.com/jamband/*.

Preparation

Build Time

▶ About 30 minutes to an hour, depending on the toy and amount of modification

Tools

▶ A standard soldering kit (See page 340.)

▶ A selection of screwdrivers (If you don't already have them, I suggest a set of *jeweler's screwdrivers*, which usually includes #0 and #1 Phillips and 1.4 mm, 2 mm, 2.4 mm, and 3 mm flathead drivers.)

▶ A small flathead screwdriver for prying open plastic cases

▶ Several sets of jumper clips (You can use *insulated test leads*, which look like tiny jumper cables. Pricier *mini-clip hook-style jumper wires* are even better.)

▶ An electric drill with bits (This is for modifying the toy's case to accommodate new jacks and switches.)

Supplies

▶ A battery-operated toy or instrument you'd like to vivisect

▶ Two *normally on* momentary pushbutton switches, also called *normally closed SPST* switches (For more on switches, see "Switches" on page 338.)

▶ A 1M ohm variable resistor (Variable resistors are also called *potentiometers* or *pots*; see "The Gory Details: Audio Taper vs. Linear Taper" on page 327.)

▶ A fistful of other resistors, potentiometers, switches, brass screws, metal knobs, and so on

▶ A 1/4-inch mono phone jack, also called a *guitar jack*

▶ 24-gauge insulated hook-up wire (22-gauge speaker wire is fine, too. It's stranded like hook-up wire, only slightly thicker.)

▶ Small boxes to accommodate the extra jacks, switches, and pots you'll add to instruments (Any thin-walled plastic box will do; hobby shops often carry a variety of "enclosures" and "project boxes," like the one attached to the keyboard in Figure 7-1. See "On Enclosures" on page 214 for more information.)

▶ Two alligator clips (The ones shown in Figure 7-12, often called *mini-hook clips*, are much easier to use when clipping into commercially produced circuits. Those pictured are American-made "E-Z-MINI-HOOK TEST CONNECTORS" from E-Z-HOOK, Digi-Key parts #461-1013-ND and #461-1014-ND.)

▶ 22- or 24-gauge bare bus wire (This is uninsulated solid core wire. Because you need only a few little snips of bus wire for this project, you can get away with using a scrap of wire, a leftover bit snipped from a component lead, or even a piece of a paper clip.)

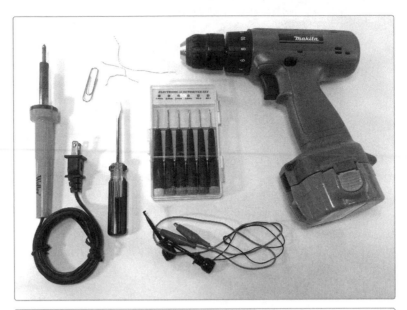

FIGURE 7-2: *Tools and supplies*

Three Basic Circuit Bends

Once you have a few toys picked out, we'll modify the output by adding a 1/4-inch jack and optional momentary mute button, use resistor-based bends to safely monkey with the circuit's internal clock, and add a power reset to set the stage for advanced circuit exploration. We'll finish up with a brief discussion of how to package a finished circuit-bent project.

Most toys, especially newer ones, have little spare space inside to accommodate new jacks, switches, and variable resistors, so the most important preparation is to consider how to package all that when you're done adding new functions. Final packaging can be tricky, but it's far from impossible; think of it as another outlet for creativity. Also, keep in mind that there's nothing stopping you from adding off-board expansion modules or entirely repackaging the toy's electronics in a larger case.

While you're pondering your packaging options, pull the batteries out of your toy and remove all the screws you can find. Remember to look for screws hiding under stickers and in the bottom of the battery compartment. Then, pry open your toy's case and determine whether you're dealing with an old or new toy. If you're unsure, check out Figure 7-3. The left panel shows a portion of the guts from a 1980s electronic toy keyboard; the right shows the entire circuit board from a newer McDonald's Happy Meal freebie toy (specifically, a 2010 *Alvin and the Chipmunks* talking Theodore).

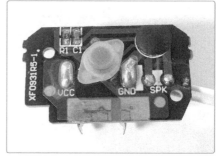

FIGURE 7-3: *The older circuit board (left) has many large, easy-to-identify components. The modern toy's circuit board (right) is much smaller (the entire board is smaller than my thumb) and has few components, which are harder to distinguish.*

The older toy is comprised of many generic, discrete components[1] that you'll recognize from building your Droid Voicebox (Project 6) or other electronics projects: resistors, capacitors, variable resistors, diodes, LEDs, and integrated circuits (ICs). The new toy, on the other hand, has few recognizable components and instead sports that mysterious black blob (in this case, it's in the upper-right corner). That blob hides a single, highly specific microprocessor that replaces most of those discrete parts. Newer toys usually rely on such black-blob chips. To add to the challenge, the few discrete components in these newer toys will usually be very small *surface-mount* components, such as resistor R1 and capacitor C1 in Figure 7-3, as opposed to the standard components you'll buy for projects in this book. Surface-mount components are harder to tell apart than standard components and, owing to their small size, are more of a pain to work with. That said, even on newer toys

1. Even integrated circuits, while specific to a given manufacturer, were fairly generic and used in multiple products. For example, the CPU in the Casio PT-10, shown in Figure 7-21 on page 111, was used across several units and lines. This is why my PT-10 has the same built-in rock beat that the German band Trio made famous with their tune "Da Da Da," even though Trio used a Casio VL-1—a keyboard noted for having a built-in calculator.

with small circuits composed of surface-mount components, changing up the output remains a straightforward operation that greatly expands the toy's audio palette.

Modify the Output

I strongly advise adding 1/4-inch jacks to every toy. I've regularly been delighted by the breadth and depth of sounds even the junkiest toy has to offer once it's properly amplified or pumped into commercial or homebrew effects. I also strongly advise using a high-quality jack: the jack is going to get jerked around a fair bit, and something nice, like a Switchcraft, will put up with that for many years. The cheapest of cheap jacks—which tend to be standard on consumer-grade electronics and even midgrade guitar amps from big-name manufacturers—can start to get loose and noisy after just a few months of regular use. Very annoying. A Switchcraft will stay sure and steady for decades.

In this section, you'll work with two kinds of mono phone jacks: normal and switch jacks. The normal mono jacks, shown at the far left of Figure 7-4, will be familiar if you've ever seen or handled a guitar, amplifier, or stomp box. The switch jacks, shown in the middle and on the right, are a little special. When nothing is plugged into a switch jack, the signal goes to the device's built-in speaker. When you plug something in, the internal speaker is disconnected, and the signal is routed instead to the external amp. If you're planning to replace the toy's original speaker altogether, use a normal mono jack. If you want to keep the original speaker and don't care that the built-in speaker makes sound even when the toy is plugged into an amp or effect, then a normal jack is still fine. But if you want the toy to automatically mute the built-in speaker when you plug into your amp or effect, then you want to install a switch jack.

FIGURE 7-4: *A standard mono jack (left), a generic switch jack (center), and a Switchcraft 112AX switch jack (right)*

Step 1   Choose a jack. For plain old jacks, I'd go with the Switchcraft L11, shown at the far left of Figure 7-4. It's Digi-Key part #SC1085-ND or Mouser part #502-L-11. (Always check both, as there can be a pretty big price difference, especially once you work in shipping and handling.) The middle jack in Figure 7-4 is a Switchcraft 12A, which is the switchable version of the standard Switchcraft mono jack. This open-style switch jack is nice because its internal mechanism is entirely visible, making it a little easier to understand if you're new to electronics. Its only drawback is that it looks an awful lot like a standard mono or stereo jack. If you have a jumbled box of mixed jacks (as I do), it's easy to miscount and get halfway through a project before realizing you don't have the jack you thought you had. For that reason, my preferred switch jack is the partially enclosed Switchcraft 112AX, shown at the far right of Figure 7-4. It's the same jack used in the Plasti-Pickup (Project 2), Digi-Key part #SC1316-ND or Mouser part #502-112AX, and it's very easy to distinguish from mono and stereo jacks.

✳ **NOTE:** *Steps 2 through 4 describe how to add a plain old jack, while Step 5 describes how to add a switch jack.*

Step 2 Decide whether you want to keep your toy's existing speaker or amputate it. If you choose to lose the existing speaker, then this mod requires only the two steps shown in Figure 7-5: snip off the speaker, solder one wire to each of the two lugs on a standard 1/4-inch jack, and you're done. The black speaker wire traditionally goes to the jack's ground lug and the red wire to its tip, but that rarely makes any difference in these situations. (If you have no idea what the *tip* and *ground* are on a mono phone jack, flip to "Quarter-Inch Phone Plugs and Jacks" on page 337, where they're illustrated. If this is all new to you, see "Soldering" on page 346.) If you've decided to keep the speaker, move on to Step 3. If you'd like to add a *momentary mute*, skip to Step 4.

FIGURE 7-5: *The unmodified toy with speaker in place (left); an amputated speaker replaced with a jack (right)*

Step 3 If you want to keep the original speaker, the quick-and-easy solution is to jumper a jack in, as shown in Figure 7-6. To accomplish this, just cut two lengths of insulated wire, strip and tin both ends of each, and solder one wire to each lug on the jack. Now, solder one wire to each of the two terminals on the speaker. On the upside, this approach is quick and easy, and you retain the ability to play anywhere while adding the ability to harness amps and effects. On the downside, even when you're plugged into an external amp, you'll still have the built-in speaker making noise, which might prove annoying, especially if you use mics to isolate your various instruments for recording. If this sounds annoying, flip to Step 5. Otherwise, you're done!

FIGURE 7-6: *A "jumpered" auxiliary output jack*

Step 4   Adding a pushbutton momentary mute allows you to quickly cut the signal to the speaker and output jack. Just snip the positive speaker wire, which is usually red and is often connected to a + sign on the circuit board, the speaker itself, or both. Then, install a normally closed—that is, normally on—pushbutton SPST switch between the circuit board and the speaker and output jack (see Figure 7-7). When you hold the button down, the output is silenced. Release the switch, and the speaker immediately comes back online.

* **NOTE:** *If the board and speaker are unmarked and the wires are otherwise identical, don't sweat it. Just choose one wire to consider positive and stick with that. "Positive" and "negative" are almost always arbitrary distinctions when it comes to small speakers in battery-powered toys.*

FIGURE 7-7: *An auxiliary output with a momentary mute*

A mute offers a lot of neat sonic possibilities, allowing you to chop and punctuate the toy's noise. For example, rhythmic momentary muting is a key to the stuttery *gate* and *transform* effects popular in electronic dance music. Even if you don't choose to add an output jack, a momentary mute is still a fun mod for any electronic noise toy.

You should be done installing your plain jack at this point, so either move on to the next mod, "Explore Resistor Bends" on page 104, or read on through Step 5 to learn how to add a switch jack.

Step 5   If you don't want the internal speaker to play at all times, use a switch jack. On the partially enclosed Switchcraft jack I've specified, the corner lug[2] is the *ground* connection, and it makes contact with the barrel, rather than the tip, of the plug. The lug farthest from the ground lug connects to the jack's *tip* and carries the audio signal. The middle lug connects to the internal switch.

To wire this switch jack, first snip the positive speaker wire in half. Solder the lead connected to the board to the tip lug of your jack. Then solder the lead connected to the speaker to the middle lug. Solder a separate insulated wire from the jack's ground lug to the other lug on your speaker, as shown in Figure 7-8. Note that this switch jack layout—with the ground and tip opposite each other and the switch connection in the middle—is likely not universal. To determine which lug is which on your jack, either check the datasheet (the site where you ordered the part should have it posted in the online product listing) or do a little guessing and checking. You can also add a momentary mute here, if you like. The procedure is exactly the same as it was with the plain old jack, outlined in Step 4.

FIGURE 7-8: *A finished switch jack output*

----

2. That's the lug mounted at an angle on the corner of the plastic housing; see Figure 7-4 for a clearer picture.

Explore Resistor Bends

Resistor bends are a good place to start because they're relatively easy to find and install and have a low likelihood of destroying a circuit.

Step 1   Now that we're getting down to the nitty-gritty of circuit bending, select a toy to explore for good resistor bends. Anything that makes a noise will do. Pop open a few candidate toys, and you'll immediately notice the difference between new and old electronic toys, as I described in the introduction of "Three Basic Circuit Bends" on page 98.

Step 2   Start by finding the resistors. With older toys, the resistors may be a touch larger or smaller than those you buy at your local hobby shop, but they'll still look a lot like what you've seen in this book so far: an oblong body (usually tan, sometimes blue or green) marked with four stripes, with wire leads protruding from either end. You may see small traditional resistors in newer electronics, too, but more often, the resistors will present themselves as flat black rectangles with a white stripe at either end; sometimes these surface-mount resistors are numerically coded, and sometimes they're just plain black. Either way, you'll almost always find their positions on the circuit board marked with an *R* followed by a number (see Figure 7-9 for an example of both old-toy and new-toy resistors).

FIGURE 7-9: *On the left, a portion of an older toy's circuit board featuring four resistors. On the right, a close-up of a newer toy's circuit board, high-lighting a surface-mount resistor next to a surface-mount capacitor. Note the* R *labels on both circuit boards, denoting the placement of resistors.*

Step 3   Once you've found some promising resistors—the ones nearest the ICs or adjacent to trim pots (see Figure 7-13) are often good ones—start poking around for fun bends. Load the batteries back into your electronic toy and power it up. Get the toy making noise and then use your fingers to bridge the leads on a single resistor.

   ✳ **WARNING:** *Poking the circuit board while it's powered is safe only if the toy is battery powered. This is* **never** *safe with anything that is, or can be, plugged into the wall!*

   I've shown several techniques in Figures 7-10 and 7-11. Some folks, especially older folks, may need to lick their fingers for this to work (older skin tends to offer more resistance than younger skin). Repeat until your poking makes something interesting happen to the sound: it might get louder, softer, faster, slower, higher pitched, or lower pitched. You might also hear rhythmic palpitations, a crazy squeal, or looping; a sample might unexpectedly play or repeat; and so on. Usually, pressing harder will make it happen more.

For smaller resistors, just one finger should work well. If the resistor is larger and your hands are small, you may need two fingers, as shown in Figure 7-11.

You've gotten a noise! Rad! But what the heck just happened? As I described in the Droid Voicebox (Project 6), your skin is a resistor. In this case, that skin resistor is being used to create a short circuit, temporarily overriding the factory-installed resistor soldered to the circuit board—that is, a *resistor bend*. If you press lightly, it will offer around 3M ohm of resistance. As you press harder, you lower the resistivity down to just a few hundred ohms. Using your skin as a variable-resistance probe is both cheap and easy, allowing you to quickly move through a circuit and find fun sounds.

If you're having trouble making good contact in a given spot, hold a pair of bare wires between your damp thumb and forefinger, making sure that the wires don't touch; you need to force them to use your skin to complete the circuit (see the right image in Figure 7-10). Squeezing the two wires will decrease resistance, just as pressing harder on the board does.

FIGURE 7-10: *Poking around for a fun bend: my target is the small resistor labeled R16, shown in the far left panel. Try poking small resistors with one finger (center) or a pair of bare wires (right). Note that the pair of bare wires—which are two straight pieces of paperclip— are not touching: my thumb is acting as a resistor bridging the gap between the wires.*

  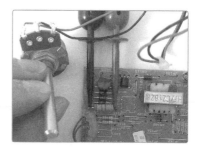

FIGURE 7-11: *Poking around for a fun bend on a larger resistor: using two fingers in two different positions (left and center) or a clip-on pot (right)*

Of course, sticking a wet finger into a live circuit is a decidedly maladaptive practice, so if you're not cool with trying it, just rig up a little clip-on variable resistor, as shown in Figure 7-12. A 1M ohm resistor is a good size here. All you need to do is solder one clip lead to lug 1 of the resistor and the other to lug 2. With the pot twisted fully counterclockwise, it will offer 1M ohm of resistance. As you turn it clockwise, you reduce the resistance. The clip-on variable resistor is a nice tool because it allows you to dial in precise resistances. If you hit a sound you really love, you can unclip the resistor and use your multimeter to measure the resistance currently dialed into the 1M ohm pot. When you go to make your bends permanent in Step 5, you can wire in a resistor of that value and thus rig up pushbutton access to that new sound you discovered.

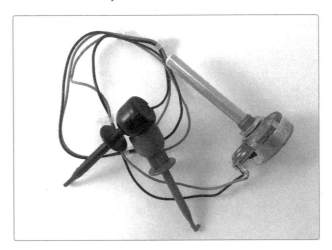

FIGURE 7-12: *A clip-on variable resistor*

Poking around your circuit board while it's on will generally uncover one of three types of resistors:

▶ **A volume-controlling resistor.** Bending these resistors will either raise or lower the volume of some aspect of the toy's sound. Zeroing in on these resistors sometimes make it possible to separate a voice from a background rhythm or to make some part of the mix in prerecorded music more prominent.

▶ **A resistor that sets the clock speed for the toy's IC.** *Clock speed* determines how often a chip repeats a function or how fast it executes an instruction, and finding a clock-speed resistor is the jackpot. Bending it will change the speed or pitch (or both!) of some aspect of the sound, possibly for multiple sounds or even all sounds the device makes.

▶ **A resistor that causes a glitch.** In this case, a *glitch* is any sort of crazy sound that you like that isn't part of the machine's usual repertoire.

Step 4   Check for board-mounted variable resistors, like those shown in Figure 7-13. Variable resistors are another very promising source of bends, especially on older toys. These small potentiometers are called *trim pots* or *trimmers*, and they are often used to fine-tune, or *trim*, a toy's speed or pitch. Trim pots are usually so small that they don't even have knobs (see Figure 7-13, left). Instead, they're adjusted using a small screwdriver or Allen wrench. Fiddle with them! You'll likely get effects like those enumerated earlier. You can remove these trim pots and replace them with large potentiometers mounted on the toy's case so that you can easily tweak them while playing. You bought that toy, so break it however you deem fit. You can also probe these trimmers with the same methods I described in Step 3. Because a trim pot frequently has three legs instead of just two leads, there are even more possibilities to explore.

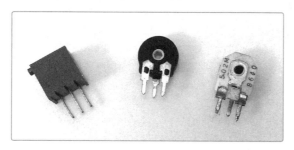

FIGURE 7-13: *Several styles of trim pots (left) and an in situ trim pot (right)*

Now it's time to make your resistor bends permanent by adding a switch, knob, or some other extension that'll allow you to easily tweak the resistance whenever you want.

Step 5   Start by adding leads to the resistor you want to bend. Just cut a pair of insulated wires long enough to reach from the circuit board to wherever you're going to mount the switch or knob on your case. Then, strip both ends of each wire and tin both ends. Now, solder one wire to each side of the target resistor, as shown in Figure 7-14.

FIGURE 7-14: *Adding leads to the resistor—before (left) and after (right)*

Step 6   Add the control for your bend. You have three options: body contacts, a potentiometer, or a switch. *Body contacts* are circuit bending's most emblematic control, formalizing your temporary "finger in the circuit" maneuver. They can be as simple as the "deely-boppers" sticking out of Theodore's head in Figure 7-19, which are just the bare leads from Step 5. Usually, however, they're a pair of conductive pads or studs bolted to the outside of your bent toy (see Figure 7-15). A quartet of body contacts are prominently displayed on the add-on enclosure bolted to the left side of my keyboard in Figure 7-21 (right).

FIGURE 7-15: *A set of body contacts viewed from the inside (left) and outside (right)*

To make body contacts, I like to use #6 brass acorn nuts mounted on #6 machine screws. A couple of washers on the machine screw will help ensure a good contact between your lead from Step 5 and the nut. Body contacts offer very expressive bend control and, because different individuals' skin will put up different amounts of resistance on different days and in different conditions, generally make for unique improvisations.

On the other hand, if you'd like the reliability and convenience of dial-in control, you could connect your leads to a potentiometer instead of body contacts. Select any variable resistor; 1M ohm is a good place to start, as it approximates the resistivity of your skin, but experiment with several values to see what results in the sounds you want. Solder one of the two leads you added to the toy to the middle lug of the variable resistor and the other to either of the two outer lugs (see Figure 7-16).

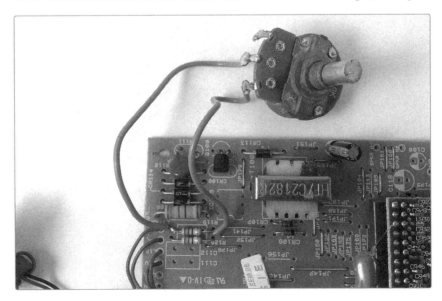

FIGURE 7-16: *A potentiometer bend controller*

Finally, some bends, especially glitches, are most fun at full warp with zero resistance. In these cases, just add a small switch—either pushbutton or toggle, whichever you prefer. Solder one wire to each lug, and you're ready to go (see Figure 7-17). If you want to insert a specific resistor value in series here—in order to capture the just-right squeal you found with your clip-on variable resistor in Step 3 on page 104—do so by soldering one resistor lead to the switch lug and the other to one of the leads.

FIGURE 7-17: *Controlling a bend with a momentary pushbutton or toggle switch*

Of course, circuit bending doesn't have to be limited to resistors or even to single components. The next logical step is to use your fingers to test for bends on capacitors or diodes, or even to create bends between adjacent components. Bends that run resistor-to-resistor, resistor-to-cap, IC-to-IC, or anything-to-ground or sprawl among a slew of different points on the board are all possible—and sometimes pretty rad.

* **WARNING:** *Messing with ICs is more likely to fry your circuit. Proceed with caution!*

If you're going to branch out beyond the simple single-resistor bends you've done so far, you want to add a power reset.

Add a Power Reset

More advanced bends have a higher likelihood of causing the circuit to freak out and lock up, and it'll take a hard reset to get it running properly again. Adding a reset switch to the power supply lets you avoid constantly popping out the batteries to get the toy working again. A reset switch also makes it possible to create some neat effects by stuttering the power supply.

Step 1   Find the positive power connection in your toy. This will almost always be red and very often connects to a + sign on the circuit board.

Step 2   Cut this lead, strip both ends, and tin them, as shown in Figure 7-18 (center).

Step 3   Solder one lead to each of the two lugs on a normally closed pushbutton SPST switch (just like the one you used as a mute switch in Step 4 on page 102).

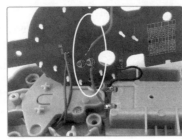

FIGURE 7-18: *Installing a momentary power reset switch. Find the power connection (the long wire in the middle of the left image), cut it (center), and solder in the switch (right).*

Step 4 Ta-da! You now have a power reset switch. If the toy locks up or goes nutsy, push the switch for a few seconds to disconnect the power from the circuit. Then, release the reset button and carry on. You can also occasionally use the power reset musically, perhaps in combination with the mute button, to chop up and recue sounds, samples, and rhythms in new ways.

Package Your Project

The final step is always to put your toy and its new circuitry all together in some usable format. Some examples of finished circuit-bent toys are shown in Figure 7-1. Let's look at two of these in detail, starting with lil' Theodore, shown before and after in Figure 7-19.

FIGURE 7-19: *Talking Happy Meal Theodore, before and after bending (You can see his unmodified brain on the right of Figures 7-3 and 7-9.)*

As you can see in Figure 7-3 (right), talking Theodore is a thoroughly modern toy: his brain is a highly specialized custom IC supported by just two off-chip parts: a capacitor and a resistor. But that resistor sets his clock speed. Removing the original resistor and replacing it with a pair of body contacts, which stick out of his head like antennae, gives you full control of the lil' guy's pitch and rate. With the body contacts, you can make him groan like a creeping glacier or buzz and chirp at hyperdrive superspeed (even by chipmunk standards). I added an auxiliary jack, too, so I can pump his vocals through effects pedals. Theodore's output was so loud I added a 1M ohm resistor to the auxiliary output (see the detail in Figure 7-20). It's not unusual to have to add a buffer resistor when you bend freebie promo toys because they tend to be very loud, even on their tiny built-in speakers.

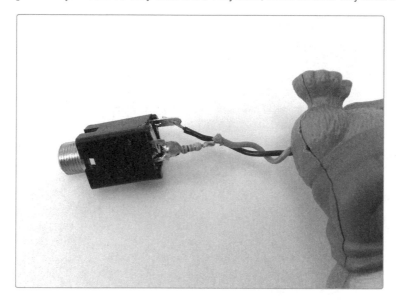

FIGURE 7-20: *Detail of Theodore's output jack, which needed a pretty large buffer resistor to tame his terrible cackles*

By way of contrast, the Casio PT-10 keyboard, shown in Figure 7-21, is a more traditional circuit-bending subject. Dinky little 1980s toy keyboards of all stripes are the circuit-bending gold standard, in my humble opinion.

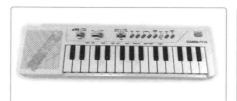

FIGURE 7-21: *A vintage Casio keyboard untouched (left) and fully bent (right)*

To accommodate the keyboard's new electronics, I bolted an add-on enclosure to the left side. The add-on enclosure sports a 1/4-inch auxiliary jack at the bottom, just like Theodore, so the keyboard can easily be amplified or fed to effects. This one didn't need a buffer resistor, as the volume level was reasonable to begin with.

Directly above the output jack are two switches: a pushbutton and a toggle. The pushbutton is a mute, and the toggle switch is a *hyperdrive*. The hyperdrive bridges the keyboard's main clock resistor, in essence reducing the resistance to zero at that point in the circuit. This sends the keyboard's preset voices and rhythms into whirring paroxysms of chiptunery. Above the mute and hyperdrive switches, I added four brass acorn nuts. These are body contacts connected to the resistor and trim pot that regulate the clock speed. Touching them individually or in combination changes both the pitch of the notes being played and the tempo of the built-in drum machine.

Of course, these are just a couple of ways to package these projects. I could have gutted lil' Theodore and stuffed his electronics into an old teddy bear, a decorative gourd, or a candy tin. With a little care, I probably could have mounted all of my jacks, switches, and contacts into the PT-10's original plastic housing; pulled the whole thing apart and installed it in an antique jewelry box; or wired it to the keys of a broken-down upright piano. However you choose to mount your electronics is awesome, because it's yours. Now go out and share that awesome with other folks!

Playing Your Circuit-Bent Instruments

If you've gone so far as to make one or more of these instruments, then I'm sure you've already started to discover the ways in which they can make the sorts of glitch-punky jazz-noise you crave. But I do have two quick notes.

First, if you haven't already done so, flip to "Playing the Scratchbox" on page 67. The advice there on using a momentary mute to sculpt sound—including transforms and gate effects—holds here and should help you start working your mute to musical effect.

Second, in all this prodding and soldering, it's easy to lose track of the strictly mechanical bends you can do while playing. Almost any toy instrument can be made infinitely more interesting by using your cupped hand to muffle, wah-wah, and warble its built-in speaker. You could even go one step further and employ an improvised horn-player's mute, like the *plunger mute* I talked about in the Elephant Trumpet's "Tips, Tricks, and Mods" on page 35.

Along the same lines, consider converting the toy's small internal speaker into an ad hoc compression driver, like the one we built for our Droid Voicebox (Project 6). This drifts a bit past playing style and into an actual mod, but it can make your hacked toys really sing.

Tips, Tricks, and Mods

Circuit bending is entirely an art of tricks and mods, so the one tip I have to add here is: *don't feel limited to just making new short circuits within an existing toy.* Consider grafting your own homebrew circuits onto the factory-made ones. The Mud-n-Sizzle Preamp (Project 12), Twin-T Phaser/Wah (Project 14), and the filters and effects included in Appendix B are all good candidates for such surgery.

The Universal LFO (Project 13) is a good option, too: its two control leads are the legs of an automated variable resistor. Solder one control lead to each of the leads you added to your toy's target resistor, and you can automate that bend or glitch, creating a wide variety of looping backup rhythms, warbling tones, slow-sweeping filter effects, and not-even-God-knows what else.

If you've enjoyed this brief excursion digging into abandoned technologies and making them do new and unintended things, then do yourself a favor and invest in a *multi-bit security driver set.* This is a full set of screwdriver bits that includes all of the most popular odd-shaped bits that manufacturers use to discourage you from opening things up: triangles, asterisks, Torx, "tamper-resistant" Torx Plus, and so on. I own the generic 100-piece security bit set, shown in Figure 7-22, and have never regretted that $16 investment.

FIGURE 7-22: *A generic 100-piece security bit set*

Resources

If you want to learn more about circuit bending, there are two books I'd recommend. First, Reed Ghazala's *Circuit-Bending: Build Your Own Alien Instruments* (2005) is a fun read for those interested in one man's experience of the development of modern electronic music, and it includes detailed instructions on bending a set of popular musical toys. Second is Nicolas Collins's excellent *Handmade Electronic Music: The Art of Hardware Hacking* (2006), which includes a section on circuit bending as well as tons of other great projects and ideas. It's a vital resource for any electronic tinkerer drawn more to exploration and expression than rigorous electrical engineering.

8 JUNKSHOP PERCUSSION

The other projects in this book focus on building or modifying an object—a synth, a uke, a garage-sale toy megaphone. This project focuses a bit more on building up your skills—tuning your ear and limbering up your limbs—so that you can coax tasty licks and rhythms from the plain old, unimproved junk cluttering up your basement, your garage, or the cheapest aisles of your local resale shop.

There are three fundamental habits that contribute to building a good junkshop percussion kit and developing solid junkshop percussion *chops*:[1]

► Listening

► Exploring surfaces

► Exploring techniques

----

1. *Chops* are a musician's technical skills and fluidity at applying these skills.

This chapter has no significant building instructions. Instead, it's a collection of tips, tricks, and techniques for finding the good sounds that lie dormant around you.

Meet Vince: Professional Washboardist

For this project, I've enlisted the aid of percussionist Vince Russo, a man uniquely qualified to speak on the topic of coaxing fun sounds out of cast-off objects. Vince tours with the Appleseed Collective, an Americana roots band that's often described as a post-folk Dixie-Gypsy blend of ragtime, bluegrass, and Django Reinhardt–Stéphane Grappelli–style "hot club" jazz.

Vince has played percussion for most of his life, working in a variety of styles on a traditional drum kit. In Vince's words, the drum kit is an inherently "rigid, biased form": different limbs have different jobs in order to produce a set of prescribed rudimentary patterns. These patterns are the building blocks for wonderfully complex rhythms, and they can be liberating once you acquire them—but the form itself can be constraining. Vince had never touched a washboard before he joined the Appleseed Collective, and the washboard was a revelation: "There's no orthodoxy in washboard," he says. As he traveled, he saw washboard players from vastly different backgrounds—traditionally trained drummers, accomplished percussionists, folks who inherited the tradition from their communities, the completely self-taught—using "wildly different playing styles."

"There are rhythmic things that carry over from the drums, but a lot of it, in terms of muscle motion, is different," Vince says. "Because there's little structure here, your mission is really to impose structures so that you can later move out of them." For example, Vince discovered that he could fall back on his old drumming rudiments, and by moving them across the many surfaces of the washboard, he could "create the same sound five different ways." It was a wild, improvisational effect that was still solid and controlled.

"This is about exploration," he explains, "about being able to explore on the washboard rather than being taught how to play a drum kit. I'm so much more of a fluid communicator on the washboard."

Credit: Luke Sass, Gold Coast Entertainment

Vince Russo, fluid percussionist

See Vince Russo demo a little junkshop percussion at *http://www.nostarch .com/jamband/*. Figure 8-1 shows a few examples of junkshop percussion instruments.

FIGURE 8-1: *A variety of junkshop percussion instruments*

Listening to Music

Vince would be the first person to point out that he's far from the world's most accomplished percussionist. Many drummers have more inborn talent, begin formal lessons earlier in life, and dedicate more hours to molding their natural talent into formidable technique. Nonetheless, he's able to make a living as a gigging musician because of something he learned early on: "There's not a lot of percussionists who listen." Listening will trump technical talent and years of study every time.

There's a broad assumption that the beat is the core of a song—that the percussionist's job is just to hold down the beat and keep everyone in step. But if you just want perfect time, you can build the basic 555 timer metronome in Appendix B (see "555 Timer Oscillator and Metronome Circuits" on page 362) or buy a great metronome for under $15. Any percussionist worth their chops knows that the drummer's job—every musician's job—is to serve the song itself. Sometimes this means leaning back into the groove and keeping the percussion tight to allow other musicians to dance over the beat. Other times, the percussionist needs to expand with the song, contributing to its energy and sense of forward momentum, which is, more or less, the musical definition of *finding the groove*.

Finding the groove—or, as drummers often say, "finding the pocket" or "staying in the pocket"—is a quasi-spiritual, you'll-know-it-when-you-feel-it sorta thing. But there are several listening practices that'll bring you closer to the groove. Top among these are catching the song's *stops* and *dynamics.*

Stops are the breaks in the forward energy of the song. Either a *rest* is formally added, or the players emphasize particular beats (i.e., shift the *accent*).[2] Shifting the accent so that the music seems to "tread water" for a few bars is often called *stop-time,* a name that carries over from ragtime music. Listen for breaks, drops, or anything else that gives a feel of moving the music into a holding pattern and then popping back out of that hold.

Dynamics are often thought of as just being about how loud notes are relative to each other and how sections of a song contrast in terms of volume. But dynamics aren't just about volume, even though that's an important aspect. The dynamics are better thought of in terms of how full or sparse the song is in various sections and how much of that sound-space is being filled by each instrument.

As a quick exercise, play your favorite song and focus your attention on the rhythmic elements, which are often carried primarily by the percussionist and bassist. Listen for stops, rests, and shifts in dynamic, the ways the song swells and recedes, and your emotional response to these shifts.

Then, listen to *my* favorite song: the original commercial recording of "I Want You Back" by the Jackson 5. This is the most common version of the song, the one that's in hundreds of movie soundtracks and has regular radio play anytime I'm on a summer road trip with my kids. Most of the songs that came from Motown Records in this period are wonders of graceful composition, flawless performance, and inspired audio engineering. In this particular song, listen to the bass and percussion: it begins with simple hand claps, adds tambourine and subdued congas, and then progresses to a simple drum kit. At the B section (roughly two-thirds into the song), all of the rhythmic elements drop away and are then reconstructed, this time focusing on the congas. Each part is remarkably sparse (the hand claps are only once per measure, and the drum beat is a straightforward beginner's lesson pattern), but taken together, these simple elements, gracefully timed, entirely direct your heart's response to the song. This is a real-world example of what we mean when we talk about a song's groove.

Exploring Surfaces and Sounds

But listening is just listening unless you have an instrument in hand. Go find something to beat on. There's nothing special or magical about the washboard in general or about Vince's washboard in particular. Rather, it's a tool that provides a wide variety of percussive surfaces in an easy-to-carry package. You could get just as far playing a mixing bowl, or plastic bucket, or elementary school desk.

Percussion, at its simplest, can be thought of in terms of heavy and light sounds: stomping on the bleachers versus clapping hands, snapping the snare versus tapping the closed high hat, beating the kick drum versus hitting a rim.

---

2. See Appendix C for more about music theory and terminology.

As you explore surfaces to percuss upon, look for heavy and light sounds that you want to use as your building blocks. Here are some ideas.

Boom

These are your kick-drummy sounds. Try boxes—cardboard and otherwise— plastic 1- and 5-gallon buckets, large plastic water-cooler bottles, the middle of the kitchen table, a clean trash can, an old filing cabinet, an empty dumpster, a closed garage door, and so on.

Ting and Ring

You can get ringing sounds from cymbals, bells, wind chimes, wine glasses, dog dishes, mixing bowls, and so on. Bowls offer a world of options for exploring all on their own; I'm especially fond of metal dog dishes. Try metal, ceramic, or glass bowls. Use them right side up, upside down, and with or without water. Play them with your hands, drumsticks, chopsticks, pencils, metal or wooden spoons, thimbles, your wedding ring, pieces of PVC, old whisk brooms, or even BBQ skewers.

You can improvise wonderful chimes by hanging the metal platters from inside a discarded hard drive on lengths of string (see Figure 8-2). You may have recently torn apart a hard drive to get magnets for the Playing-Card Pickup (Project 9) or just because you were angry with it.

FIGURE 8-2: *A dead desktop computer hard drive with cover torn off (left) and the three platters ready for their new life as chimes (right)*

Click Sounds

Think of a metronome or the sound you get by striking the rim of a snare drum, a stroke percussionists call a *cross stick*[3] or *rimshot*.[4] You can often find good clicks

---

3. Also called a *rim knock* or *rim click*

4. You comedians in the back row probably think of a rimshot as the "ba-dump-*ching!*" that accompanies a Borscht Belt zinger. Drummers call that maneuver a *sting*; many variations don't even include a cross stick or rimshot stroke. No one seems to know why anyone started calling a sting a "rimshot," but here we are, with a confusing lexical overlap.

on mixing bowls and dog dishes. Flip one over and set it on a towel to muffle the reverb. Or you can lay a stick across the top of the bowl and strike the rim by lifting one end of the stick while the other end mutes the bowl, as shown in Figure 8-3; this technique is actually very similar to playing a cross stick on a snare drum.

FIGURE 8-3: *Playing a cross stick on a dog dish*

You can also find nice clonks and clicks in glass soda, juice, and milk bottles. Try tapping on a bottle right side up or upside down. Play it on its side on top of a towel, in your lap, or gripped between your knees. Leave the cap on, take it off, or intermittently mute the open mouth with your free hand while playing.

Bones and spoons, as shown in Figure 8-4, are a traditional source of clicks and clacks in American folk music. Bones were once made from sections of beef rib and are now mostly made from wood. They're a bit tricky to find these days and have a steep learning curve. But spoons are readily available and are very much a "minutes to learn, lifetime to master" sort of instrument. Figure 8-4 shows a classic spoon grip, and a quick YouTube search will bring you more spoon-playing instructional videos than you ever thought necessary.

FIGURE 8-4: *Bones and spoons (left) and a classic spoon grip (right)*

Another traditional click—one that's very important to Latin, and especially Cuban, music—comes from the *claves* (pronounced *CLAH-veys*).[5] These are a set of 1-inch-thick, 9-inch-long hardwood dowels that can, for example, be cut from an old shovel handle (see Figure 8-5).

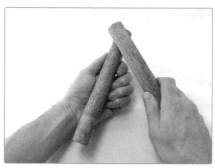

FIGURE 8-5: *A set of claves (left) and a basic clave grip for a right-handed person (right). Note that my left-hand fingertips are* below *the surface of the clave, which prevents me from inadvertently hammering my fingers between the two heavy dowels.*

The claves are traditionally played by letting one dowel, called the *hembra*, rest lightly in the nondominant hand while striking it with the other dowel, called the *macho*.[6] This should permit the resting hembra to resonate freely, a bit like the wooden bar of a marimba. Experiment with where to strike your hembra to get the best tone. My best tone is either at the far tip of the hembra or between the base of my thumb and tip of my middle finger.

Shake and Rattle

The sound of a shaker depends on both the body of the shaker and its contents. Try plastic Easter eggs, pill bottles, soda cans, soup cans, coffee cans, cardboard mailing tubes, cigar boxes, plastic or glass bottles, and jars of all sorts, and fill them with ball bearings, BBs, gravel, plastic beads, dried beans, rice, sand, wire scraps, snips of aluminum pie pans, metal bottle caps, screws, washers, or bolts. I recommend a glass jar with a metal lid, charged with a handful of dried beans. This configuration gives several tonal options: shake side to side for an all-glass shaker-sound, shake end to end for a hard metal smack countered by a softer glass pitter-pat, mute the metal lid occasionally with your fingertips, and so on.

Flattened bottle caps sound really neat inside mailing tubes, and also make excellent *zils* (finger-cymbalesque jingles) for improvised rattles and easy tambourines, like the one shown in Figure 8-6.

5. Claves are so important that there's a basic rhythm—one traditionally played on this instrument—that bears their name: a *clave rhythm* is a syncopated pattern of three hits, a pause, and then two—something like *Pow Pow Pow. Pow-pow*. It's central to Cuban *son* music. Search YouTube for *clave son 3-2* to find plenty of examples.

6. Fun fact: these names translate as *female* and *male*, respectively. As with bullfrogs, the female clave can be noticeably larger.

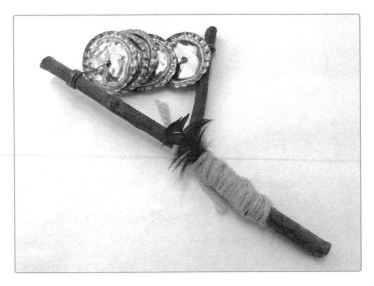

FIGURE 8-6: *A bottle-cap rattle tambourine*

Sticks

When you start playing percussion instruments with sticks instead of hands, consider using either chopsticks, which are light and easy to finesse, or something heavy, like claves or sections of a broom handle. For many newcomers, these thicker sticks will be easier to grip and play steadily than traditional drumsticks.

You can also fashion excellent mallets by drilling a small hole about halfway through a high-bounce ball, squirting in some silicone-based glue, and sticking in a piece of dowel or chopstick (see Figure 8-7). You can get a solid bass drum sound playing an overturned 5-gallon bucket using a pair of these.

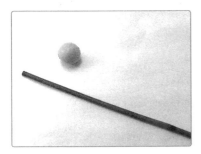  

FIGURE 8-7: *Making a bouncy ball mallet*

Traditional drumsticks are amazingly versatile, but using them to good effect requires quite a bit of practice—it ain't hammering a nail. If you want to explore traditional drumsticks, pick up any old pair of all-wood sticks and invest some quality time on YouTube watching the videos that come up when you search *traditional grip*, *matched grip*, and *traditional vs. matched grip*. You can learn to play a great drumroll with nothing more than a pair of decent sticks and a hardwood coffee table or hardcover book.

Your Hands as Percussion Instruments

For many beginning junkshop percussionists, the hands are a more serviceable playing tool than sticks or mallets, especially when those hands have been slightly enhanced. Take your lead from the washboard players of the world and explore the sounds you find in surfaces by augmenting your fingertips with these gaffs, gimmicks, and gizmos:

- Thimbles (metal or plastic)
- Rings
- Fingerpicks (the kind used by banjo players and bluegrass guitarists, which are available in a variety of styles in metal and plastic)
- Frosting-bag tips (for example, those made by Wilton and sold as *decorating bag tips*)
- Used shotgun shells
- Test tubes
- Cigar tubes
- Work gloves modified with bottle caps or other metal scraps

If these don't strike your fancy, dive into the junk drawer, stick some stuff on your fingers, and see what you hear.

Multisurface Improvised Instruments

You might have noticed some repeat appearances in this chapter, and that's because different surfaces in the same object can make different sounds. Mixing bowls can serve in different modes for tings, rings, and clicks; claves can also be drumsticks; glass bottles might serve as shakers or as cowbells. But the most famous multisurface junkshop percussion instrument is the washboard (see Figure 8-8).

FIGURE 8-8: *Vince's antique washboard (left) and a homemade mini-washboard crafted from wood and an accordion-folded "tall #3" soup can—the sort of can that stands about 7 inches tall (right). Craig W. Van Otteren and Jim Jett developed these mini-washboards as a kid's project for the 2014 Wheatland Music Festival in Remus, Michigan.*

Despite the name, Vince finds that many folks don't immediately make the connection that the washboard is an early 20th-century tool for scrubbing clothes. Brand-new washboards are increasingly common in music shops, but Vince doesn't necessarily suggest getting one of these—they are cheap-o novelties that, in his experience, are poorly constructed and tinny sounding. He finds it much easier to work with finesse on an old board. While a new made-for-music washboard runs about $30, an antique washboard is only about $50 and will survive much longer. Some folks prefer to play washboards free of any additional add-ons, while others completely encrust their boards in service bells, cowbells, wood blocks, cymbals, cans, lids, chimes, and so on. Vince has added two accoutrements to his board: a service bell bolted to the soap saver and a small splash cymbal mounted to the left-hand rail.

Another great improvised, multisurface instrument is the humble hardcover book. Badly out-of-date encyclopedias—which tend to have thick, glossy pages, ripe for riffling—are an excellent example. A hardcover can be drummed like bongos with hands, fingers, knuckles, or any combination thereof. You can play a drumroll on it using standard drumsticks. The pages can be riffled with the book open or shut, and you can slam the book itself for an accent. And don't forget about your nails, which can drum and rasp the fabric of the cover or skritch-skratch up and down the edges of the pages.

For an example of an improvised multisurface instrument that's made the transition from ad hoc make-do to "real" instrument, check out the *cajón*, as shown in Figure 8-9. What started out as a drum improvised out of packing crates by Peruvian slaves in the 1600s is now a commercially mass-produced instrument, an all-in-one tunable drum with bass and snare in a single package. You can buy cajóns in stores, take cajón lessons, and even hear the cajón in songs on Top 40 radio.

Credit: Gluckstadt, Wikimedia Commons

FIGURE 8-9: *A commercial cajón*

Exploring Techniques

The trick to drumming is in breaking up the rhythmic movement of your limbs in new and counterintuitive ways. When I'm walking down the street, I'm moving rhythmically and largely symmetrically: my legs take equal strides at a steady pace, and my arms swing back and forth. But when I sit down at a drum kit, I assign my limbs jobs that aren't at all symmetrical. For example, to play a simple 4/4 rock beat, my right hand taps the closed hi-hat eight times, while my right foot stomps only on the 1st and 5th beats; my left hand hits the snare twice but only on the 3rd and 7th beats; my left foot holds down the hi-hat pedal, but I might lift it on 1 and 5 so that those beats are accented.

This is confusing to read and can feel absolutely mind bending to do at first, but anyone can learn to do it. Remember, you learned to walk, skip, and hop—all of which are insanely complicated rhythmic activities that you now do all the time, completely on autopilot.

Practice Rudiments with Your Hands

Drummers start retraining their limbs by practicing *rudiments*. These are simple patterns originally developed in Northern Europe so that drummers could be used to coordinate complex battlefield maneuvers during the early Renaissance. These patterns were later refined to serve as the codified building blocks of really complex marching band drumline patterns. Today, rudiments are still used to form the basis of the formal drumming vocabulary.

The best rudiment for breaking up your limbs is the *paradiddle*. This is a steady pattern of *RIGHT-LEFT-RIGHT-RIGHT-LEFT-RIGHT-LEFT-LEFT*, repeated ad nauseam. Get a metronome ticking and pat the pattern out on your lap using your open palms. If you don't want to buy or build a metronome, just search *metronome click track* on YouTube for a virtual alternative. Make a point of practicing this with a metronome; maintaining an even beat is just as challenging as using the right hand at the right time, so you might as well work both those mental muscles at once.

You'll note that this pattern forces your hands to take turns leading; you very much feel this in your body as a sense of passing the beat from left to right. Work on playing this pattern evenly and slowly at first, and then bump up the speed. Finally, try steadily speeding up and slowing down as you play.

Once the paradiddle gets comfortable, you can further blow your mind by doing *flam paradiddles*. A *flam* is when you precede your drum stroke with a little tap[7] from the alternate hand; think of the "pa-rump" you hear in The Little Drummer Boy's "pa-rumpa-pum-pum." That "pa" tagged to the front of the "rump" is a flam. A flam paradiddle might be written out like this, where the lowercase words are the tapped grace notes: *leftRIGHT-LEFT-RIGHT-RIGHT-rightLEFT-RIGHT-LEFT-LEFT*. An even more mind-bendingly limb-liberating hand alternator with flams is the *pataflafla*, which goes like so (where, again, lowercase words indicate flams): *leftRIGHT-LEFT-RIGHT-rightLEFT-leftRIGHT-LEFT-RIGHT-rightLEFT*.

---

7. This little tap is called a *grace note*.

Once you've gotten the hang of these, the *paradiddle-diddle* will be no sweat. It's a paradiddle with an extra diddle—that is, an extra double stroke: *RIGHT-LEFT-RIGHT-RIGHT-LEFT-LEFT*, repeat. Because the paradiddle-diddle has that extra diddle, it doesn't alternate hands. You can address this by dropping half of a normal paradiddle in between two paradiddle-diddles. This will force the paradiddle-diddle into the other hand. If this sounds confusing, start tapping it out on the table, and you should hear it immediately.[8]

Try playing these patterns on different bits of your improvised percussion kit or using different objects in your hands. Vince, for example, plays washboard with thimbles, fingerpicks, and pastry-bag tips on his right hand and a heavy beer-bottle opener in his left. In this way, he can make a lot of music with very little. "You don't need to have a multitude of surfaces to have a multitude of sounds," he says. "I'll play a countertop or my knees." He laughs, "One time I jammed with a dulcimer player, and I didn't have my washboard, but I had my washboard picks and a glass juice bottle. The bottle had a little bit of that waffle pattern, so I clinked on it with my picks."

Free Your Fingers and Feet

But wait, there's more! Once you've freed your hands from working in lockstep synchronization, the next logical step is to liberate your fingers. Vince notes that "using your fingers—and not just thinking, *I have right and left*—really opens it up and expands the possibilities." This is the place where junkshop percussion really pushes beyond the drum kit: your hands don't have to be wrapped around sticks, nor your feet dedicated to pedals controlling the hi-hat or thumping the big bass drum. Try practicing paradiddles by tapping your toes,[9] alternating between the index and ring finger on your right hand, or alternating between your left heel and left hand.

The more you practice automating these movements and breaking up your limbs, the more free mind-space you'll have for listening to the song as it forms around you and feeling for the groove. And, as you do all this practicing, don't stop listening. Pay attention to what's fluid and comfortable as you play, but also listen for sounds and patterns you like—little tasty licks that you can keep handy and that might fit a gap you hear in the next song you play.

8. Still confused? Search YouTube for *Vic Firth Rudiment Lessons: Paradiddle-diddle*, which is nice and clear. He mentions the half-paradiddle switch about three minutes in.

9. You can mod and enhance your feet just as you did with your fingers. Try gluing bottle caps to the bottoms of some old flip-flips to give a sharp tap-shoe zing to your toe tapping.

PART II

Weekend Projects

None of us start our musical lives by saying, "How do I play a dominant seventh on this thing?" or "How many sharps are there in F major?" But most of us pretty quickly start asking, "How do I play 'Camptown Races' or the intro to 'Iron Man' on this thing?" If that's where you're at musically, then Part II is calling (as are the appendices, especially Appendix C).

The projects in Part II are more advanced than those in Part I. Some call for more tools or trickier construction techniques, while others simply take longer to complete, with more sawing, drilling, sanding, soldering, and so on. But the scope is what really differentiates these projects from those in Part I. Here, you'll make full-fledged instruments and musical tools. That said, despite the greater complexity, nothing here really requires an excessive amount of preexisting skill—just some patience and determination.

9 THE PLAYING-CARD PICKUP

The Playing-Card Pickup is a true magnetic pickup, based on the same basic design principles as the pickups built into professionally made electric guitars. Although this pickup was developed for the next two projects (the Robo-Tiki Steel-Stringed Ukulele and the Twang & Roar Kalimba), it's well suited to amplify any number of homemade stringed instruments, such as cigar-box guitars and "canjos," as well as smaller store-bought steel-stringed instruments. Much like the Plasti-Pickup (Project 2), the Playing-Card Pickup is a fun, experimental noise toy in its own right, useful for quickly electrifying ad hoc percussion kits composed of cookware, scrap metal, playground equipment, and so on.

The finished Playing-Card Pickup is shown in Figure 9-1. Hear samples at *http://www.nostarch.com/jamband/.*

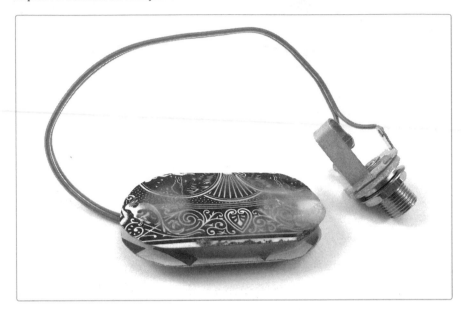

FIGURE 9-1: *The finished Playing-Card Pickup*

Preparation

Build Time

▶   Under an hour

Tools

▶   A standard soldering kit (See page 340.)

▶   A ruler or straight edge

▶   A hobby knife, utility knife, or pocketknife

▶   Scissors

▶   A quarter

▶   An amp and instrument cable for testing the pickup

▶   (Optional) A multimeter (See "Using a Multimeter" on page 349.)

Supplies

▶   A playing card

▶   Four small neodymium disk magnets (Available in most hardware stores, these are often called *supermagnets* or *rare earth magnets*. I used disk magnets that are 8 mm in diameter and 3 mm tall, but others should work fine.)

- A few hundred feet of 43-, 42-, or 30-gauge enameled winding wire (30-gauge wire—also called *magnet wire*—is by far the easiest to find on the spur of the moment. 42- and 43-gauge wire can be ordered online; see "Resources" on page 141 for details.)

- A chunk of beeswax roughly 1 1/2 inches by 3 inches, or about 2 ounces by weight

- 24-gauge insulated hook-up wire (Stranded wire is best.)

- A 1/4-inch mono phone jack, also called a *guitar jack*

- Control-gel cyanoacrylate glue (Any brand of control-gel formula superglue will do. I especially like Loctite Ultra Gel.)

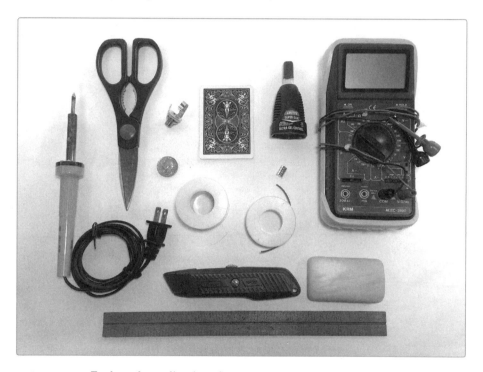

FIGURE 9-2: *Tools and supplies (not shown: amp and instrument cable)*

Building the Playing-Card Pickup

Step 1 We'll start by cutting a pair of identical body plates from an old playing card or a similar thin, rigid cardboard. Place a quarter on the card, trace a circle around it, and then move it down so that it's *tangent to* (barely touching) the circle you just drew (see the left image of Figure 9-3).

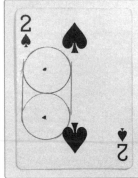

FIGURE 9-3: *Cutting the matching front and back body plates*

Trace a second circle, mark the centers of both circles, and draw a pair of tangent lines joining their outer edges, as shown in the middle of Figure 9-3. Cut this oblong shape out and use it as a template to create a matching plate (as in the rightmost image in Figure 9-3). With a little care, you should be able to make both plates from a single card.

Step 2   Now we'll mount the magnets. Because they'll want to stick to each other, you can make this step a bit easier by having a nice, flat hunk of magnetic metal handy (an old toolbox or metal filing cabinet is great). Start by putting a dab of control-gel superglue at one of the center points you marked in Step 1. Place one of your neodymium magnets in the puddle of glue and squeeze for 5 or 10 seconds to let the glue begin setting. Repeat with a second magnet at the other center point, taking care not to let magnet #2 attach itself to magnet #1. Now snap your partially finished pickup body, magnets up, on your metal work surface. Control-gel superglue takes a few minutes to firm up, but you can speed that up by blowing on it.

Once your magnets are glued down, stack another small neodymium magnet atop each of the ones you just glued (see the leftmost image in Figure 9-4). Put a bead of glue atop each magnet stack and then glue the other body plate in place (see the rightmost image in Figure 9-4).

＊ **SUPERGLUE FIRE WARNING:** Cyanoacrylate (CA) *glues have an* exothermic *reaction with cotton and some wool blends. In other words, CA tends to get hot when it makes contact with many fabrics. CA may even react with cotton balls or swabs (some brands are all or mostly natural fiber), and the reaction can be rapid enough to cause burns or catch fire. This risk is minimal but real. Be prepared. If you squirt some glue on your clothing, just strip down in a hurry and risk embarrassment rather than ending up with a smoldering T-shirt glued to you.*

＊ **PROTIP:** *If you glue your fingers together, remember: superglue binds fast and hard, but it has relatively poor* shearing *strength. In other words, it holds tight if you're pulling straight up, but it often releases if you work from the edge and peel the glued surfaces apart. Dip glued-together fingers in olive oil and then rock them back and forth; don't try to pull them straight apart.*

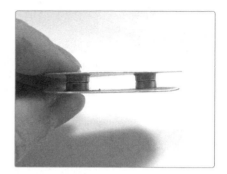

FIGURE 9-4: *Completing the pickup body*

Step 3   Next, we'll wind the coil. Leave a 6-inch tail of enameled magnet winding wire dangling and begin wrapping your pickup as you'd wind up the string on a yo-yo. Even though 30-gauge enameled wire is thin, it's pretty durable, so you should be able to wrap quickly without snapping the wire. Wind until the body is filled with wire along its long edges, and then leave another 6-inch tail and snip the wire. Many pickup builders speed this up by sticking their pickup body to the exposed flywheel on an electric sewing machine and then running the machine at a slow-ish setting, carefully feeding the wire to the pickup as it spins. Using this method, you can have the coil wound in just a few minutes. Finish this step by using sandpaper to scrape the enamel off the last half-inch of each free end of the coiled magnet wire.

However you arrive at it, the final result should look like Figure 9-5 and, if you've used 30-gauge wire, will probably consist of a few hundred windings, totaling less than 200 feet. If you've used a thinner-gauge wire, you'll have used a greater length of wire, resulting in more windings. That's great, and it will work just as well—if not better!

FIGURE 9-5: *The wound pickup*

Step 4   If you have a multimeter, now is a great time to check that your pickup is in good shape. The coil should have *continuity* (that is, electricity should be able to flow through the coil from one end to the other), and it should offer around 14 ohms[1] of resistance. Resistance as low as 8 ohms is fine, and higher is also okay.

1. If this jargon—multimeter, continuity, resistance, ohms—is throwing you off, read the "Components Primer" on page 324.

Guitar Pickups Demystified

The pickup we're building is a passive, single-coil magnetic pickup with a relatively flat winding. It's modeled on the original soapbar pickups used in '40s and '50s Gibsons and Fenders, and it retains something of their bright, crisp twang. A single-coil design like this is a bit more susceptible to interference than a dual-coil pickup, but given our junkyard-punk hillbilly aesthetic, that wash of warm background hum is a feature, not a bug.

That paragraph you just read is a dandy little spiel to rattle off if a gearhead asks you about the electrified cigar-box instrument you're showing off, but probably sounds like a bunch of in-group jargon to you. Don't fret: guitar pickups are pretty mysterious to most folks, including many a road-weary musician. What the heck are pickups? How do they do their voodoo?

From an electrical engineering standpoint, the pickups on an electric guitar are simple magnetic *transducers*: they transform the vibration of the metal guitar strings into a matching electrical signal suitable for electronic amplification.

A magnetic guitar pickup is made of one or more magnets surrounded by many coils of fine wire. When you pluck a guitar string, that steel string vibrates near the pickup, disturbing the standing field created by the magnet. This moving magnetic field induces a moving electrical current in the coil, and your amp amplifies this fluctuating current. The fluctuations in this electrical current are identical to the fluctuations in the magnetic field, which are themselves identical to the fluctuations of the vibrating guitar string. This is why the note coming out of the amp matches the note the string is playing: same frequency = same note.

When you think about it, your guitar amplifier isn't really amplifying your guitar, per se: it's amplifying an electrical current that matches a magnetic ripple that matches the vibrations of the guitar strings themselves. That's why an electric guitar sounds distinctly different from its acoustic cousins. With an acoustic instrument, the design and composition of the body has a profound impact on the instrument's voice because you hear the entire body of the instrument vibrating. With an electric instrument, the body has almost no impact on the sound because the pickup only detects the vibrations in the metal string itself. Body composition has almost no impact on how those metal strings vibrate. It could be fine Adirondack spruce or an old wood toolbox; it's gonna sound about the same. If you really think about it, when you're listening to Jack White or Joan Jett tear it up, you aren't even hearing the strings, not really; you're hearing the electricity.

Most electrified ukes and cigar-box guitars floating around the Internet don't have a true magnetic pickup. Instead they rely on an acoustic piezo pickup, like the ones we built for the Slinkiphone and the Plasti-Pickup (Projects 1 and 2). That's why they don't sound like electric guitars: strictly acoustic pickups amplify the entire instrument—including every bump and scrape on the fingerboard—just as much as the notes you're playing. You don't get the same tone as an electric guitar because that true electric-guitar tone arises from directly amplifying the string itself, rather than the wood it's strapped to.

If you don't have continuity, then the coil is broken; unwind it and try to find that break. You can probably repair this break by scraping off the enamel on the two broken ends and resoldering the wire, although that might prove a pain. (Unwinding the broken coil and entirely rewrapping with brand-new wire is also an option.) If the pickup's resistance is less than 8 ohms, you probably have a short in the coil. This will definitely make the pickup sound quieter and could lead to it not working at all. Proceed with caution, or play it safe by rewinding the pickup and testing again.

Step 5 Let's add the insulated leads and 1/4-inch jack. Cut two 6-inch lengths of 24-gauge insulated wire. Strip the ends, tin them, and solder one to each end of the pickup coil, as shown in Figure 9-6. Wrap at least one of these connections in a snip of electrical tape or shrink tubing in order to prevent annoying shorts in the pickup.

FIGURE 9-6: *Leads and jack connected*

Finally, tin the lugs of the jack and solder one lead to each lug so you can test the pickup easily in the next step. It doesn't matter which lead goes to which lug. (Need a refresher on soldering? Flip to "Soldering" on page 346). Keep in mind that this is just a temporary connection; you'll need to remove the jack when you install the pickup later. Going light on the solder now will make desoldering later a breeze. If you're *not* planning to install this pickup in an instrument, then feel free to go a bit heavier on the solder now.

Step 6   It's time for a sound test! You could skip this step if your pickup had continuity in Step 4, but if this is your first pickup, I strongly advise testing now so you'll know how much the sound improves after potting the pickup in Step 7. Carefully plug a guitar cable into the jack, connect it to an amp, power up, and crank the volume. Tap the pickup with your finger, and you should hear some scraping and thumping. You should also hear a nice loud *thump* when you drop a small magnetic screw or nail on the pickup. If you don't hear anything but your tests in Step 4 confirmed that the coil has 8 or more ohms of resistance, then you have a short in your leads or jack. Check for crossed wires or solder bridges, fix them, and test again.

Once your pickup checks out, wrap the remainder of the enameled wire around the body of the pickup and glue the two leads to the center of the back of your pickup (see Figure 9-7).

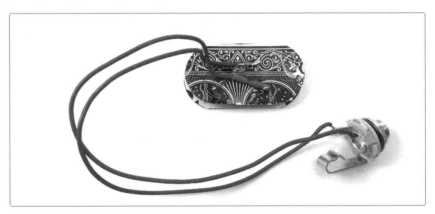

FIGURE 9-7: *Fully functional pickup, ready for potting*

Step 7   Let's *pot* the pickup, which means soaking it in molten wax. Remember how the pickup thumped when you poked it with your nonmagnetic finger? That's because the enamel coils moved, and even tiny shifts of the wire in the fixed magnetic field will generate enough current for the amp to pick up. Potting the pickup fixes the coil in place and eliminates this unwanted nonmusical noise.

To begin, place your beeswax in a clean glass jar, place the jar in a pot with a few inches of water, and place this over medium heat until the wax has just melted.[2] Then, remove it from the heat. Dunk the pickup in the molten beeswax for one minute. Tiny bubbles should drift to the surface as the wax soaks into the little crevices between the windings. Pull out the pickup, and let it drain and cool for 30 seconds. Repeat a few more times. After the third or fourth dip, you shouldn't see any more bubbles when you soak it. Hang it over the edge of a jar to cool, as in Figure 9-8, so you don't get wax all over your table. Then, carefully check the outside of the coil for loose loops of wire that can still wiggle. If you see any, try a couple of quick *candle dips*: five seconds submerged, five seconds to drip, five more seconds submerged. Candle dips will build up a thick outer layer of wax pretty quickly.

When no wires move, set aside your pickup. You're ready to install it in your instrument or to start exploring the electromagnetic soundscape around you.

FIRE DANGER!

＊ **HOT WAX FIRE WARNING:** *Never leave beeswax unattended on the stove! If the beeswax smokes, it's too hot; cut the heat immediately. A beeswax fire is a lot like a grease fire: never throw water on it! Flaming wax will splash and spread the fire when doused with water. To put out such fires, cover the pot with its lid (if possible) or smother the fire in baking soda, salt, or sand. Do not smother a fire with flour! Flour burns very well and can be explosive when thrown at a grease fire. According to US Army improvised munitions manuals, a 5-pound sack of flour— if properly initiated—can easily level a small building.*

FIGURE 9-8: *Drip-drying the potted pickup*

---

2. This arrangement is called a *double boiler.* It's also handy if you're melting chocolate—perhaps to drizzle over popcorn bars, which are essentially identical to Rice Krispie treats, but with air-popped popcorn in place of cereal. Mix in a handful of dried cranberries and you have a concoction I call "Cranberry Floozies," superior to Rice Krispie treats by any metric.

The Facts on Wax

Beeswax is available at most health/organic food stores and farmers' markets, as well as many craft, hardware, and art supply stores. Although beeswax (or a 20/80 blend of beeswax/paraffin) is traditionally preferred by many luthiers, straight paraffin wax has gained traction in professional circles over the last several years. I've talked with makers who've used paraffin candle wax instead of beeswax to pot their playing-card pickups and gotten good results, although I haven't tried it myself.

Paraffin wax is an industrially produced petroleum product with very consistent composition; you can be sure that pretty much any hunk of paraffin wax is basically about the same as any other. Beeswax is made by bees with low standards for quality control. I've had batches of beeswax that contained minute magnetic particles in suspension—you could see this magnetic dust gather like iron filings over the magnets on the potted pickups. How does this metal end up in the beeswax? It's what's left of shooting stars. Meteorites burn up as they enter our atmosphere, in most cases leaving nothing behind but a dusting of iron powder. Tons of this stardust rains down on Earth each year, and some gets inadvertently collected by bees as they gather their sweet, sweet nectar. I've personally never noticed any audible difference between my "high iron" beeswax pickups and those with less obvious accumulations of meteorite dust, but my pickups are not precision devices to begin with. Nonetheless, this variable iron content may contribute to the "dark" tone that veteran luthiers sometimes attribute to 100 percent beeswax-dipped pickups.

As for real and verifiable differences between paraffin wax and beeswax: paraffin wax has a lower melting point, which some folks find easier to work with in general. That makes paraffin wax preferable for some applications, and it should pot your pickup just fine. Beeswax, however, is more pliant at room temperature. This makes the pickup more durable over time. Because durability trumps all for me (I regularly hand these homebrew instruments to excitable children), I stick with beeswax. It doesn't hurt that I've already got a big hunk of beeswax on hand, and I love the faintly honey-dipped smell of the finished beeswax pickup.

Finally, note: *paraffin wax is just as flammable as beeswax*. Please take the same precautions when handling it as you would for beeswax.

Playing the Playing-Card Pickup

The primary use for this pickup is to amplify the next two projects: the Robo-Tiki Steel-Stringed Ukulele (Project 10) and the Twang & Roar Kalimba (Project 11). It's likewise a great design for amplifying a homemade cigar-box guitar, diddley bow, or any number of smaller steel-stringed instruments, such as lutes, mandolins, strum-sticks, mountain dulcimers, and so on.

You can also use this magnetic pickup to amplify and record many harmonicas. The wooden or plastic comb you blow into and the brass reeds that vibrate to make the harp sing aren't magnetic, but most harmonicas have a chrome body, which is. You can stick the Playing-Card Pickup right to the side of your Marine Band Special 20 and blow with abandon—no worries about picking up pops and 'plosives.

Like the Plasti-Pickup (Project 2), this homespun magnetic pickup is a great tool for sonic exploration. For example, stick this guy to the side of anything magnetic that you might want to percuss and amplify—such as a flagpole, handrail, garbage can, swing set, tire rim, and so on. If you have metallic bowls or dog dishes that aren't magnetic enough to hold the pickup tight, just set the bowl on top of the pickup and drop in a couple of magnetic ball bearings, washers, or screws to keep the pickup in solid contact with the bowl. This is a *great* way to capture the sound of Tibetan singing bowls for recording, amplification, or effects processing, and it even works with totally nonmagnetic vessels like "singing" wine glasses or large glass mixing bowls. Check out Junkshop Percussion (Project 8) for more ideas.

You can even use the Playing-Card Pickup to safely listen to the 120-volt AC lines in your walls. That's not terribly useful musically, but it can come in handy if you're hanging a picture and worried about hitting a live wire, or just curious about what path your wires take behind the plaster.

Tips, Tricks, and Mods

More windings mean a *hotter* pickup—one that's more powerful, sensitive, and louder. By using a thinner-gauge wire (one with a higher number; see "Wire" on page 336 for a quick explanation of wire gauges), you'll be able to fit more windings into the same size pickup.

Commercial pickups use 42- or 43-gauge wire. If you've successfully made a pickup with 30-gauge wire, making another with 42- or 43-gauge wire is well within your skill set and definitely worthwhile. This thinner wire is more delicate, so wind slowly and consider doing it entirely by hand. Everything else with the design is otherwise the same. Just use the thinner-gauge wire and add windings until the coil is within 1/8 to 1/16 inch of the long edge of the body plate. The resulting pickup should have a DC resistance around 2.5k ohms—that's 2,500 ohms, 170 times more resistance than what you measured on your 30-gauge pickup in Step 4.

With magnetic pickups, higher resistance corresponds to greater amplification; 2.5k ohms is at the lower end for fairly cool (i.e., low-powered) commercial pickups. Such a pickup will still work with a lot of different homebrew and commercial effects.

Resources

Sourcing 30-gauge wire is pretty easy: it's readily available at some hardware stores and most electronics shops, hobby shops, and science supply stores. It can also be salvaged from discarded motors and transformers. The thinner 42- and 43-gauge wires are easier to find online than locally, though a *very* well-appointed guitar repair shop *might* have a spool to sell.

Stewart MacDonald (*http://www.stewmac.com/*), a popular online luthiery supplier, sells large half-pound spools of pro-grade 42- and 43-gauge pickup winding wire, both new and vintage. Either gauge and style is fine for this project. A half-pound spool is enough winding wire to make at least a dozen pickups like this one. Antique Electric Supply (*http://www.tubesandmore.com/*) sells smaller, and thus cheaper, spools of 42-gauge wire that also work well for this project. Both stores are good sources for other musical instrument and effect hardware, such as switches, jacks, tuning pegs, and so on.

As for appropriate magnets, many hardware stores stock the small neodymium magnets I call for here, but they can be a tad pricey, up to a buck each. If you order online, you can get magnets significantly cheaper. K&J Magnetics (*http://www.kjmagnetics.com/*) is a popular, well-stocked source for rare-earth magnets. The 8 mm × 3 mm discs I favor are the same as K&J's 5/16 × 3/32 disc magnets. As of this writing, K&J Magnetics sells them for $0.38 each.

You can also salvage neodymium magnets from old computer hard drives (as shown in Figure 9-9). Each drive contains two arc-shaped bar magnets. You might need to modify the size and shape of your playing-card body plates in Step 1, but it's a neat opportunity to upcycle. I'd suggest a chunky 2-inch by 2 1/2-inch pickup, which is a bit more like a bulky soapbar pickup. If you're already taking apart a hard drive, hold on to the platters; they make really great chimes for your Junkshop Percussion setup (Project 8).

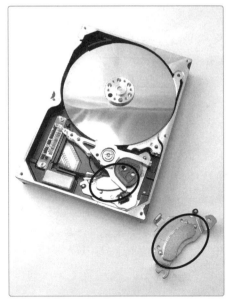

FIGURE 9-9: *A broken hard drive with the cover removed (left) and the same drive partially dismantled to reveal the pair of curved neodymium magnets (right). The circles indicate the location of these wonderfully powerful magnets.*

10 THE ROBO-TIKI STEEL-STRINGED UKULELE

There are several reasons the ukulele is an ideal first instrument, especially for adults who never learned to play anything as a child. First, ukes are cheap and highly portable. Second, they have low-tension strings, which means they're easy on fingers new to holding down strings while strumming. Third, the fingering for all the basic chords is exceptionally easy. But the main advantage of the ukulele is that it's often dismissed as a toy. Anyone playing a recognizable song on the uke gets disproportionate accolades based on the *talking dog effect*.[1] Mark my words: had electrified ukes dominated the market early on, the vast majority of punk bands would have lead ukulelists, not lead guitarists.

---

1. A talking dog doesn't have to say anything wise, funny, or helpful; folks are delighted just to meet a dog that can talk.

But guitars won out, and while electric guitars are plentiful and diverse, commercial ukes are almost exclusively acoustic, with nylon strings. This significantly limits the sound palette, genres, effects, and recording options for ukulelists. The Robo-Tiki Steel-Stringed Ukulele fills those gaps on a shoestring budget of about $25—less than you'd pay for an entry-level acoustic uke that likely won't even stay in tune, let alone rock a house party at the drop of a hat. You can source the parts from most hardware stores and hobby/electronics shops. If you wake up Friday morning with a hankering to get weird and make a jazz noise, you can build a playable electric-acoustic uke with good intonation by Sunday night.

If you're thinking of going either all acoustic or all electric, check "Tips, Tricks, and Mods" on page 164 before you head to the hardware store.

The Robo-Tiki Steel-Stringed Ukulele is shown in Figure 10-1. Hear samples at *http://www.nostarch.com/jamband/*.

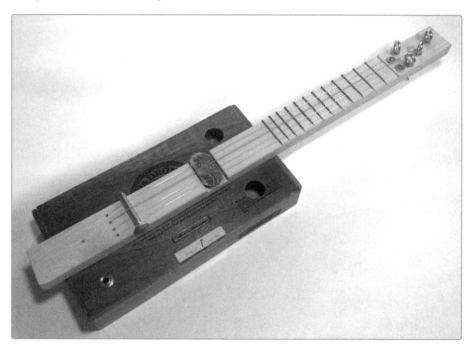

FIGURE 10-1: *The finished Robo-Tiki Steel-Stringed Ukulele*

Preparation

Build Time

▶ About 3 hours, including pickup build (plus 4 to 24 hours of drying time if you use Gorilla Glue)

Tools

▶ A standard soldering kit (See page 340.)

▶ A ruler marked in both inches and millimeters[2]

▶ An electric drill with bits (You'll definitely need 1/8-inch, 3/16-inch, and 3/8-inch bits; a 1/16-inch bit might also prove handy for drilling guide holes.)

▶ A 1-inch spade drill bit, also called a *paddle bit* (See Figure 10-11.)

▶ A framing square, carpenter's square, or combination square (The top photo in Figure 10-2 shows a combination square at the lower left.)

▶ A miter box

▶ A standard wood saw

▶ 100-grit sandpaper (You might want to use an electric sander if you have one available.)

▶ A chromatic tuner (A dedicated tuner is handy, but there are plenty of iOS and Android apps for this and lots of nice free tuners online. I especially dig *http://www.tunerr.com/*.)

▶ A table vise or several clamps

▶ A hardwood scrap about the size of a bar of soap

▶ A triangular file (A file with sides that are about 1/4-inch wide is a nice general-use file and good for this project.)

▶ A hobby knife, utility knife, or pocketknife with a small, sharp blade

▶ Scissors

▶ A quarter

▶ An amp and instrument cable for testing the pickup

▶ (Optional) A tapered half-round file (This is very handy for cleaning up drill holes, but you can also do so with a pocketknife or a little scrap of sandpaper rolled into a tight tube.)

▶ (Optional) A multimeter (See "Using a Multimeter" on page 349.)

---

2. I wouldn't normally mix English standard measuring units with the metric system, but as you'll see when you're laying out the frets, getting proper intonation on a multistringed instrument calls for a level of precision that would drive a sane person crazy if expressed in inches.

Supplies for the Pickup

▶ A playing card

▶ Four small neodymium disk magnets (Available in most hardware stores, these are often called *supermagnets* or *rare earth magnets*. I used disk magnets that are 8 mm in diameter and 3 mm tall, but others should work fine.)

▶ A few hundred feet of 43-, 42-, or 30-gauge enameled winding wire (30-gauge wire—also called *magnet wire*—is by far the easiest to find on the spur of the moment. 42- and 43-gauge wire can be ordered online; see "Resources" on page 141 for details.)

▶ A chunk of beeswax roughly 1 1/2 inches by 3 inches, or about 2 ounces by weight

▶ 24-gauge insulated hook-up wire (Stranded wire is best.)

▶ A 1/4-inch mono phone jack, also called a *guitar jack*

▶ (Optional) Gorilla Glue

Supplies for the Neck

▶ A finish-grade poplar 1×2 at least 24 inches long (Other hardwoods are also acceptable.)[3]

▶ A dozen 1 3/4-inch straight finishing nails (Get nails with smooth shafts—a style sometimes labeled *bright finish*.)

▶ Control-gel cyanoacrylate glue (Any brand of control-gel formula superglue will do. I especially like Loctite Ultra Gel.)

Supplies for the Body

▶ A cigar box or other materials to make your uke's body

▶ Four 3/4-inch-long wood screws

▶ A bamboo skewer roughly 1/8 inch thick

▶ A 5/16-inch-ish coarse-threaded carriage bolt

Supplies for the Pegs and Strings

▶ A set of the lightest steel ball-end electric guitar strings available (I used Ernie Ball 2225 Extra Slinky Electric Guitar Strings, which are less than $5 at most guitar shops.)

▶ Four 2-inch-long 3/16-inch eyebolts (also called #10 eyebolts)

▶ Four 3/16 wing nuts (also called #10 wing nuts)

▶ Ten 3/16 washers (also called #10 washers)

✳ **NOTE:** *If you prefer, you can use storebought tuning hardware rather than building your own pegs.*

---

3. Note that a 1×2 board actually measures 3/4 by 1 3/4 inches. These reduced dimensions are the result of how lumber is milled and dried, but also happen to be a better size for a stringed-instrument neck. Win-win!

FIGURE 10-2: *Tools and supplies (not shown: soldering kit, amp, and instrument cable)*

Building the Robo-Tiki Steel-Stringed Ukulele

It may look daunting, but this project breaks down into four parts: building a pickup, building a neck, drilling some holes in a cigar box, and then screwing it all together. After that, it's just stringing and tuning, as with any store-bought instrument. Let's get started!

Build the Pickup

Flip back to the Playing-Card Pickup (Project 9) for complete instructions on building and testing the uke pickup using the parts listed in "Supplies for the Pickup" on page 146. Although the 30-gauge version works great, if you think you might want to use this uke with effects—such as the Twin-T Phaser/Wah (Project 14)—and have a little patience for handling the more delicate 42- or 43-gauge winding wire, you should seriously consider building the hotter pickup. It's well worth the extra effort!

Build the Neck

Step 1   The first step is to prepare the neck of the ukulele. You'll want a length of hardwood 1×2, about 22 to 24 inches long. Poplar, white maple, or sapele are all good wood choices. You can leave this board a little longer for now and trim it later based on the dimensions of your cigar box or how you want to anchor a strap. (The neck in the illustrations is 24 inches long, as the lovely cigar box I used for the body was slightly wider than average.)

In these instructions, the neck's *B-side* is its right side, as shown in Figure 10-3. This is the side of the neck where a right-handed player's fingers rest when holding the uke in a playing position. The other side is the *A-side,* where a right-handed player's thumb might rest while playing (although that's a bad habit—try to center your thumb behind the neck as you play). These long edges are fairly sharp. As you play, you'll run your fingers up and down both back edges of the neck, as well as the front edge on the B-side. Rounding them off will make the uke significantly more comfortable to play.

Decide which 2-inch face looks nicest and declare that the *fingerboard* (which is another name for the front of the neck, because it's the board your fingers press the strings against while playing). We're going to round off both the A-side and B-side back edges of the neck, as well as the B-side edge of the fingerboard.[4] You have a few options here. With a little determination, this can be entirely accomplished with nothing more than the sheet of 100-grit sandpaper; just keep sanding until you can run your hand up and down the edge smoothly, with no worries about cuts or splinters. You can speed this up by first whittling the long edges of the neck down with a pocketknife, taking off that keen hardwood edge, and then smoothing with the sandpaper. Or, if you have an electric sander handy, you can grind that edge to silky smoothness in just a few noisy, sawdusty minutes.

---

4. Because you won't touch the A-side edge of the fingerboard while playing, you shouldn't bother sanding it now. Keeping this edge sharp can also make setting the frets a little easier in Step 9.

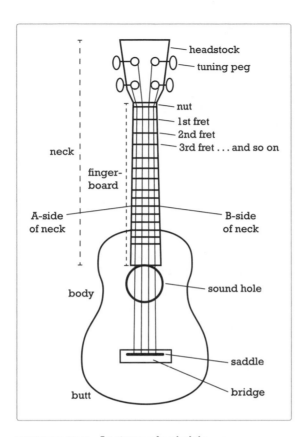

FIGURE 10-3: *Anatomy of a ukulele*

Step 2   Now add the guidelines for the nut, pickup, saddle, and bridge (illustrated in Figure 10-4). Place the sanded uke neck on your workbench with the A-side of the neck closest to you. This places the head (where the tuning pegs will be located) to your left and the butt to your right. Measuring from the left end of the neck, draw lines at 4 inches (for the nut), 15 1/2 inches (the middle of the pickup), 19 1/2 inches (the saddle), and 20 1/2 inches (the bridge). You want these to be as perpendicular to the neck as possible, so use a square when drawing them.

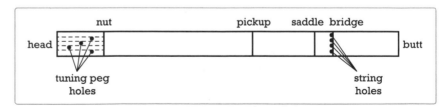

FIGURE 10-4: *This diagram shows, from left to right, the placement of the nut, pickup, saddle, and bridge. The dashed lines on the headstock indicate the guidelines for Step 3, and the dots at the head and butt show the placement of the tuning pegs and string holes, respectively.*

Step 3   Mark the headstock for drilling. The 4 inches of wood to the left of the nut are the *headstock*, which is where we'll mount the tuning pegs. Draw four guidelines perpendicular to the nut and running all the way to the end of the headstock; use the square to make them as straight as possible. Start 7 mm from the B-side of the neck and space these lines 10 mm apart. You'll thus end up with lines 7 mm, 17 mm, 27 mm, and 37 mm from the B-side edge, as shown in Figure 10-5 (left). These lines will correspond to the spacing of your strings, and we'll use the same spacing to mark the string holes at the bridge line in Step 4.

Now we'll mark where you'll actually mount the tuning pegs. Place the first mark next to the 7 mm line, 1 inch up from the nut line, as shown in Figure 10-5. The second dot goes 2 inches up from the nut, beside the 17 mm line; the third goes 3 inches up, next to the 27 mm line; and the fourth goes 1 inch up, beside the 37 mm line, mirroring the first dot. This placement is all shown in Figure 10-5 (right).

✳ **NOTE:** *If you want to do something fancy with the headstock—like cutting it at an angle or adding a little flourish—this is a good time to do so.*

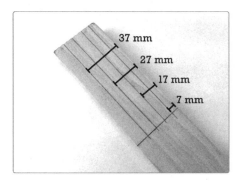 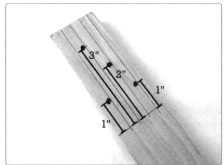

FIGURE 10-5: *The headstock, showing string-spacing guidelines (left) and markings denoting where the tuning pegs will be placed (right)*

Step 4   Next mark the bridge for drilling. Measuring from the B-side edge of the neck, place dots on the bridge line at 7 mm, 17 mm, 27 mm, and 37 mm (see Figure 10-6). These are the points where the guitar strings will be anchored to the body.

Step 5   Time to drill the headstock and bridge, making a total of eight holes. The four tuner holes in the headstock are 3/16 inch, and the four string holes along the bridge line are 1/8 inch.

Step 6   Now we'll mark where the frets will go. This is, by far, the most persnickety part of this project, so work carefully, double-check all measurements, and use a square to make these lines absolutely perpendicular to the neck and dead parallel to each other. Otherwise, you'll throw off the instrument's *intonation* (the consistency and accuracy of pitch), which will frustrate you later. Use Table 10-1 to guide you. All measurements are from the nut line.

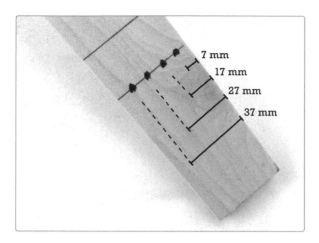

7 mm
17 mm
27 mm
37 mm

FIGURE 10-6: *Placement of the string holes*

TABLE 10-1: Distance from Nut to Each Fret

Fret number	Distance from nut
1	22 mm (2.2 cm)
2	42 mm (4.2 cm)
3	62 mm (6.2 cm)
4	79 mm (7.9 cm)
5	97 mm (9.7 cm)
6	112 mm (11.2 cm)
7	127 mm (12.7 cm)
8	142 mm (14.2 cm)
9	157 mm (15.7 cm)
10	170 mm (17.0 cm)
11	182 mm (18.2 cm)
12	193 mm (19.3 cm)

Step 7   Next, you'll slot the neck. This means cutting the initial slots at each of the 12 fret lines as well as cutting the nut and saddle slots. We aren't using commercial fret wire, so our slots will need to be a bit thicker than normal. Use a standard cross-cut wood saw here, as opposed to a hacksaw or *ryoba* (a Japanese-style pull saw with a thin, flexible blade). A standard wood saw's slightly thicker *kerf*, or cut width, should accommodate your frets' thicker profile nicely.

All of these cuts need to be parallel to each other, so use a miter box to guide the saw, as shown in Figure 10-7. Even the cheapest miter boxes will give you 90-degree and 45-degree angles, and we need only 90 degrees for this project. If you don't have access to a miter box, you can carefully clamp your square to the neck and use that as a guide for each cut.

Fortunately, while the orientation of the cuts is important, the depth is not. Aim for around 2 mm, but don't despair if you go a touch deeper.

 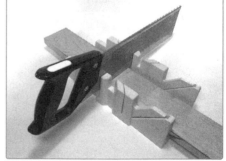

FIGURE 10-7: *A miter box is used to cut accurate, precise, reproducible angles (action shot on the right).*

Step 8   Now to notch the nut and saddle slots. Use your triangular file to widen these two slots into triangles. Each should be about 2 to 3 mm wide at the surface of the fingerboard and taper to a point as it goes into the neck.

* **NOTE:** *In the next step, you'll press the finishing nails into the fret slots so that the sharp tip of the nail rests just about level with the top of the fingerboard. If you're using sapele or a similarly dense hardwood, you may want to widen the fret slots a bit now, using your triangular file, so that each slot is a millimeter or two wide.*

Step 9   You're ready to install the frets. We'll just install one at first, to get the feel for the process. Select 12 perfectly straight finishing nails. Make sure they're all the same length and have no deformities. Next, squirt a line of control-gel superglue into the first fret slot, starting about 1/4 inch from the A-side edge of the fingerboard, where the nail head will ultimately rest, and stopping about 1/4 inch shy of the B-side edge (see Figure 10-8).

FIGURE 10-8: *Laying glue for the second fret*

Place the nail in the groove so that the sharp point doesn't quite reach the B-side edge of the fingerboard. Wipe away any excess glue with a paper towel—remembering the warning on page 134 about superglue and textiles! Finally, clamp your scrap piece of hardwood against the fingerboard, sandwiching the nail between the pieces. Tighten the clamps until the two pieces of hardwood are almost flush and the sharp tip of the nail is even with the surface of the fingerboard (see Figure 10-9). If you're working with denser hardwood, you might want a table vise for this part (although a combination of C clamps and prenotching the slots with a file will work fine, even on the densest neck wood). Once you've gotten a feel for this, you can pretty safely install frets in this fashion two at a time. After all the frets are installed, give the glue a minute or two to set before moving on to the next step, routing the neck to accommodate the pickup.

FIGURE 10-9: *Pressing the frets into place with a set of clamps*

Step 10 The pickup cavity needs to be about 1 1/8 inches wide—i.e., wide enough to accommodate the pickup you made using the instructions in Project 9—and centered on the pickup line we drew in Step 2. Draw two guidelines, one at about 9/16 inch above the pickup line and the other 9/16 inch below (obviously, if you've made a pickup with significantly different dimensions, you should adjust these measurements accordingly). Once you draw these guidelines, check them against your actual pickup rather than the ruler. If the lines are far enough apart to accommodate your pickup, then saw about 1/4 inch into the neck along each, as shown in Figure 10-10. The depth of the cuts should match the pickup's height so that the pickup's top plate will ultimately rest roughly flush with the fingerboard.

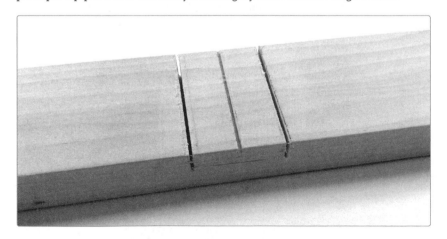

FIGURE 10-10: *Preparing to rout the pickup cavity*

Step 11   Next, we'll rout[5] the cavity with a 1-inch spade drill bit, like the one in Figure 10-11. You'll accomplish this by drilling three holes about as deep as the cuts you made in the previous step.

FIGURE 10-11: *A 1-inch spade drill bit*

The goal is to clear the bulk of the wood out of your pickup cavity but to leave enough wood for the neck to provide structural stability to the finished uke. To drill the first hole, set the tip of your spade bit on the pickup line, about 1/2 inch from the edge of the neck, so that the spinning bit will just nick the edge (you can see the results in Figure 10-12a). Drill down about 1/4 inch (the depth of your cuts from Step 10). Drill the second hole tangent to the first, as in Figure 10-12b, and finally drill out the center, as in Figure 10-12c. With the bulk routing done, it's time to use your pocketknife and triangular file to clear out the remaining nubbins of wood and square off the cavity. Once you have a cavity that your pickup will fit into, clean up the edges with sandpaper. The results are shown in Figure 10-12d. Your neck is done!

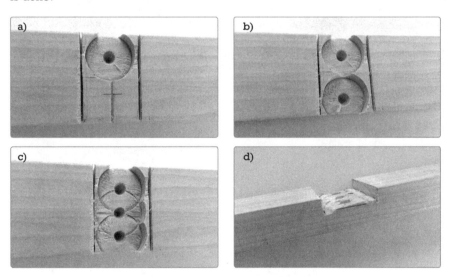

FIGURE 10-12: *Routing the pickup cavity in four steps*

....
5. We're not technically *routing* this cavity, as we aren't using a *router*, which is a power tool created for this task. Though a router does a much better job, this way works, and it's quick, easy, and cheap. If you have access to a router and know how to use it (or have a pal to show you), then you'll be able to make a beautifully routed pickup cavity with less effort. Do so!

Build the Body

Step 12 Prepare the cigar-box body. Decide which side of the cigar box you want people to see when you play. This side is the *soundboard*, and it's responsible for much of the uke's tone. Lay the neck down the center of the soundboard as a guide. Then mark the places where you want to have your uke's sound holes and jack. For optimal acoustic performance, you'll want two sound holes in the soundboard and a jack mounted on the side or butt of the uke's body. At this point, you might also want to mark the location of the neck with a pair of light pencil lines, if you're concerned about getting the same alignment later.

Drill the sound holes with the 1-inch spade bit and the jack hole with a 3/8-inch bit. Clean up the holes with sandpaper (a small rattail or half-round file is very helpful here). Figure 10-13 shows my cigar box after drilling. (I placed the jack hole on the soundboard instead of the side because my cigar box had very thick walls.)

＊ **NOTE:** *If you'd rather go all electric and replace the cigar box with a non-resonating alternative, see "Tips, Tricks, and Mods" on page 164 for some notes and ideas.*

FIGURE 10-13: *The sound and jack holes in the cigar-box body*

Step 13 Now mount the neck to the body. It's good to have the 12th fret just clear of the body so it's easy to hit those high notes. With many cigar boxes, you can mount the neck this way and still have the string holes at the butt of the neck hang clear of the cigar box, as in one of the several ukes shown in Figure 10-22. This makes for easy stringing later on. If you end up with the string holes lying atop the soundboard, as they do in Figure 10-1, don't worry: all you need to do is drill 1/8-inch string holes through the soundboard. Protip: you'll avoid headaches if you keep the string holes at least 1/2 inch from the cigar box's walls.

Once you settle on an alignment, use your two 3/4-inch wood screws to attach the neck to the soundboard from the inside (see Figure 10-14). If you want to pre-drill guide holes for this (which you probably will if you're using anything denser than poplar), use a 1/8-inch bit. (Just be very careful not to pop through the fingerboard!) If your string holes are resting on the soundboard, then this is the time to use your 1/8-inch bit to drill through the soundboard using the string holes in the neck as a guide.

FIGURE 10-14: *Anchoring the neck to the uke soundboard (the circles indicate the location of the two neck-mounting screws)*

Step 14   Now to install the pickup. Scrape off any excess wax lumps on the pickup body, and if there's a jack connected to your pickup, remove it.

Drill a 3/16-inch to 1/4-inch hole in the bottom of your pickup cavity in approximately the same location as the leads coming off your pickup. Make sure the hole goes all the way through both the neck and the soundboard, as shown in Figure 10-15. Slip the pickup's leads through the hole, and press the pickup into place. In many cases, you'll find that the pickup fits pretty snugly and will stay put thanks to the beeswax coating. If you feel like it needs a little extra anchoring, squirt some Gorilla Glue behind the pickup and let the glue dry for 24 hours.

✳ **NOTE:** *Gorilla Glue expands as it cures! A little smear can turn into an unsightly glob in 24 hours. Follow the instructions on the packaging and be tidy as you work.*

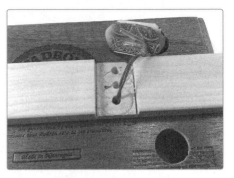

FIGURE 10-15: *The hole for my pickup's leads aligns with one of the spade bit's guide holes from Step 11. You can see that more clearly with the pickup wires installed (right).*

Step 15 Once the pickup is securely mounted and any glue is dry, mount the jack in its hole in the uke body and solder its leads (see Figure 10-16). Hook this to your amp, power up, and confirm that the pickup still works.

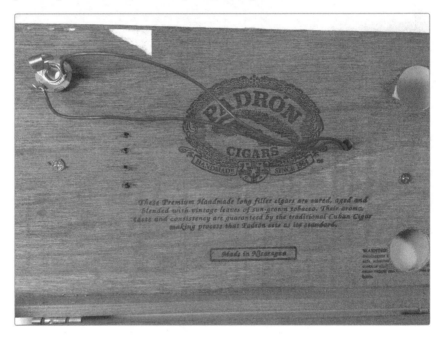

FIGURE 10-16: *The jack and pickup leads in place inside the cigar box (you can also see the string holes I drilled through my soundboard)*

Add Pegs and Strings

Step 16   Install the tuning pegs. First, clean up the headstock: erase those unsightly guide-lines you drew earlier and do any final sanding. Next, take one of the four #10 eye-bolts and slide a #10 washer on to it. Working from the front of the neck, slide the bolt through one of the holes in the headstock and add a second #10 washer behind the neck so that you end up with a washer-neck-washer sandwich. Finally, spin on a wing nut until it's tight, as in Figure 10-17. Repeat for the other three holes. (If you have commercial tuning pegs or tuning machines, you're on your own, kid; do whatever makes sense for what you bought.)

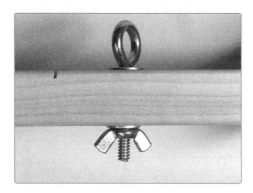

FIGURE 10-17: *A homemade tuning peg*

Step 17   String the uke. If you're familiar with stringed instruments, you might find the stringing diagram in Figure 10-18 strange. But don't worry—it's correct! Unlike other stringed instruments, the uke isn't strung lowest to highest. Its lowest note—the C—is actually the third string (counting from the B-side of the neck), and the next lowest is the second string.

Of the six electric guitar strings we bought, we'll use the ones marked 8, 11, 14, and 30 (or, as a guitarist would call them, the high E, B, G, and A strings) as our E, A, G, and C uke strings. But don't throw out the other two! The 22 can also work as a uke C string, and the 38 would make a good electric diddley bow string. (For full diddley bow build instructions, check out my first book, *Snip, Burn, Solder, Shred.*)

Start with the string for the uke's G note, the fourth string. Slide it up through the string hole closest to the A-side of the neck and feed it through the eye of the tuning peg. Pull the string tight and then back it up, adding 1 inch of slack. Bring the free end of the string toward you (counterclockwise), loop it under the length of the string, and pull the end of the string up and back, crimping it where it crosses under itself. These three steps are demonstrated in Figure 10-19.

Finally, wind the peg counterclockwise several turns (you may need to tighten the wing nut as you do so), but don't make the string taught yet. Finish by clipping off the dangly tail. Repeat this process, adding the C, E, and A strings in that order.

String number on uke	4 (top string)	3	2	1 (bottom string)
Open note on uke	G	Middle C	E	A
Open note on guitar	G	A	High E	B
Ernie Ball Extra Slinky string thickness (in thousandths of an inch)	14	30	8	11

FIGURE 10-18: *The diagram on the left shows the open notes on a uke with standard tuning. The chart on the right indicates which electric guitar string serves for which string on our uke.*

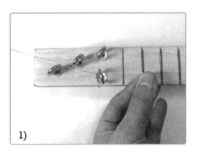
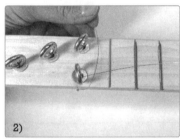
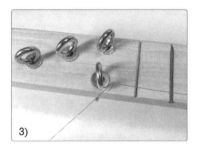

FIGURE 10-19: *The three steps to properly stringing any steel-stringed instrument: (1) give 1 inch of slack, (2) pass the free end under, and (3) kink it up.*

Step 18   Let's add the nut, saddle, and string trees. Grab your bamboo skewer and 5/16-inch coarse-threaded carriage bolt, which are your nut and saddle, as well the last two #10 washers and the last two 3/4-inch wood screws, which we'll use to fashion a pair of *string trees*. Slide the carriage bolt into its notch at the saddle line, making sure that the strings lie along threads that line up nicely with the string holes. Avoid odd angles and tensions, as they make tuning a pain. Then slide the bamboo skewer nut into its notch at the head. Cut it down so its width matches the neck of your uke. Tighten the strings a touch to hold everything in place. It's highly unlikely the strings are firmly resting on the nut right now, so we'll add a set of string trees. These redirect the tension of the strings, forcing nice solid contact with the nut. You'll find a similar feature in classic straight-headstock guitars, like the Fender Telecaster.

Our string trees are just 3/4-inch wood screws with #10 washers. Drill 1/8-inch guide holes not more than a 1/4 inch deep between the first and second string. Put the washer on the screw and then drive the screw into the headstock, as in Figure 10-20. Leave a gap between washer and wood; you want to press firmly against the string but not lock it down (the string needs to be able to slide under the string tree as you tune the uke). Repeat this for the second string tree but place it between the third and fourth string.

FIGURE 10-20: *Placement of the string trees*

Tune and Test

Step 19   Time to tune 'er up! The strings of the uke are G, C, E, and A (see Figure 10-18). For your initial tuning, you should get those strings as close as possible to their proper pitches. Use a good electronic or software tuner. Free tuner software and mobile apps abound.

New strings will need to stretch, and new tuning pegs are often pretty loose, so this first tuning will be a bit repetitive. Tune everything up, starting with the C and then tuning the E, G, and A. Then, go back to check the C; it'll probably be flat (i.e., a touch lower pitched than it should be). Tune it back up, tightening the pegs as you do so, and repeat until everything is in tune. Let the uke rest for an hour or two and check it again. Retune, tighten the pegs, and you should be good, but don't be surprised if it drifts out of tune overnight. After a day or two, the strings should finish stretching, and the uke should be rock solid. If not, try tightening your pegs further.

Tuning by Ear

If you don't have a tuner handy, you can still get your uke in playable condition by tuning it relative to itself. Assuming that the third string (the thickest) is at middle C (often labeled C_4 on electronic or computerized tuners), hold that string down at its fourth fret and pluck it to hear the proper pitch for the E string (which is the second string). Tighten the second string until it matches this pitch. Now, if you hold the second string at the third fret and pluck it, you'll hear a G. Tune the fourth string to that note. Finally, once you have that fourth string tuned to G, pluck it while holding the string down at the second fret to hear an A, the note of the final string.

Learning to tune a uke to itself is useful when you don't have other equipment, but it also offers three big advantages for the non-pro musician. First, tuning this way trains your ear to pick out what "sounds right" on the uke. In music, especially folky music, "sounding right" trumps "being right" every time. Second, your uke will sound good even if it's technically out of tune because the strings are tuned at the appropriate intervals. Melodies and chord progressions will be recognizable (although they might technically be in the "wrong" key). Third, you can tune your "C" string to something that sounds good with your singing voice and then tune the rest of the uke from that baseline (technically *transposing* the uke to a more favorable key). That way you don't have to force yourself to attempt to sing in a key that doesn't work for you. To heck with the arbitrary standards of those fat-cat music teachers!

Step 20   Test each string, working up each fret and listening for clarity. Do you hear buzzing when you play a particular string at a particular fret? If so, the next fret down from the one you are stopping the string against is a touch too tall. Try giving it a little squeeze with the clamp at the point where the buzzing string crosses the fret. If that doesn't solve the problem, loosen the strings, file down the top of the fret, and then recheck. It's not uncommon to have a fret or two buzz. If they're all buzzing, then you might need more distance between the fingerboard and the strings. Check the slots for the nut and saddle. If these are crazy deep, consider replacing the bamboo skewer nut with a small bolt.

Step 21   As you test the strings, also listen for points where you move up a fret but the note doesn't change. That usually means that the next fret down from the missing note is so tall that it's prematurely stopping the string. Treat this like a buzzing fret and see whether that fixes the problem. Otherwise, the fret you're trying to stop against is probably too low. Remove the strings carefully, pop the fret out with a hammer and flathead screwdriver, and install a new one. (When you press the new fret into place, you may need a smaller hardwood block than you used during the initial installation to avoid disturbing the neighboring frets.) If you have lots of misses, check the uke for structural problems. Are your nails from two different batches (and thus different diameters)? Is the neck warped? If you see such issues, you may need to rework parts of the uke.

Step 22 Once you've eliminated any misses and buzzes, check the intonation. First, find the *harmonic* on each string. Measure the distance between the nut and saddle and divide this in half to find the center of the section of the string that can freely vibrate. Mute the center of each string with a light touch of your left hand (don't press hard enough to touch the fingerboard or fret; just lay your rigid finger across the strings). Then, pluck each string with your right as you would when playing normally (that is, in the general vicinity of the pickup or sound holes). A high, pure version of each string's open note should quietly sing out. If you just get a "thunk" (as you would when lightly stopping most other points along the string), try sliding your hand up or down the neck a few millimeters and pluck again. (This is very much a "one video is worth a thousand words" situation; look online for a video on *ukulele harmonics* to see what I mean.)

You should find that your center harmonic is right over the 12th fret. If it hovers around the 10th or 11th fret instead, make sure the saddle and nut are both secure in their notches (the saddle is most likely the culprit; it can hop out of its notch during the tuning and retuning process). If the saddle and nut are fine but the harmonic still isn't in the right place, you may have messed up the measurements in Step 6. The distance from your nut notch to your saddle notch should be within a millimeter or two of 392 mm.

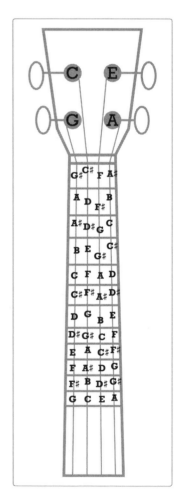

* **NOTE:** *If you find the harmonic but it's way up around the 5th fret, then don't worry: you've just found the second easiest to play harmonic. Try muting around the 12th fret and plucking again.*

Step 23 Once you have the harmonics over the 12th fret, it's time to check the intonation of each fret. Get out your chromatic tuner, check out Figure 10-21, and work down the neck, fretting each note on each string. Check the pitch that you're getting against the pitch you *should* hear (according to the chart). Ideally, they'll be spot on; the intonation in this design is pretty good. Fretting at the 12th fret, each string should be one octave above its open note (the C string, for example—which the computer calls C_4—should be C_5 at the 12th fret).

FIGURE 10-21: *This diagram of the ukulele's neck shows where all the notes are.*

If the octaves are sharp or flat, check your saddle bolt. If the saddle bolt keeps popping out of its groove, then you may need to cut the groove in the neck a tad deeper with your triangular file. It's common to find that all of the notes are sharp at the first few frets closest to the headstock but are in tune farther down the neck. Sharp notes here mean that the nut is too high; remove the nut, file the nut notch deeper into the neck, replace the nut, retune, and check the notes again. On the other hand, if all of the strings are sharp at the 10th, 11th, and 12th frets, then the saddle is too high. Loosen the strings, remove the saddle, and file the saddle notch a little deeper into the neck. Then, put it all back together, retune, and check the intonation again. If most of the strings are fine but one is still a touch too sharp (the second string—E—is often the culprit), then use your pocketknife to cut a little slot into the nut for the string to rest in. Once the string is in its slot, tighten its string tree a tad. This lowers the tension on the string when you're playing to slightly flatten the note and bring it into alignment with the uke as a whole.

All this tinkering—called *setup* by guitarists and ukulelists—is usually necessary only when an instrument is brand-new or has been neglected for a long time, especially in an environment with poor temperature and humidity control, like a hot attic or damp basement. Once you get a little practice setting a uke up, you'll have a powerful tool in your musical arsenal, because you'll be able to pick up cheap or beat-up ukes worth salvaging and coax them into playable condition.

Playing the Robo-Tiki Steel-Stringed Ukulele

The Robo-Tiki Steel-Stringed Ukulele is a legit musical instrument, so I'm going to largely defer to *actual* uke instructional materials here. I'll offer one piece of concrete playing advice: *keep your thumb behind the neck!* It is very common for ukulelists to allow the neck to rest in the crook of their left hand while playing, with the thumb lightly (and uselessly) curved over the upper edge. Guitarists already know this is a bad habit because it robs your hand of the strength it needs to get a good sound out of taut steel strings. Ukulelists can get away with bad hand posture because most ukes have soft nylon strings and commercially built ukulele necks are fairly narrow. Playing the Robo-Tiki, however, requires full hand strength, so do yourself a favor and keep your thumb centered behind the neck.

One final tip: nylon-stringed ukes sound basically the same strummed with fingers or a pick (during the first uke craze, in the 1920s, fat felt picks were all the rage). Steel strings, however, offer a much broader palette, especially when amplified. Invest in picks of various weights and materials and find a sound that suits you.

Tips, Tricks, and Mods

Before we talk tips, consider our uke's design. The Robo-Tiki uke is a true steel-stringed, electro-acoustic ukulele, which is rare in the open market. It's a *concert* uke, which means it's a slightly larger version of the standard *soprano*[6] ukulele and thus offers greater volume and sustain. This uke's total *scale*—the length of string that actually vibrates to produce a given note or chord—is 392 mm. That's a bit long, even for a concert uke, but going with a longer scale means the strings have less tension than a conventional soprano or concert uke, making the Robo-Tiki uke a bit more forgiving to build and a bit easier to play.

Mass-produced ukes have bolted-on necks because stringed instruments are cheaper to mass-produce that way. Ours has a *neck-thru*—that is, a *neck through body*—design, just like a high-end electric bass guitar. This purportedly offers better sustain and undeniably makes for an easier build. By anchoring both ends of the strings to the same beam, we totally eliminate the need for cross-bracing in the body and thus can use relatively fragile or inconveniently shaped items as soundboards and bodies. Neck-thru ukes are also sturdy instruments; you can bludgeon a wild beast with it and still be in tune. The only downside is that we've anchored the neck to the outside surface of our soundboard. This somewhat mutes the purely

---

6. There are also tenor and baritone ukes out in the big wide world; a tenor uke has a somewhat different tuning from the standard soprano, and the baritone is somewhat more akin to the guitar.

acoustic volume (i.e., when you aren't plugged into an amp) and can prove awkward to handle if you choose a deep resonator. (Takeaway lesson: consider using a relatively shallow cigar box.)

That said, this design gives you a lot of options, both aesthetically and functionally, as you can see in Figure 10-22. On the strictly decorative end, you could show a little flair on the headstock and tailpiece. If you decide to swap in store-bought tuning pegs, please note that our homemade ones will probably perform *better* than the cheaper factory-made friction tuning pegs. Either invest in the nicer geared tuning pegs—called *tuning machines*—or save both cash and frustration by sticking with the pegs you made yourself.

FIGURE 10-22: *A variety of acoustic and electric ukes riffing on this project's core design*

One very quick and easy aesthetic improvement is adding *fret markers*. These little dots along a stringed instrument's neck aren't just an opportunity for individualization; they also make the fingerboard easier to navigate. On a uke, markers are traditionally placed at the 3rd, 5th, 7th, 10th, and 12th frets. You can keep it simple—dots of marker or paint work fine—or get fancy by burning in little chevrons with the tip of a hot soldering iron. For something more like the inlaid pearl dots commonly seen on guitars, consider drilling shallow holes in the fingerboard and gluing in disks of contrasting wood cut from a dowel. You could also buy decals that convincingly approximate real pearl inlays. Jockomo, a Japanese company that sells these online (*http://inlay-stickers-jockomo.myshopify.com/*), is a popular brand that offers a variety of ukulele-specific inlay decals for under $10.

Swap Out the Cigar Box

Different bodies offer different tones and amplification when playing acoustic-only. Cigar boxes are nice because the thin, rigid wood makes them decent resonators. Cookie tins and other metal containers are less consistent but can give you a bright banjo voice. Steel mixing bowls and dog bowls have a wonderful singing tone, although they can also make your uke awkward to play. As long as you're experimenting, consider buying some thin, attractive veneer wood to use as a soundboard and then attaching that to a heavy-duty plastic food-storage container to form the body. It sounds cheap and janky but can have a nice effect, both musically and aesthetically (think of the famous Ovation round-body acoustic guitars, which have wooden soundboards and fiberglass bodies). Depending on your uke body's dimensions and your playing style, you may discover that portions of the upper edge of the fingerboard near the pickup and tailpiece, which we didn't sand, wind up rubbing you the wrong way as you strum. Sand these down if they bother you.

Straight Acoustic or All Electric

This design can be easily adapted to a straight acoustic ukulele, simplifying the build in a number of ways. Not only will you skip building the pickup, but you also won't need to rout the neck to accommodate the electronics. Because you aren't using a magnetic pickup, you won't need steel strings. If you go with nylon strings, then you can use much softer material for the frets: replace the nails with flat-sided toothpicks superglued directly to the fingerboard, no slotting required.

One tip for nylon strings: avoid the terrible "crystal clear" strings installed on most entry-level ukes. The widely available GHS H-10 black nylon uke strings cost only about $3, are more comfortable to play, and sound better. If you want a pro-quality string, Ko'olau Gold Strings are pricey but really nice. While our uke is technically concert sized, many strings for a standard—that is, soprano—uke should fit. The only trick is that, because our bridge isn't designed for non-ball-end strings, you'll need a way to secure the loose ends to anchor them beneath the bridge. Tying them off on a length of bamboo skewer, matchstick, scrap wood, or even a few shirt buttons will work just fine. (If you have trouble with the knots coming undone, seal each knot with a little bead of superglue.)

Conversely, you could go all electric. To build a flat body for your all electric uke, take a 1-inch-thick board, cut it to any shape you choose, finish it to your liking, and bolt it to the neck. Traditionally, with a neck-thru design, the body of the instrument is actually a set of "ears" or "wings" that are bolted to either side of the neck. You can make a really slick solid-body uke using the bolt-on-ears technique. Because you aren't relying on the body for any of the instrument's sonic character, you can even go with a totally minimal *uke stick* that's nothing more than the electrified ukulele neck, with the body reduced to an Altoids tin housing the pickup wiring and jack (there's a uke stick at the bottom of Figure 10-22).

The pickup itself can also be tweaked for better performance. The closer the pickup is to the strings, the more volume and presence you'll have. To raise the pickup, glue a few pennies behind it. Truly cunning builders might even fabricate a lever-and-hinge arrangement so they can vary the pickup-to-string distance while playing, a sort of Doppler-effect whammy bar. The pickup's proximity to the saddle also affects tone, which is why most modern guitars have several pickups mounted at different distances from the neck and bridge. You could move your single pickup closer to the saddle if you prefer the bridge tone. We've left plenty of room on the neck to add a second pickup (there's an example of a two-pickup uke in Figure 10-22). A two-pickup configuration can be controlled using a single SPDT switch wired as shown in Figure 10-23 or with a pan-pot arrangement akin to the one in the Twang & Roar Kalimba (Figure 11-12 on page 182).

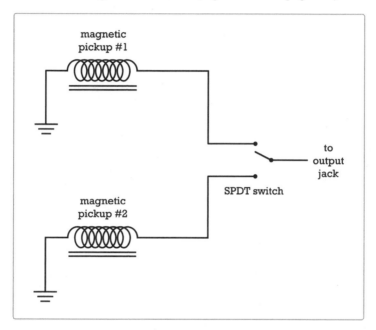

FIGURE 10-23: *Wiring diagram for a dual-pickup uke*

Finally, as mentioned in the Playing-Card Pickup instructions, this pickup design works just as well with true guitar-pickup winding wire, which is 42 or 43 gauge—significantly thinner than the 30-gauge magnet wire sold for kids' science-fair project electromagnets. Pickups wound with this thinner wire are significantly *hotter*—that is, *louder*—than 30-gauge pickups. These hotter playing-card pickups made with 42- and 43-gauge wire generally perform at the low end of commercially available guitar pickups. Subsequently, they tend to get along better with recording equipment and effects pedals (both homemade and store-bought) than the standard 30-gauge pickup does. Working with 42-gauge wire can be frustrating—it's hair thin and breaks much more easily than 30-gauge wire—but the results are generally well worth the added frustration. If you're digging your Robo-Tiki uke, I *strongly* urge you to make one with a hot-rod pickup.

Resources

The Internet is awash in homebrew (and occasionally dubious) chord transcriptions of every song imaginable. Just search for a song's title followed by the word *chords*, and you're on your way. Chord breakdowns of songs are fairly instrument agnostic. If you're strumming a uke, guitar, mandolin, or lute, Bob Dylan's "Quinn the Eskimo" is still going to be A-D-A-D-A-D-A-D-A-D-A-D-A-E-D-A (with a A-A-E-A-A-E-D-A played twice for the chorus)—repeated ad infinitum.

You'll find plenty of free chord breakdowns online, but there's still a soft spot in my heart for the ukulele books by Jumpin' Jim Beloff. They offer excellent uke arrangements of popular songs at a variety of skill levels. My first uke-specific music book was *Jumpin' Jim's '60s Uke-In*, which is a little Beatles heavy but has great versions of Otis Redding's "Dock of the Bay" and Roger Miller's "King of the Road." As an added bonus, each of Jumpin' Jim's books includes an extensive ukulele chord chart in a two-page spread at the beginning. The charts alone are invaluable for the beginner, though I'm sure you could find similar charts online and print them out yourself.

Mel Bay—that old standby for mandated after-school music lessons—continues to offer an almost mind-numbing array of materials for all instruments, including inexpensive entry-level books filled with no-frills arrangements of royalty-free folk and traditional tunes. *Fun with Folk Songs* and *Songs and Solos for Uke* have given me literally thousands of hours of enjoyment, and neither cost more than $10 with shipping.

As your uke skills improve, I'd also recommend the venerable *fake books*. These thick, spiral-bound tomes are available at any guitar shop, with titles like *Classic Rock Fake Book: Over 250 Great Songs of the Rock Era*. Each entry in a fake book includes the sheet music for the melody, complete lyrics, and a breakdown of the basic chords for that song coordinated with the appropriate lyrics. Musicians use these books to learn songs they're vaguely familiar with, allowing them to "fake it" with minimal preparation. Like the chord transcriptions you'll find online, fake book lead sheets apply to all instruments, so there's no need to chase down the *Ukulele Classic Rock Fake Book*—the one the guitarists use will suit you just fine. You'll be shocked at how upbeat you can rock Pink Floyd's "Brain Damage" on four strings and a box.

11

THE TWANG & ROAR KALIMBA

The kalimba, or *thumb piano*, has a sweet, harmonically complex music-box voice—when you think "generic African music" you are likely imagining lots of hand drums and a kalimba. Despite being so immediately recognizable, this instrument is really distinguished by its diversity: it has a dozen names throughout sub-Saharan Africa, and it has been crafted from every material imaginable—from scrapyard remnants to now extinct hardwoods, with countless idiosyncratic tunings. The Twang & Roar Kalimba is made from common materials and has a Westernized tuning that's easy to improvise on, and with the addition of just two extra tines, it's ready to play lots of familiar folk tunes—a mod discussed in detail in "Expand Your Scale" on page 193.

While our kalimba is a perfectly usable acoustic instrument that's great for hikers and campers, it's also kitted out with a pair of pickups—one magnetic, the other piezo—so that you can plug in and crank up the twang and roar. The two pickups connect through a simple pan-pot mixer, which allows you to smoothly transition between them, with lots of tonal options in the middle.

The piezo pickup amplifies any contact made with the instrument, bringing out the aggressive twang of the tines as well as the thump, rasp, and rumble of the player's hands on the body of the instrument. The piezo is also loud and distorts wonderfully.

In contrast, the magnetic pickup amplifies only the vibration of the metal tines themselves, for a more natural tone. Because those tines are fairly massive pieces of vibrating metal (by guitar-pickup standards), they can really rattle that magnetic field. Play hard, and you'll get a great Rhodes electric piano–style fuzzy bark tone.[1]

FIGURE 11-1: *The finished Twang & Roar Kalimba*

---

1. The Twang & Roar Kalimba's magnetic pickup arrangement is inspired by the internals of the classic Fender Rhodes Stage Piano.

There's a *ton* of room to improvise and expand on this kalimba design: you could build an all-acoustic or all-electric kalimba, or use a big booming oil drum or a tiny candy tin for the body. The first kalimbas, built 3,000 years ago on the west coast of Africa, had bamboo tines instead of metal ones. Today, folks make neat kalimbas with all sorts of upcycled tines: bike spokes, popsicle sticks, plastic sporks, and so on.

The following instructions assume that you're building a six-note kalimba using a box less than 5 inches wide, with metal tines tuned to the C pentatonic scale (see "Scales and Chords" on page 373 for more on the pentatonic scale). If you want to build something larger or with an alternative tuning, check "Playing the Twang & Roar Kalimba" on page 190 and "Tips, Tricks, and Mods" on page 191 before proceeding.

Hear the Twang & Roar Kalimba in action in the samples at *http://www .nostarch.com/jamband*.

Preparation

Build Time

▶ About 2 hours for a full electric-acoustic model, plus 4 to 24 hours of drying time for the glue (An all-acoustic kalimba can be built in about half that time.)

Tools

▶ A standard soldering kit (See page 340.)

▶ An electric drill with bits (You'll need 3/16-inch, 1/4-inch, and 3/8-inch bits. Depending on the exact dimensions of your potentiometers, a 5/16-inch bit might be handier than a 1/4-inch bit.)

▶ A 1-inch spade drill bit, also called a *paddle bit* (See Figure 10-11 on page 154.)

▶ A ruler marked in both inches and millimeters

▶ Any wood saw

▶ 8 1/2-inch bolt cutters (Larger bolt cutters are fine, too; a hacksaw will do in a pinch, but it's a pretty miserable alternative.)

▶ A screwdriver that matches your #8 machine screws (You'll almost certainly want a Phillips screwdriver.)

▶ Medium-grit sandpaper

▶ Pliers

▶ (Optional) A hacksaw (This tool is mandatory if you've skipped the bolt cutters.)

▶ (Optional) A tapered half-round file or other metal file with a flat face

▶ (Optional) A multimeter (See "Using a Multimeter" on page 349.)

▶ (Optional) An amp and instrument cable for testing the pickups and mixer

Supplies for the Body and Tines

▶ A cigar box (A narrow, deep box will be easiest to work with and play. For best results, get a box with a wooden bottom.)

▶ A 7/16-inch dowel, at least 4 inches long (Anything thinner will prove hard to drill.)

▶ A 1/4-inch carriage bolt, roughly as long as the width of your cigar box (The one shown in Figure 11-2 is 5 inches long. You could also get a 1/4-inch threaded rod and cut it to your desired length.)

▶ Two 1-inch-long #8 machine screws

▶ Two 1/2-inch-long #8 machine screws

▶ Two #8 wing nuts

▶ Two #8 nuts

▶ Six #8 washers

▶ 3 feet of 1.4 mm, or .055, music wire (See "Buying Music Wire" on page 174 for details.)

▶ A few rubber bands (The thick ones used to secure lobster claws or bunches of broccoli are great.)

Supplies for the Pickup and Mixer

▶ Control-gel cyanoacrylate glue (Any brand of control-gel formula superglue will do. I especially like Loctite Ultra Gel.)

▶ A playing card

▶ Four small neodymium disk magnets (Available in most hardware stores, these are often called *supermagnets* or *rare earth magnets*. I used disk magnets that are 8 mm in diameter and 3 mm tall, but others should work fine.)

▶ A few hundred feet of 30-gauge enameled winding wire (Also called *magnet wire*, this is regularly stocked by any hobby or science supply shop that carries electronic supplies, and it can also be found at many hardware and craft stores.)

▶ A chunk of beeswax roughly 1 1/2 inches by 3 inches, or about 2 ounces by weight

▶ 24-gauge insulated hook-up wire (Stranded wire is best.)

▶ A 1/4-inch mono phone jack, also called a *guitar jack*

▶ A piezo element, such as Digi-Key part #102-1126-ND (Any piezo element will do, but getting one with pre-soldered leads will save your sanity.)

▶ A 10k ohm audio potentiometer (This is a variable resistor with an audio taper; see "The Gory Details: Audio Taper vs. Linear Taper" on page 327.)

- A 100k ohm variable resistor with a linear taper
- Two control knobs that fit your variable resistors
- Silicone-based household glue, sometimes called *room-temperature vulcanizing rubber* or *RTV-1*

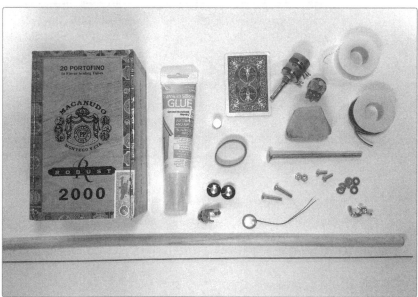

FIGURE 11-2: *Tools and supplies (not shown: multimeter, amp, instrument cable, and control-gel cyanoacrylate glue)*

Building the Basic Twang & Roar Kalimba

Similar to the Robo-Tiki Steel-Stringed Ukulele (Project 10), this project is a matter of building several subunits—an acoustic kalimba with a few extra holes for wires and hardware, a pickup, and a very simple mixer—and then bringing them together. We'll start by doing all the acoustic bits—that is, everything shown in Figure 11-3—so that those uninterested in going electric can be on their merry way after the final step on page 181.

Prepare Your Tines and Bridge

Step 1   Cut the music wire into six pieces, each 6 inches long. These will be the tines shown in Figure 11-3. Music wire is durable, precision-drawn, high-carbon, tempered spring steel, but you can snip right through it with a little muscle and a decent pair of bolt cutters. The cut ends will be pointed but not terribly sharp. You can easily smooth them with a metal file if they prove too prickly for your fingers.

You *can* use a hacksaw to cut the tines, but it's incredibly frustrating. Steel this thin is hard to secure while cutting; it springs around a lot, and you'll definitely need to file the playing ends down afterward. Sawing is not recommended, but it is possible. However you cut your tines, finish them by using pliers to bend over the last 1/4-inch of each tine (see Figure 11-4).

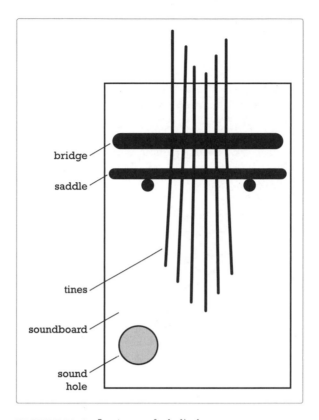

FIGURE 11-3: *Anatomy of a kalimba*

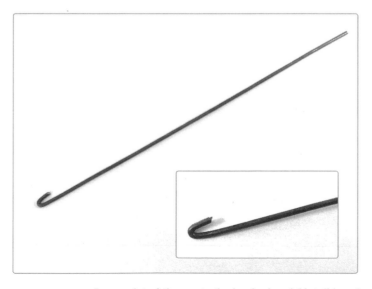

FIGURE 11-4: *A completed tine; note the hooked end (detail inset)*

Step 2   Prepare the bridge. This is a 4-inch length of 7/16-inch dowel with two 3/16-inch holes, 1 inch from either end of the dowel. (If that string of numbers makes you dizzy, then a quick look at the far-right image in Figure 11-5 will likely clear your head). Take your 4-inch dowel and draw a centered guideline down its length, as in the far-left image of Figure 11-5. Measure 1 inch in from each end and mark your drill holes, as in the middle image of Figure 11-5. Finally, drill those holes.

> * **WARNING:** *Drilling holes in the side of a cylinder can be tricky. Before pulling the trigger, ask yourself where that drill bit is going to end up if you slip. If you want to be extra safe, you can use a hammer and nail to pop little guide divots into the dowel before drilling.*

Finish the bridge off by sanding down any rough edges. Don't worry if the drill holes look bad; they'll be hidden behind washers when you're done.

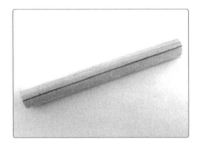  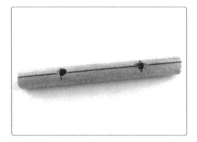

FIGURE 11-5: *Preparing the bridge*

Step 3   Prepare to drill the body of your kalimba. The bottom of the box will be your piano's soundboard. Flip the box over so that its lid is down and bottom up. Peel off any unwanted price tags or pesky surgeon general's warnings. Pencil in horizontal guidelines 2 1/2 inches from the top (for the *saddle line*) and 1 1/2 inches from the top (that's the *bridge line*). Next, add two *X*s along the saddle line, each 1 inch from either edge of the box (see Figure 11-6).

> * **NOTE:** *The bottom is a cigar box's sturdiest construction. Using it for the top of your instrument makes for a better resonator overall. Lids, on the other hand, are often made of inferior wood and are almost always poorly attached. Using such lids as soundboards results in a muffled tone, poor amplification, and annoying buzzes.*

Drill the Holes

Step 4   Drill the body of your kalimba. This will take several steps, starting with the acoustically mandatory sound hole. Using the 1-inch spade bit, drill a hole in either of the lower corners of the soundboard; in Figure 11-7, I chose the lower left. Be sure to leave a 3/8-inch to 1/2-inch margin around the hole so you don't bore into the sidewalls of the cigar box. If the hole looks rough, take some time to sand it smooth now.

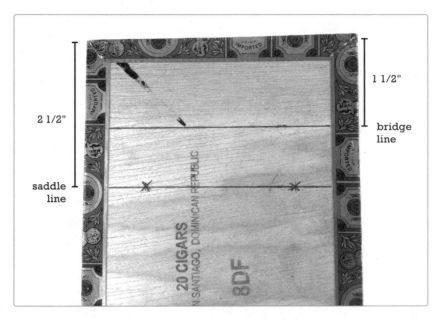

FIGURE 11-6: *Guidelines for drilling the kalimba's soundboard. The saddle line is the lower of the two lines, and it has a pair of Xs on it.*

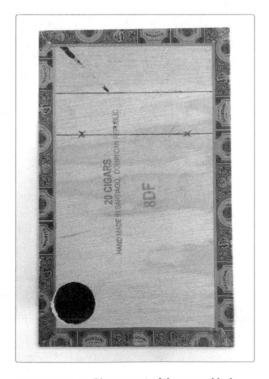

FIGURE 11-7: *Placement of the sound hole*

Step 5   Add the saddle-block holes. Using a 3/16-inch bit, drill holes at the two *X*s you marked on the saddle line. Then, center the bridge you prepared on top of the bridge line and use it as a guide to drill two more 3/16-inch holes (see Figure 11-8). If using the bridge as a guide in this fashion is too tricky—as it might be for those with small hands or heavy drills—you can use it as a guide to mark the placement of the holes and then remove the dowel and drill the holes on your marks.

FIGURE 11-8: *Using the bridge as a guide for drilling the bridge anchor holes*

Step 6   Now drill holes for the jack, pan pot, and volume control pot. If you're not going to electrify your kalimba, then you can skip this step. Otherwise, drill two 1/4-inch or 5/16-inch holes for the volume and pan pot, respectively (if you're using only one pickup, then omit the pan pot).

Next, drill one 3/8-inch hole for the jack. I placed all three holes on the tail-piece, as shown in Figure 11-9, but you could drill them almost anywhere. Just don't mount them on the soundboard, as that will take a real bite out of the kalimba's sweetness.

* **NOTE:** *The hole sizes you need for this step depend on your potentiometers' dimensions. If you're unsure of the exact size, use the smaller bit and ream out the hole until your hardware fits.*

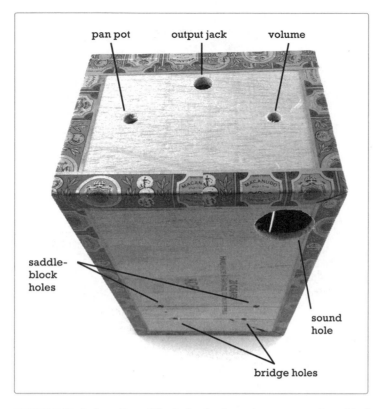

pan pot   output jack   volume

saddle-
block
holes

sound
hole

bridge holes

FIGURE 11-9: *Location of the holes for the volume, pan pot, and jack*

Install the Saddle Blocks and Bridge

Step 7   Erase any guidelines you drew, unless you dig the "sketchy construction lines" aesthetic. Once the kalimba body looks presentable, install the *saddle blocks*, which are screws that keep the saddle from sliding around. Start by placing a washer on a 1/2-inch #8 machine screw. Thread that screw out through either of the saddle-block holes so that the screw head is inside the kalimba body. Then snug a #8 nut onto the machine screw from the outside. Repeat with the other 1/2-inch screw, #8 washer, and #8 nut.

Step 8   Add the mounting hardware for the bridge. Place a washer on each of the 1-inch #8 machine screws and thread each through the soundboard from the inside. Next, slide the bridge onto the two machine screws, put a final washer on each, and spin on the wing nuts. Leave them fairly loose for now. You just want to keep everything from falling apart when you set the kalimba down. Your kalimba should look like Figure 11-10 at this stage.

FIGURE 11-10: *The installed mounting hardware, bridge, and saddle blocks (for an interior view of the mounting hardware, see the right image in Figure 11-17)*

Install the Tines

Step 9   Complete the acoustic construction of the kalimba by adding the six steel tines. As shown in Figure 11-11, the carriage bolt—which serves as the instrument's saddle—will rest against the tops of the two hexagonal saddle-block nuts. Because we want the saddle to rest evenly against these and run parallel to the top edge of the kalimba, we'll want to make sure these nuts are twisted around so they are identically oriented, with flat sides parallel to the top and bottom edges of the soundboard (see Figure 11-11a). Next, lay the 5-inch-long, 1/4-inch-diameter carriage bolt snug against the aligned nuts (see Figure 11-11b). You want the saddle to rest flat on the soundboard, so either saw off its head or let it hang off the edge of the box.

Step 10   Add the first tine. Working from the top, slide the straight end of the tine under the center of the bridge and over the center of the saddle. Levering the end of the tine up with a screwdriver helps (see Figure 11-11c). You'll want about 3 inches of tine to extend below the saddle. Add the remaining five tines, distributing them evenly across the middle 1 1/2 inches of the saddle (as in Figure 11-11d).

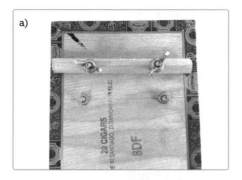

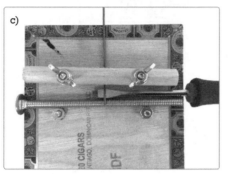

FIGURE 11-11: *Installing the saddle and tines*

Step 11   Once all of the tines are in place, use pliers to adjust their lengths according to the chart in Figure 11-21 on page 190. Make sure none of the hooked *tails* are touching (these are the ends that stick out past the top edge of the box, above the bridge), and then tighten the wing nuts. You may need to use a screwdriver to hold the machine screws in place from the inside of the body in order to keep them from impotently spinning while you're trying to tighten.

Step 12   If you're building a nonelectric kalimba, you're finished. Flip to "Tuning the Twang & Roar Kalimba" on page 189 and "Playing the Twang & Roar Kalimba" on page 190, and get plucking! Even if you're going electric, now is a good time to tune up, as doing so will make the next few steps significantly less annoying for anyone who shares living arrangements with you.

Time to Go Electric!

We'll be building and testing the dual-pickup mixer shown in Figure 11-12, with the caveat that you'll at least partially unbuild this after testing in order to install the magnetic pickup.[2] Get out both pots, your jack, your piezo element, the supplies you need to build the magnetic pickup, your insulated 24-gauge hook-up wire, and your soldering kit.

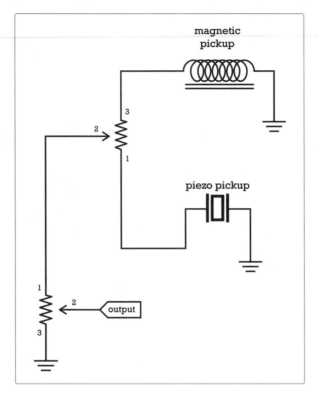

FIGURE 11-12: *The circuit diagram for the Twang & Roar Kalimba's pan-pot mixer*

Build the Electronics

Step 13 First, you need to build the magnetic Playing-Card Pickup (Project 9). Please do so now.

Step 14 Prepare to wire the pan pot. Look at your 100k ohm pot. I like to talk about pots as they appear when viewed from the front with their lugs up like a crown (see Figure 11-13). You'll need to add two leads to the pan pot: one will run from pan pot

---

2. If you're confident with your electronics skills and want to avoid resoldering, then you can use alligator clips to connect the magnetic pickup for testing rather than soldering it. But if you're newer to electronics, I highly suggest soldering everything now. If you run into a snag during the testing phase, it'll be easier to tell whether the problem is a bad connection, a bad pickup, or something else entirely.

lug 2 to the volume pot, and the other will go from pan pot lug 1 to the piezo pickup (which you'll mount on the inside of the soundboard, likely below the tines). A 6-inch wire will almost certainly be able to make either of these connections, but take a second to eyeball the layout of your specific kalimba and be sure. Maybe you need a slightly longer wire to make that run from pan pot to piezo? If your pan pot and volume pot are very close, you might prefer a shorter lead between them. Keeping these wires on the shorter side is a good policy (anything over 8 inches in an audio application like this starts to run the risk of picking up strong AM radio signals). Once you've settled on appropriate lengths, cut your wires, strip the ends of each, and tin them (this handy skill is discussed further in "Tinning Wires" on page 348).

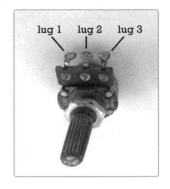

FIGURE 11-13: *A pot viewed from the front. Doesn't it look like a little metal-nosed Pinocchio wearing a crown?*

Step 15   Let's solder the pan pot! Grab the magnetic pickup you built in Step 13 and solder either of its insulated leads to lug 3. Now solder one of the freshly trimmed and tinned wires from Step 14 to lug 1. As indicated in Figure 11-14, this will be the input from the piezo pickup. Take the other length of insulated wire from Step 14 and solder it to lug 2. This lead is the pan pot's output (it will go to the volume control). Finally, solder the red lead from your piezo element to the free end of the lead running to lug 1. You can use either piezo lead, but it's good practice to always respect the convention of running black wires to ground.

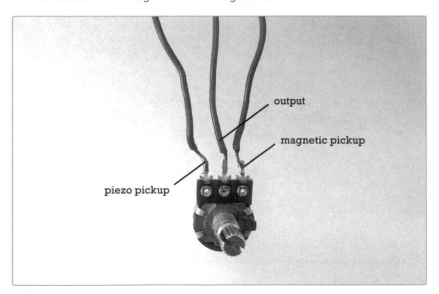

FIGURE 11-14: *The wired pan pot*

Step 16   Prepare the 10k ohm pot, which we'll use for volume control. Cut two 4-inch lengths of insulated wire and strip and tin both ends of each. Solder one wire to lug 2 of the 10k ohm pot and the other to lug 3, as shown in Figure 11-15.

FIGURE 11-15: *The prepared volume control*

Step 17   Complete the electronics. Start by routing your pan pot to your volume control. You should have one free lead on the pan pot; it's connected to the middle lug. Solder the free end of this lead to the first lug on the volume pot.

Now solder the jack, which has two lugs. One lug connects to the hooked metal tongue that will make contact with the *tip* of the 1/4-inch guitar cable; this lug carries the signal from the instrument. The other lug connects to the cable's *sleeve*, thus completing the ground connection.[3] Solder the lead connected to the middle lug on the volume pot to the tip lug of the jack. Then, take all three ground connections—the black lead on the piezo, the free lead on the magnetic pickup, and the remaining lead on the volume pot—and solder them to the ground lug on the 1/4-inch jack. Confused? Check Figure 11-16.

---

3. "Quarter-Inch Phone Plugs and Jacks" on page 337 includes clarifying illustrations of 1/4-inch instrument plugs and jacks.

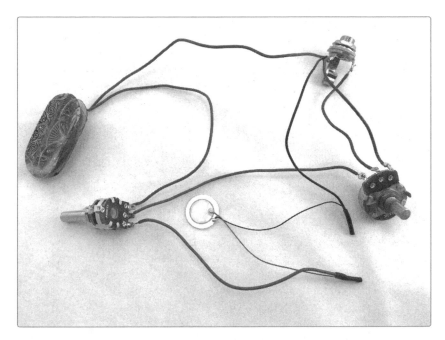

FIGURE 11-16: *The wired electronics. I've extended the piezo's ground lead and added shrink tube to both the piezo leads' solder joints.*

Test the Electronics

Step 18   Twist both the pan pot and the volume pot counterclockwise until you can't twist anymore. If you've followed the instructions correctly so far, this should correspond to the lowest volume setting, with only the signal of the magnetic pickup—which is quieter than the piezo—running to the jack. Connect the jack to your amplifier, turn the amp on, and set its volume control to a comfortable level.

Step 19   Test the piezo element. Run the pan pot fully clockwise and bring the volume control up halfway. Poke the piezo; you should get all manner of scrapping and thunking. If it isn't feeding back, raise the volume until it does or until the volume pot is fully clockwise. If you start getting feedback, then bring down the volume on your amp and go back to nudging the volume up on the kalimba's electronics until the volume pot is fully clockwise.

Step 20   Once the volume control is sorted out and you know the piezo works, twist the pan pot fully counterclockwise. You should now hear nothing when you tap the piezo element, and you should also hear nothing when you tap the magnetic pickup with your nonferrous finger.[4] Tap the magnetic pickup with an iron nail or steel screw, and you should hear clunking. Success!

Step 21   Set the volume pot back to full counterclockwise, shut down your amp, unplug everything, and move on to final construction. Problems? Check your work against Figure 11-16 and the schematic in Figure 11-12. If you see no obvious mismatch between your build so far and those images, flip to "General Troubleshooting" on page 355.

Install the Electronics

Step 22   Find and mark the best locations for each pickup. This is trickiest with the magnetic pickup, which has a very limited range. Start by placing the magnetic pickup on top of the soundboard beneath the tines. Face up or down makes no difference, but I prefer facedown, as shown in Figure 11-17, because it's important that the magnetic pickup lie flat to get an accurate sense of what the tone will sound like on the finished project. Slide the pickup around while plucking each of the kalimba's tines until you find the position where it gets the cleanest, most even tone. This will generally be centered beneath the tines, with the pickup's top edge about 3/4 inch below the saddle line you drew in Step 3. (If the two end tines are within about 1/4 inch of the rounded ends of the pickup, they'll be *much* quieter than the others. Tighten the spacing between tines a thread or two if you need to.) Once you've found your sweet spot, mark this location and then mark where you'll want to drill the hole for the magnetic pickup's wires.

   Next, flip open the body of your kalimba and press the piezo pickup to different locations on the underside of the soundboard while plucking various tines. Find the spot with the sound you like best and mark it. I like my piezo pickup tone to be as aggressive as possible, so my sweet spot tends to be centered about 1 inch below the saddle block screws on the inside of the body (see the right image in Figure 11-17).

Step 23   Prepare to mount the pickups. While your soldering iron heats up, drill a 3/16-inch hole for the magnetic pickup's leads, as shown in Figure 11-18. Depending on how you've finished your magnetic pickup, this hole may or may not be visible when the project is finished. If it's going to be visible, take a second now to sand and file it into a presentable state. Then desolder the leads on your magnetic pickup from the jack and pan pot.

---

4. If you have a steel, iron, nickel, or cobalt finger, use a different finger for this test. If your entire body is magnetic, use a hot dog or carrot or ask an adult you trust for help.

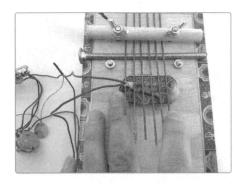

FIGURE 11-17: *Finding the sweet spot for the magnetic pickup (left) and piezo pickup (right)*

FIGURE 11-18: *The location of this kalimba's pickup-wire hole. (Different hand-made pickups will have their leads in different locations, so adjust accordingly.)*

Step 24   Mount the pickups, starting with the piezo pickup. Flip the kalimba over, open up the box, and squirt a pea-sized dab of silicone-based household glue on the piezo pickup sweet spot you marked earlier. Press the piezo into place brass side down and hold it firmly for a good 30 seconds. It should stay in place, but avoid bumping or jostling it for the remainder of this step. Turn the kalimba back over and carefully feed the leads of the magnetic pickup through the 3/16-inch hole you drilled in the previous step (see the left image in Figure 11-19). Smear a generous dollop of glue on the soundboard where you want to mount the magnetic pickup. Press the magnetic pickup into the adhesive, adjust it so that it's centered beneath the tines, lash it in place with a thick rubber band, and double-check the alignment and centering of the pickup (see the right image in Figure 11-19). Leave everything to dry for 30 minutes or so.

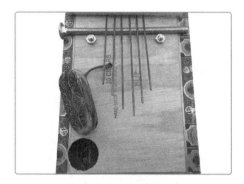

FIGURE 11-19: *Mounting the magnetic pickup*

Step 25   When your 30-minute wait is up, carefully remove the rubber band that's holding the magnetic pickup in place, as it's also holding the box shut. Now, mount one pot in each of the smaller holes shown in Figure 11-9 and then mount the jack in the larger hole. The washers go on the outside of the box, between the wood and the nut. If a little post protrudes from the face of either pot's casing and that post starts getting in the way, just snap it off with a pair of needle-nose pliers, as discussed in "Resistors: Fixed and Variable" on page 325.

Step 26   Solder the leads from the magnetic pickup back to the pan pot and jack; it doesn't matter which wire goes to the pan pot and which goes to the jack. Install the knob covers on the two potentiometers and then retest the electronics, following the process from "Test the Electronics" on page 185. If all is swell, then finish off the kalimba by smearing an insulating layer of silicone-based glue over the piezo (see Figure 11-20). This will drastically reduce—and perhaps even eliminate—feedback. When that's finished, close the kalimba up, check to confirm that the magnetic pickup is still situated to your liking, replace the rubber band, and leave the whole thing to dry overnight.

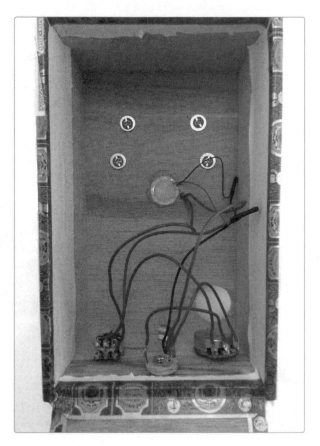

FIGURE 11-20: *The guts of the finished Twang & Roar Kalimba. Note the silicone insulating the piezo element.*

Tuning the Twang & Roar Kalimba

First, make sure your tines correspond to the initial lengths shown in Figure 11-21. Your kalimba might sound okay after adjusting the lengths according to the chart, but it's unlikely to *actually* be in tune at this point.

You'll need two tools to fine-tune your kalimba: pliers and a chromatic tuner. If you don't own a tuner, you can download one. Several free smartphone/tablet apps have chromatic tuners, or, if your computer has a mic and an Internet connection, you could just go to Tunerr.com.

Part of the aural charm of the kalimba comes from the sympathetic vibration of adjacent tines: when you pluck one, its neighbors vibrate too. This characteristic is musically rich, but annoying while tuning, so you'll want to mute all the tines you *aren't* tuning in order to focus on the one you are. Do this by pressing down on the other tines with your free fingers or by lightly securing the other tines with a rubber band.

Let's tune the longest tine first. That's the middle one, which will provide the lowest note. Mute the other tines, pluck the middle tine, and watch the tuner's readout. You want this tine to be middle C. (See Appendix C for details about musical terms.) It'll almost certainly be a touch sharp or flat at this point. In fact, don't be shocked if it's pretty far off, as sharp as C#/Db or D, or as flat as B. If the tine is sharp, then it's too short; use pliers to pull it out a millimeter or two. If it's flat, then it's too long. Use the side of your pliers like a little hammer to tap the tine in a millimeter or so. Check the pitch again; if you are dead on centered at middle C, then rejoice and move on to the D tine, which is one to the left. If not, keep fiddling until you get as close as you can to a perfect middle C; you need to do this only once in a blue moon, so get it right and lock it down.

Repeat these tuning steps for each tine. It's easiest to tune them in order from lowest to highest pitch—that is, longest to shortest tine. This means working out from the middle and alternating left and right. Periodically recheck the tines you've already tuned because it's easy to accidentally knock something out of whack. Once your kalimba is all tuned up, tighten the wing nuts, double-check the tuning, and make any final adjustments.

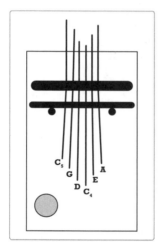

Note	Length (measured from front edge of saddle)
C_4*	81 mm
D	76 mm
E	71 mm
G	64 mm
A	60 mm
C_5	55 mm

* The subscript 4 denotes *middle C*, which is the C more or less smack in the middle of a standard 88-key piano keyboard. It's around 261.6 Hz—an ugly number, but a pretty universally recognized note in Western music. C_5 is thus the next C up, pitched one octave higher than middle C.

FIGURE 11-21: *A diagram showing which tine corresponds to which note (left) and a chart giving the initial lengths for the six tines on your pentatonic kalimba (right)*

Playing the Twang & Roar Kalimba

This kalimba is now tuned to the C-major pentatonic scale. This is a great place to start because there are no "wrong notes" in a major pentatonic scale: you can noodle around in basically any sequence, and you won't hit a sour note. For example, the signature riff in The Temptations' "My Girl" is basically just the major pentatonic played from lowest to highest. Try it! On the kalimba, playing the scale lowest to highest just means starting with the middle tine and working your way out, alternating between left and right. To construct your own riff, pick a few notes, and build them into a repeating pattern. Divert from that pattern by just a note, creating tension, and return to your original pattern. Congratulations: you're a composer!

While many tunes use the pentatonic scale as their basis, there are relatively few still-common tunes that confine themselves entirely to a single octave of it. One exception is good ole "Camptown Races":[5]

The	Camp-	town	la-	dies	sing	this	song,
G	G	E	G	A	G	E	

Doo-	da,	Doo-	da
E	D	E	D

The	Camp-	town	race-	track's	five	miles	long
G	G	E	G	A	G	E	

Oh,	the doo-	da	day
D	D E	D	C_4

Goin'	to run all	night,	Goin'	to run all	day
C_4	C_4 E G	C_5	A	A C_5 A	G

Bet	my	money on a	bob-	tailed nag
G	G	E E G G	A	G E

Some-	body	bet on the	bay
D	D D	E D	C_4

Yes, it's a hokey tune, but you can still make it swing. Hit YouTube to search for Dave Brubeck's 1959 recording of this minstrel show classic to see what I mean.

Tips, Tricks, and Mods

There are tons of ways you can mod your kalimba to fit your own style: adding tines, changing the tuning, building around a much bigger or smaller body, or using very different materials. But first, you'll want to sort out any little annoyances that could be keeping your kalimba from operating at its full potential.

Troubleshooting

Acoustically speaking, there isn't a ton that can go wrong with this project. One common issue is slight buzzing on some or all of the notes, especially when plucked hard. There are a couple of possible sources of buzzing.

Most commonly, the tails of two or more tines turn out to be touching—sometimes just barely. They may buzz only when the body of the instrument is really vibrating, as with an especially loud riff. Check to be sure that all of the tails are well clear of each other.

....

5. In an early draft of this project, I was on cold medicine and inadvertently typed *Compton Ladies* here. Such a song, to my knowledge, doesn't yet exist—but I invite readers to compose this for the kalimba and send me a video. Extra points if you can get Snoop Dogg, or a reasonable facsimile, to sing it.

A more subtle problem is that the tines may occasionally rattle against the soundboard. A quick solution is to run a rubber band or strip of felt beneath the tails of the tines, as shown in Figure 11-22.

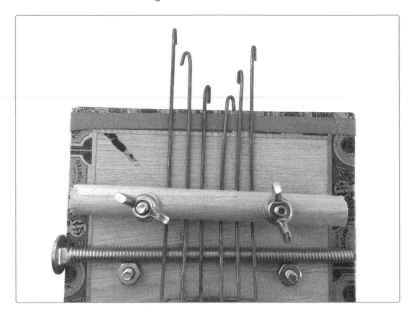

FIGURE 11-22: *Padding the tines to eliminate buzz*

Or, it could be the lid of the box itself that's buzzing; this is especially common if you've used the top of the box as your soundboard. Sometimes this can be addressed by running a thick rubber band, like the one shown in Figure 11-22, snugly around the box.

A more permanent and aesthetically pleasing solution is to add *blocking* to the interior corners to secure the lid. Cut a couple lengths of wood long enough to run from the bottom of the cigar box to the underside of the closed lid, and glue these into the two corners opposite the hinge. Once the glue dries, screw the lid down by driving screws through the lid and into these blocks. You'll probably want to use relatively thin screws and to predrill these holes so you don't split the lid or blocks.

As far as electronic troubleshooting goes, provided you haven't used a metal box as the body of your kalimba, there are no special pitfalls here. The most common problem is going to be a wrong connection or short circuit. Flip to "General Troubleshooting" on page 355 for a list of things to check.

Acoustic Mods

Sound hole placement is a much-debated art and science unto itself. Because of the small soundboard and resonator, a centered guitar-style sound hole isn't a great idea here; it consumes your sweet spot, making for a much duller instrument. But you could certainly try violin-style *f-holes* or replacing the single large corner hole with a spray of small holes in various sizes, as in the style of Ovation guitars. (Google *ovation sound holes* for inspiration.)

If you have a good ear and chromatic tuner, you could consider a different material for the tines themselves. I've specified music wire because it's a precision material that's easy to source. If you live in a town that uses street sweepers, the lost bristles—which are flat-rolled spring steel—make *excellent* kalimba tines. They have a rich, mellow tone and are easy on the fingers. These bristles are about a foot long, twice as wide as the music wire I specified, and half as thick. At the lengths on the charts in Figures 11-21 and 11-23, bristles are about an octave lower in pitch, so start with lengths half that in the chart and tune from there. I've never seen bristles for sale, but street sweepers seem to lose one or two on every block. Take a walk within a day or two of the sweeper's passing, keeping an eye on the gutter, and you should come up with the four bristles you need in no time.

Electronic Mods

As for electronic mods, you might choose to use only one of the pickups and thus eliminate the need for the pan pot. Or, instead of panning between and mixing the magnetic and piezo pickups, you might want to switch between them, as one does with a guitar. You can accomplish this by replacing the dual-ganged pot with an SPDT switch. (For an explanation of various switch types, flip to "Switches" on page 338.) To do this, take the output lead—that is, the one that would normally run from the middle lug of the pan pot to the volume pot—and connect it to the central lug on the SPDT switch. Then, wire one lead from each pickup to each of the outer lugs of the switch, similar to what's shown in Figure 10-23. If you want finer control when mixing the signals from the two pickups, you could give each its own 100k ohm audio-taper potentiometer.

You might also choose to add more pickups. There's no real benefit to adding another magnetic pickup because the instrument has only one spot where a magnetic pickup works—directly under the metal tines—and your first pickup is already there. But there are vast aural possibilities that can be explored by placing piezo pickups elsewhere on the instrument. The one caveat here is that piezo pickups can be really loud and aggressive. The volume control, as wired in this design, does a great job of taming this single piezo so that it doesn't overwhelm the magnetic pickup. If you start tossing in more piezos, you may need to experiment with adding a single resistor to the red lead of each of the additional piezos in order to tame the tone further.

We've already discussed going solely acoustic (just skip the "Time to Go Electric!" section), but you can also go all electric. Just as an electric guitar is pretty darn unobtrusive if you feed it into headphones—but still aggressive when amplified—you can build a resonator-less kalimba, replacing the hollow box with a plain old board. Use something a bit on the heavy side—like a stray hunk of scrap wood from a 2×4 or 1×6—and the piezo pickup on your all-electric model will be less prone to feedback.

Expand Your Scale

Probably the most tempting mod is to add more tines, thus expanding the scale of your instrument and increasing the number of tunes you can play. It's relatively easy to add two tines to the pentatonic Twang & Roar Kalimba for a full eight diatonic notes, which is perfect for a full major or minor scale. If you use the lengths described in Figure 11-23, your eight tines will be tuned to match all of the white keys on the piano between middle C and the next highest C, giving you the eight-note C major scale. (See Appendix C for a bit more explanation of musical terms.) I strongly suggest starting with C major, as so many American folk tunes can be adapted to this scale. You'll add only two tines to get that diatonic scale, but that'll rearrange how the notes are placed on the kalimba, making it much easier to play chords and, thus, to accompany other musicians.[6]

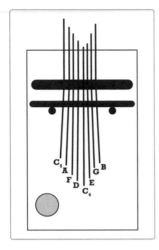

Note	Length (measured from front edge of saddle)
C_4	81 mm
D	76 mm
E	71 mm
F	70 mm
G	64 mm
A	60 mm
B	57 mm
C_5	55 mm

FIGURE 11-23: *A diagram showing which tine corresponds to which note (left) and a chart giving the initial lengths for the eight tines on your diatonic kalimba (right)*

To add extra tines, just scooch the existing tines in one or two threads toward the center. Slide your additional tines in at either edge, adjust all the lengths, and retune. On the C major diatonic kalimba, you'll ultimately have the eight tines spaced between 1/8 inch and 1/4 inch apart, taking up a total of roughly 1 1/2 inches, centered over the pickup. Figure 11-24 gives a side-by-side comparison of pentatonic and diatonic kalimbas built on this design.

The straight pentatonic and diatonic versions I've discussed are probably the best options for beginners, because you're going to have trouble cramming more than eight or nine tines into the 1 1/2-inch zone where the magnetic pickup works. If you forgo the magnetic pickup and just rely on the piezo pickup, then you'll have a bit more room to work within this design. You *may* even be able to bump this up to a full 13-note C_4–C_5 chromatic scale.[7] That would let you play even more scales, including major, harmonic minor, natural minor, blues, and so forth. But having that many tines creates a more cramped playing space and a much more challenging

---

6. A few popular chords are C major (C_4EG), G7 (FGB), F major (FAC_5), and G major (GBD).

7. A full 13-note C_4–C_5 chromatic scale corresponds to all of the black and white keys on the piano between middle C and the next highest C.

layout (you'll no longer have the chords conveniently grouped as you do on the diatonic kalimba).

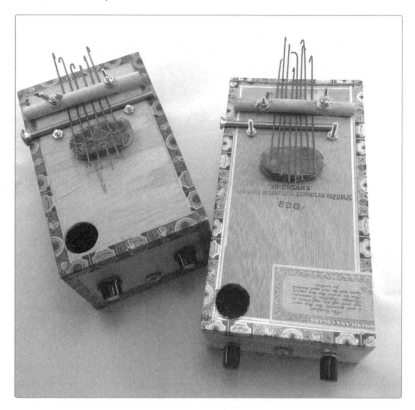

FIGURE 11-24: *A pentatonic Twang & Roar Kalimba (left) and a diatonic one (right). Note the tighter tine spacing on the diatonic kalimba.*

If you're going fully chromatic, consider busting completely out of the Twang & Roar mold. The 5-inch-wide cigar box is pretty constrained, and the 1.44 mm music wire has a relatively limited acoustically usable range. For an optimal, fully chromatic, electro-acoustic kalimba, you'll need a wider cigar box, a longer saddle and bridge (anchored at either end and in the middle), and a whole new magnetic pickup design. The pickup should be longer, with several magnets evenly spaced along its length, more like a modern multi-magnet guitar pickup. A chunky "hard drive salvage" pickup, like the one described at the end of the Playing-Card Pickup chapter (Project 9), might work here. You'll also want to go with beefier tine material, such as those street-sweeper bristles or sections cut from an electrician's flat-steel *fish tape* (a tool sold at most hardware stores and used to pull wires through walls and conduit).

Resources

I was first introduced to the notion of electrifying a thumb piano by the work of Konono N°1, a band from Kinshasa that performs traditional Congolese music on traditional instruments that they've augmented with their own homebrew electrical amplification systems. Their debut album, *Congotronics*, was very popular with fans of electronic music and inspired a series of albums. I especially enjoy their original album and the second release in the series, *Congotronics 2: Buzz 'n' Rumble from the Urb'n'Jungle* (an anthology album featuring eight bands in addition to Konono N°1). If these tracks fit your ear especially well, you might also dig the *Éthiopiques* series of anthology albums (there are now almost 30), which showcase electrified traditional and jazz/pop-influenced Ethiopian and Eritrean music.

12 THE MUD-N-SIZZLE PREAMP

When you pump up the volume, you have more sound to work with. A preamplifier boosts your instrument's volume, allowing a quieter instrument to cut through the mix when it's being played with louder instruments. This is also handy if you have an *instrument-level* signal—like the one produced by an electric guitar or by the Playing-Card Pickup (Project 9)—that you want to feed into effects or other gear that expects a hotter *line-level* input (the sort generated by synths, drum machines, powered mics, and so on).

Even in a stripped-down solo application, a preamp can be your best friend: boosting the volume brings out subtle sounds and tones, lengthens the decay of your notes, and gives you access to a broader dynamic range.

But the Mud-n-Sizzle Preamp doesn't just boost your signal; a beefy built-in filter lets you shape your tone. At one extreme, you get a muddy, mellow throb; and at the other, you get a sharp sizzle that, at full volume, crests into a warm fuzz tone.

Building this project will also give you practice using a generic *printed circuit board (PCB)* that has no predefined connections between soldering pads. Learning to run your own traces (connections between components) is a great skill that gives you much more flexibility. Now when you find a neat schematic online, you'll be able to build the project, even without step-by-step instructions.

Hear the Mud-n-Sizzle Preamp in action in the samples at *http://www.nostarch .com/jamband/.*

FIGURE 12-1: *The finished Mud-n-Sizzle Preamp*

Preparation

Build Time

▶   Under an hour

Tools

▶   A standard soldering kit (See page 340.)

▶   A ruler that shows 1/4-inch increments or smaller

▶   A fine-point Sharpie or other permanent marker

▶   An electric drill with bits (You'll probably need 3/16-inch, 1/4-inch, 5/16-inch, and 3/8-inch bits.)

- ▶ (Optional) Other tools to work your enclosure (If you're using a wooden enclosure, you'll want sandpaper and possibly files to clean up drill holes.)

- ▶ (Optional) Foam-backed double-sided tape or hardware to mount the circuit in its enclosure

- ▶ (Optional) A hacksaw for trimming down potentiometers with long shafts

Supplies

- ▶ A 2N3904 NPN transistor, such as Mouser part #512-2N3904BU

- ▶ A red LED

- ▶ A 1.5 µF electrolytic capacitor

- ▶ A 10 µF electrolytic capacitor

- ▶ A 0.1 µF capacitor (marked *104*)

- ▶ Two 1M ohm resistors (brown-black-green stripes)

- ▶ A 10k ohm resistor (brown-black-orange stripes)

- ▶ A 10 ohm resistor (brown-black-black stripes)

- ▶ A 470 ohm resistor (yellow-violet-brown stripes)

- ▶ Two 10k ohm audio potentiometers (These are variable resistors with an audio taper; see "The Gory Details: Audio Taper vs. Linear Taper" on page 327.)

- ▶ Two control knobs that fit your potentiometers

- ▶ A 9-volt battery

- ▶ A 9-volt battery clip

- ▶ 24-gauge insulated hook-up wire (Stranded wire is best here.)

- ▶ 22- or 24-gauge bare bus wire (This is uninsulated solid core wire, which we'll use to run the ground connection among the potentiometers and jacks.)

- ▶ Two 1/4-inch mono phone jacks, also called *guitar jacks*

- ▶ A pad-per-hole printed circuit board (Figure 12-2 shows an old piece of RadioShack part #276-148, but you can use a 10 hole × 10 hole piece of any generic PCB, which is about 1 square inch. The SparkFun PRT-08811 is a perfect alternative, and the Jameco 105100 is good, too.)

- ▶ A small SPST toggle switch (You can use Mouser part #108-MS550K, which is shown in Figure 12-2, but any similar switch will work.)

- ▶ A small, sturdy enclosure (You can buy a suitable project box at many hobby shops, especially those appealing to electronic tinkerers, or you can use something you find around the house or garage. See "On Enclosures" on page 214 for more information.)

- ▶ (Optional) A 9-volt battery holder clip, such as Digi-Key part #71K-ND or Mouser part #534-071

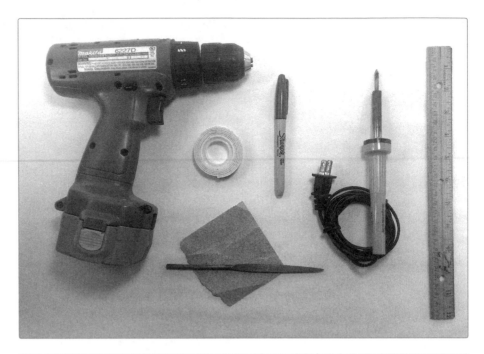

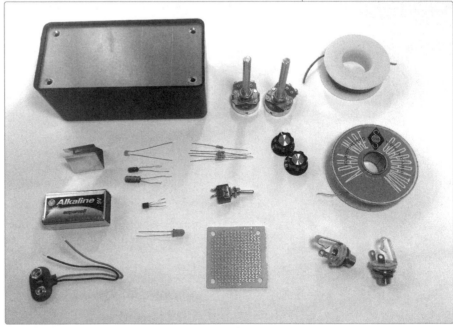

FIGURE 12-2: *Tools and supplies (not shown: hacksaw)*

Building the Mud-n-Sizzle Preamp

First, peek at the diagram for this circuit, as shown in Figure 12-3. If this schematic means nothing to you, don't sweat it: that's why there are illustrated step-by-step instructions.

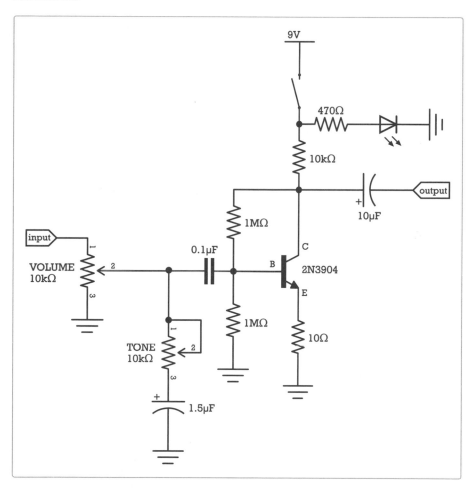

FIGURE 12-3: *The circuit diagram for the Mud-n-Sizzle Preamp*

Keep in mind that a circuit diagram shows functional relationships between components; it's not an image of how the pieces are physically laid out but instead illustrates how the pieces connect. Connections to the ground—that is, the negative battery terminal—are a perfect example: this circuit has quite a few ground connections, each represented by a little triangle made from three parallel lines. Five of these are shown in Figure 12-3, peppered around the circuit. The input and output jacks will also each be grounded, for a total of seven ground connections. The schematic shows ground connections all over the place, but on the physical board, they're all clustered in the same row.

Now that you've seen the circuit, we'll prep some components, build the circuit, and then put it all together.

Prepare the Hardware

We'll prepare the potentiometers and jacks first, starting with the volume and tone controls, so you can practice soldering with the components that are the most difficult to damage.

Step 1   While your soldering iron warms up, grab your two 10k ohm potentiometers, or *pots*, and your marker. Mark the back of one pot with a *V* (for volume) and the other with a *T* (for tone).

Step 2   Cut a 4-inch length of insulated wire. Strip about 1/4 inch of insulation off either end and tin each end of the wire. (If this is your first soldering project or if you just need a refresher, see "Soldering" on page 346.) Solder one end of this wire to the V pot's middle lug, as shown on the left in Figure 12-4, and then set it aside.

Step 3   Cut another 4-inch length of insulated wire. Strip 1/4 inch of insulation from one end and 1/2 inch of insulation from the other. Then tin both ends. Thread the tinned 1/2-inch end of the wire through lugs 1 and 2 of the T potentiometer (see Figure 12-4; we're talking about the pot on the right). Solder this wire to those lugs.

Step 4   Finally, solder the positive leg of the 1.5 µF electrolytic capacitor to lug 3 of the T pot. Most electrolytic caps (including this one) are *polarized*, which means they have a positive and negative leg. The negative leg is marked with a stripe running down the capacitor's body, so in this case you'll want to solder the stripeless leg to lug 3 (see Figure 12-4, right).

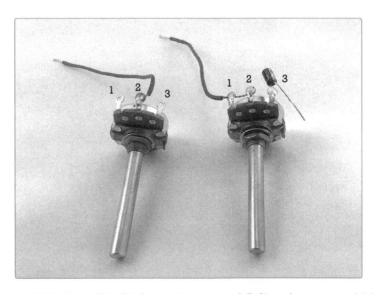

FIGURE 12-4: *The finished volume control (left) and tone control (right)*

* **NOTE:** *Long potentiometer shafts—like the ones shown here—are a pain to work into most small enclosures. If you have long-shafted pots like these, now is a good time to shorten them. For details, see "Resistors: Fixed and Variable" on page 325.*

Next, we'll build the power supply, beginning with the indicator LED.

Step 5   Solder the 470 ohm resistor (marked with yellow-violet-brown stripes) to the positive LED lead, which is the longer one. The shorter, negative lead is the one closest to the flat section of the LED's mostly round casing. The finished indicator LED is shown on the left of Figure 12-5.

Step 6   Now for the power switch itself. Start by tinning both lugs of your toggle switch. Next, cut one 5-inch length of insulated wire and strip and tin the last 1/4 inch of each end. Then strip and tin the final 1/4 inch of both the red and black leads attached to the 9-volt battery clip. Solder the positive (red) lead to one lug of your switch. To the remaining lug, solder both the 5-inch length of wire and the resistor side of the indicator LED assembly (see Figure 12-5, right).

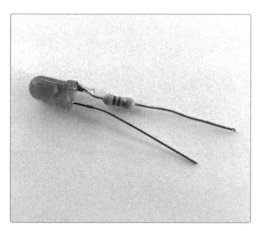 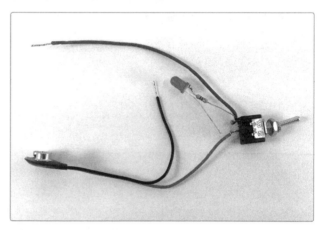

FIGURE 12-5: *The completed power supply (detail of the LED assembly on the left)*

Finally, prep the input and output jacks, which are the last pieces of hardware.

Step 7   Cut a pair of 4-inch wires, strip the ends, and tin them. Solder one wire to the tip lug of each jack, as illustrated at the top of Figure 12-6. (If you're not sure which lug is which, see "Quarter-Inch Phone Plugs and Jacks" on page 337 for an illustrated introduction to 1/4-inch jacks.) That's all for the output jack, so put one jack aside.

Step 8   Grab the volume control pot (that's the one marked with a *V*) and the remaining jack. Solder the lead you added to the jack in Step 7 to the first lug on the volume control pot, as shown at the bottom of Figure 12-6.

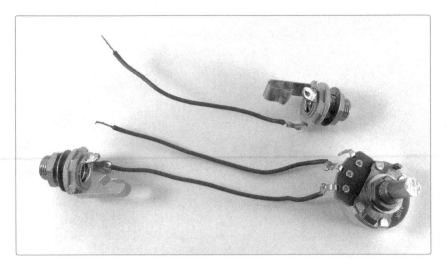

FIGURE 12-6: *The prepared jacks (output jack at the top, input jack at the bottom)*

For now, set aside the tone control pot, the power switch with LED indicator, and both jacks. We'll wire these all up at the end, once we've mounted all the components on the PCB.

Build the Circuit

Let's begin building the circuit itself, starting with the transistor.

Step 9   Mount the transistor in your PCB. Hold the PCB so that the copper pads face away from you and slide the legs of the transistor through three vertical holes at the middle of the board. You want the transistor's flat face pointing right (see Figure 12-7).

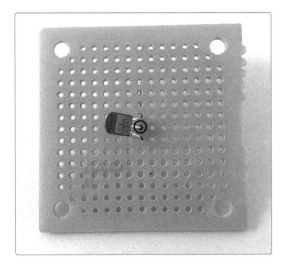

FIGURE 12-7: *The transistor is in place, and the leg you solder in Step 10 is circled.*

* **NOTE:** *If you're using a 10 hole × 10 hole scrap of PCB (the smallest you can get away with here), then make sure your transistor is positioned so that there are at least three open rows of holes above the transistor, four rows below the transistor, four open columns of holes to the right, and five open columns to the left.*

Step 10   Once the transistor is in place, bend its upper leg up and its lower leg down to hold it in place. Then solder the middle leg (also called the *base*) to the solder pad on the underside of the PCB, but *don't clip the lead yet.* For the remainder of this project, clip leads only when specifically told to do so.

Step 11   Bend one of your 1M ohm resistors (brown-black-green stripes) into a *V* shape, as shown in Figure 12-8 (left). Then, slide one lead into the hole immediately to the left of the transistor's middle leg and the other lead into the first hole immediately above the transistor's upper leg (also called its *collector*), as shown in Figure 12-8 (right). Once you insert the resistor, bend its upper leg to hold it in place and solder the leg closest to the transistor's middle leg to its pad.

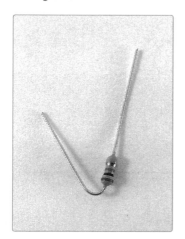 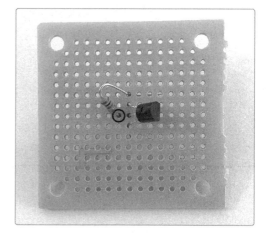

FIGURE 12-8: *The first 1M ohm resistor is in place, and the solder point is circled.*

Step 12   Bend your other 1M ohm resistor into a *U* shape, as shown in Figure 12-9 (left). Mount it on the PCB vertically so that one lead is in the first open hole to the left of the transistor's middle leg and the other lead is five rows down. When your resistor is in place, solder the top leg to its pad, as indicated in Figure 12-9 (right), but don't solder the bottom leg for now.

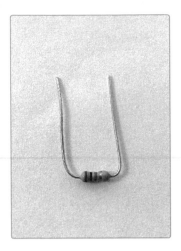

FIGURE 12-9: *The resistors are installed, and the solder point is highlighted.*

Step 13   Insert the 0.1 µF input capacitor into the PCB. Place one lead in the first open hole to the left of the transistor's middle leg. Insert the other lead two holes to the left so that there's an empty hole between the two legs (see Figure 12-10). Solder the rightmost lead to its pad. The middle PCB row should now have the following, counting from the right: the transistor's middle leg, the bottom lead from the first brown-green-black resistor, the top lead from the second brown-green-black resistor, and both input capacitor legs.

FIGURE 12-10: *The placement of the 0.1 µF input capacitor*

When you use a PCB without predefined traces, you have to connect the components by soldering bridges between pads on the underside of the PCB so that a current can run through the metal to each component. This requires a bit of practice, so don't worry if your first try looks ugly. Ugly metal conducts just as well as pretty metal.

Step 14   Begin with the four leads you installed along the PCB's middle row: the middle transistor leg, the two resistors, and the right leg of the 0.1 μF input capacitor (that is, the leg closest to the transistor). For the tidiest traces, start by trimming the excess lead sticking out of each solder point. To run a solder bridge between these four points, as shown in Figure 12-11, reheat each solder joint, add a little more solder, and carefully stretch the liquid solder across the four solder points using a smooth brushing gesture.

This prettier method is preferable because it lowers the likelihood of short circuits between rows, even on fairly tightly packed PCBs. If it's proving to be a royal pain, then read "Bridging Connections the Easy Way" (below) for an easier method that will work just as well in this case (I've made a point of spacing the project out in the PCB to minimize the possibility of short circuits). For the remainder of the project, use whichever method you prefer.

FIGURE 12-11: *The pretty method for connecting components on generic PCBs*

Bridging Connections the Easy Way

First, bend the stray end of one lead so that it runs straight along the row of holes, touching the other leads (see the left image in Figure 12-12, where I bent the input capacitor lead over to touch the resistors and transistor base). Next, heat the length of that bent lead with your soldering iron until the three other solder points remelt. Add more solder to each joint if you need to so that they make a solid connection with the bent lead. Finish by trimming the ends of the four leads once the solder has cooled. The final result is shown in the right image in Figure 12-12. The result isn't as slick, but it's just as serviceable.

 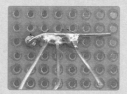 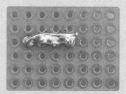

FIGURE 12-12: *The easy method of connecting components on generic PCBs*

Step 15   Now install the 10 µF output capacitor, keeping in mind that the stripe on the cap's body indicates the negative leg. Mount this cap with the positive leg in the first open hole above the transistor and insert the negative leg three columns to the right, as shown in Figure 12-13. Solder the positive leg to its pad.

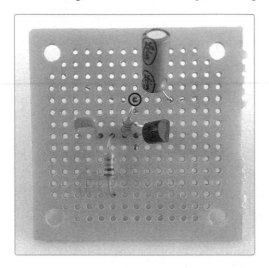

FIGURE 12-13: *The 10 µF capacitor in place*

Step 16   Insert one leg of the 10k ohm resistor (brown-black-orange stripes) into the first open hole above the positive leg of the output cap that you just installed. Insert the other leg four holes to the left in the same row, as shown in Figure 12-14.

FIGURE 12-14: *Connecting the 10k ohm power supply resistor*

Step 17   Solder the rightmost leg of the 10k ohm resistor to its pad. Then use one of the solder-bridging methods from Step 14 to connect that leg to the three directly below it: the positive leg of the output capacitor, the 1M ohm resistor, and the transistor's upper leg. You should have the four circled leads in Figure 12-14 connected. Clip the excess leads after the solder joint has cooled.

Step 18   Finally, mount the 10 ohm resistor (brown-black-black stripes). Bend the resistor into a *U* and run one lead into the first open hole to the right of the transistor's lower leg (also called the *emitter*). Insert the other lead into the fourth hole down in that same column (see Figure 12-15). Solder the resistor lead to its pad, solder the transistor's lower leg to its pad, solder the two leads together, and clip the excess wire.

FIGURE 12-15: *Mounting and connecting the 10 ohm ground resistor*

Complete the Circuit

With all of the board components mounted, it's time to add the hardware you built in "Prepare the Hardware" on page 202.

Step 19   Start with the output jack, which is the jack without a potentiometer attached. It has one insulated lead. Insert this into the first open hole to the right of the negative leg of the 10 μF output capacitor (see Figure 12-16). Solder the output lead to its pad and the negative capacitor leg to its pad, and then solder the two to each other. Clip the excess wires.

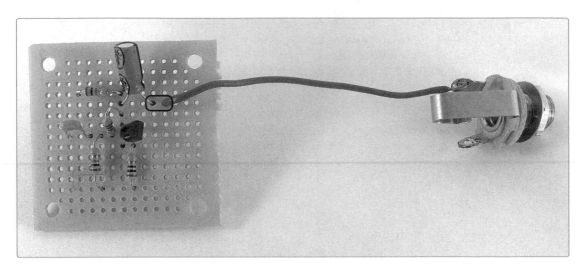

FIGURE 12-16: *Adding the output jack*

Step 20   Next, add the volume control/input jack and the tone control knob. Take the volume control pot and insert the insulated lead into the hole immediately above the 0.1 μF input capacitor. Now add the insulated lead from the tone control to the first hole below this capacitor (these wires are highlighted in Figure 12-17). Solder each lead and the capacitor leg to its respective pad, and then solder both leads and the capacitor together. Trim away excess wires once the solder joint cools.

FIGURE 12-17: *Completing the input. The upper lead connects to the volume control, and the lower one connects to the tone control.*

Step 21   Time to add power! Take the power switch assembly from Step 6 and insert the insulated wire into the first open hole to the left of the 10k ohm resistor (brown-black-orange stripes), as shown in Figure 12-18. Solder the power wire to its copper pad, solder the free 10k ohm resistor leg to its copper pad, and then connect the two. Trim the leads.

 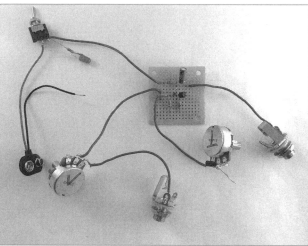

FIGURE 12-18: *Power is installed, but the ground is still incomplete (detail at left, full circuit at right)*

The circuit is nearly finished; it's just missing the ground connection (for reference, the completed, groundless circuit is shown in the right panel of Figure 12-18). Right now you have two components connected (but not yet soldered) to the row of holes that will serve as the *common ground* linking your circuit to the battery's negative terminal: a 1M ohm resistor (brown-black-green stripes) and a 10 ohm resistor (brown-black-black stripes).

In order for the circuit to actually function, we'll need to run the ground connection to all of the hardware. There are a variety of ways to accomplish this. In this case, we're using a *star ground*, where each component gets its own direct connection to the ground (this minimizes unwanted noise and interference from local AM stations, wireless devices, high-voltage appliances, and so on). You'll now add the final six connections to the ground: the input and output jacks, both controls, the indicator LED, and the negative power supply lead.

Step 22   Start with the black wire from your battery clip—that is, the ground wire itself. Place it in the hole in the ground row between the two resistors, as shown in Figure 12-19. Solder it to its pad.

Step 23   Cut a length of 22-gauge bus wire long enough to reach from the PCB to the input jack. Solder it to the open lug on your input jack and then run it into one of the open holes in the common ground row. Solder it to its pad.

Step 24   Cut another length of 22-gauge bus wire, long enough to reach from the PCB to the output jack. Solder it to the open lug on your output jack and run it into one of the open holes in the common ground row. Solder it to its pad.

Step 25   Cut two more wires long enough to comfortably run from the PCB to the controls, which are likely to go on the enclosure's lid. Solder one wire to lug three on the volume pot; solder the other wire to the open leg of the capacitor on the tone control. Insert both into holes on the ground row and then solder them to their pads.

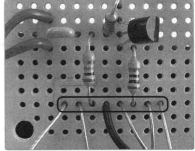

Step 26   Cut a final length of wire long enough to run from the negative leg of the indicator LED to the ground bus. Solder one end to the LED's open leg and the other to the ground row. Solder it to its pad. Now that you have all eight connections in place, solder them all together using one of the two methods described in Step 14 and trim any excess wire. When viewed from the top (that is, the side with no copper solder pads), your ground row should look like Figure 12-19.

FIGURE 12-19: *The completed common ground row. Note that the wires are all bunched together with no open holes and the negative battery lead—which is the ground—is at the center.*

Install the Circuit in Its Enclosure

Now that the circuit is built, let's mount the hardware on the case.

Step 27   Add holes to your enclosure to accommodate the switches, jacks, and controls. You'll need drill bits that correspond to the size of these components; in general, the following works: 3/16 inch for the LED, 1/4 inch for a small switch, 3/8 inch for a jack, and either 5/16 or 3/8 inch for a potentiometer (although your components may be slightly different dimensions; double-check before you drill). Decide where you want your components to go and drill the appropriate holes. I like to have indicator LEDs and controls, including power, on the face of the enclosure and jacks on the sides (see Figures 12-1 and 12-21).

Step 28   Once you've drilled the enclosure, secure your battery. If the enclosure is too large, the battery is likely to rattle around, causing short circuits or damage to sensitive components. To prevent the battery from busting up the place, I suggest screwing a 9-volt battery holder into the side of the enclosure, as shown in Figure 12-20. (If you have a fairly snug-fitting enclosure, you can often omit the battery holder.)

FIGURE 12-20: *A 9-volt battery holder clip in a finished project, with and without a battery in place*

Step 29   Mount the switch, jacks, and potentiometers on your enclosure. These come with their own mounting hardware. If that includes two washers, then put one inside the enclosure and one outside. If there's only one, put it on the outside of the enclosure. You should be able to press the LED into place and have it stay put on its own. Finally, add the control knobs to the shafts of your pots.

At this point, all of your basic hardware should be mounted, and the PCB should be floating freely, as shown in Figure 12-21—but before you close that lid, make sure your circuit works!

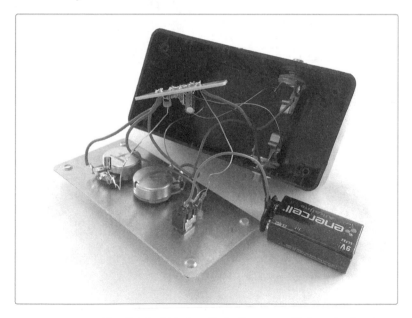

FIGURE 12-21: *The finished Mud-n-Sizzle Preamp with its guts showing*

On Enclosures

The *enclosure* is simply the box for your electronics project: drill a few holes, stuff in the batteries and wires, label your knobs and jacks, and you're good to go. You can get cheap, sturdy electronic project enclosures, like the one I've used for this preamp, at any hobby shop that specializes in electronics or online (try searching for *project box* or *project enclosure*). For something fancier, you can get the professional-grade die-cast enclosures made by Hammond Manufacturing and Velleman Incorporated. Small Bear Electronics (*http://www.smallbearelec.com/*) and Antique Electronic Supply (*https://www.tubesandmore.com/*) sell a variety of traditional stompbox-style enclosures. If you're looking for something simpler, plastic or wooden storage boxes from any office or craft supply store are perfectly suitable. Even plastic food-storage boxes sold at the grocery store will do in a pinch. On the other hand, you can also go quaint: keep your eyes peeled for old cigar boxes, pencil boxes, candy or cookie tins, lunch boxes, or military surplus equipment cases and kits.

Questions to consider when choosing enclosures:

▶ *Is it conductive or nonconductive?* A conductive metal enclosure is great for a project like a preamp—which often benefits from the extra shielding—but it can cause short circuits if the PCB brushes up against unpainted metal.

▶ *Do you have access to the tools needed to modify the enclosure?* You'll need to add holes for jacks, switches, and pots; you can cut through an old plastic pencil box with a pocketknife, but you're going to have trouble piercing a steel ammo case with anything short of a drill press.

▶ *Is the box big enough?* It needs to accommodate the battery and the jacks *with plugs inserted* as well as the circuit itself. Cramming too much into too small a case is a recipe for annoyingly intermittent short circuits.

▶ *Will the box be able to withstand the abuse?* A pretty little Chinese gift box might make an awesome Droid Voicebox (Project 6) or handheld synth but probably isn't sturdy enough for a preamp, Twin-T Phaser/Wah (Project 14), or amplifier, all of which will tend to get knocked around a bit.

▶ And finally: *does it look frikkin' awesome?*

Test and Install the PCB

Time to power up and test out the circuit! Before connecting the battery, do a quick visual check to make sure all of your soldered connections are indeed connected, with no wire left unsoldered. Figures 12-18 (which shows everything but the ground connections) and 12-19 (which gives a tight detail of the star ground) are good references for this check, as is the schematic in Figure 12-3.

Step 30   When you're satisfied that everything is wired properly, connect a battery, flip the power switch, and check to see that the LED lights up. If the LED doesn't light, make sure that your battery isn't flat and that the ground connection is complete. Then, check for short circuits. Are bare wires brushing up against the PCB? Move them so that they aren't making contact. Are there stray snips of wire or solder resting on the PCB, or stuck to a switch or pot? Remove them. Are there gobs of solder connecting adjacent components that shouldn't be connected? Remove them (see "Desoldering" on page 349 for details).

Step 31   Once your preamp has working power, turn it off, turn the volume knob all the way down (fully counterclockwise), and set the tone knob to the center of its range. Connect an amplifier to the preamp's output jack, power up the amplifier, and then power up your preamp. There should be no sound beyond a quiet hum. If you bring up the volume on the preamp, you should get a louder hum. If you're getting excess noise or AM radio, check the ground connection and the length of the leads. Any lead longer than 8 inches should be shortened. If all else fails, consider moving your innards to a partial or all-metal enclosure, which will shield the circuit from electromagnetic interference, an effect you can maximize by running a piece of bus wire from the star ground to a metallic portion of the enclosure. (This is traditionally accomplished by soldering one end of the wire to the ground and then wrapping the other end around one of the lid's mounting screws so the wire gets sandwiched between the case and lid when you screw it shut).

Step 32   Once the unwanted noise is eliminated, power the preamp down and connect a low-level instrument to the input jack, such as the Robo-Tiki Steel-Stringed Ukulele (Project 10), Twang & Roar Kalimba (Project 11), or a standard electric guitar without active electronics (guitars and basses with *active electronics* have a built-in preamp wired directly into the pickups; as a rule, if an instrument has a battery, then it has active electronics). Once the instrument is hooked up, switch the preamp back on and slowly bring up the volume while you strum, pluck, flick, tweak, or otherwise play the instrument. You should get a pretty loud signal quickly, one that's obviously louder than if you plugged the instrument directly into the amp.[1] With the volume at the extreme clockwise position, you may even get a fuzzy kind of distortion known as *clipping*.

Step 33   Now turn the volume back to the point where the signal is both loud and clear, and then begin turning the tone control knob counterclockwise. You should hear the signal make a very definite shift toward bass-heavy rumble. The filter controlled by this knob is a *low-pass filter*, meaning that as it engages, it eliminates an increasing portion of the higher frequencies in the signal while permitting the lower frequencies through to the output—that is, it lets the low frequencies pass and shunts the highs to ground. Try playing a variety of notes to get the clearest sense of what's going on. Now turn the tone knob clockwise. The tone should clear as the filter

---

1. You might hear AM radio chatter now that you've attached an instrument because the instrument or its cable can act as an antenna. You can generally eliminate such interference as described in Step 31: shorten leads, make sure the ground is as direct as possible, and move to a grounded metal enclosure. Also, try using a shorter instrument cable; having the shortest connection possible between the low-level instrument and preamp means you can have a much longer cable connecting the preamp and amplifier.

disengages. Keep turning clockwise and the tone will shift into a buzzy sizzle, emphasizing the higher-frequency components of each note. If you aren't getting an obvious effect, check your PCB for short circuits, which are often either between the tone potentiometer's input and ground connections or between some point on the PCB and ground.

Step 34   When you're confident your circuit works, install the PCB into the enclosure. Prefab PCBs often come predrilled with larger holes (usually in the corners or centered along two edges) for bolting the PCB to the inside of the enclosure. Many purpose-built enclosures are designed with slots, stand-offs, or bolt-holes for just this purpose. Just slot the PCB into place or bolt it down. For other enclosure types, you can either drill the case and bolt the PCB in place or secure it with a couple pieces of foam-backed double-sided tape.

Using the Mud-n-Sizzle Preamp

A preamp is pretty straightforward: just plug in your instrument and find the volume and tone you want, and then you're ready to rock. There are lots of possibilities to explore when you make little sounds big—for an example, check out the work of Ryoji Ikeda. You might have fun connecting this preamp directly to a stand-alone Plasti-Pickup (Project 2) or Playing-Card Pickup (Project 9) and exploring your local soundscape.

But primarily, preamps are seen as utilitarian workhorses. In addition to beefing up your sound and giving your amp and effects more to work with, a small battery-powered preamp like this can help you eliminate unwanted noise. Place the preamp early in your effects pedal chain, as close as possible to your instrument and on the shortest possible cable. This way, your boosted signal is as pure as possible, assuring that more signal—and thus less noise—reaches the effects, mixing board, recorder, or *power amp* driving the speakers in your instrument amplifier.

Tips, Tricks, and Mods

You can customize your Mud-n-Sizzle Preamp in all sorts of ways, from swapping in new components to building the preamp into an existing instrument.

Boost the Fuzz with a Darlington Pair

If you like the fuzzy overdrive sound you get when the volume is cranked all the way up, want an even more aggressive fuzz, or just need to boost a very weak signal even further, you can supercharge this preamp by replacing the single transistor with a *Darlington pair*. This is a set of matched transistors wired so that the first transistor amplifies the signal and the second transistor amplifies that amplified signal.

You can buy single-package Darlington pairs that look just like plain old transistors and will directly replace the 2N3904 shown in the circuit diagram in Figure 12-3. The 2N6426, 2N6427, and MP3A13 are all workable substitutes, with the first having the highest gain by a wide margin. Or, you can build a Darlington pair yourself by taking two 2N3904s and soldering them together, as shown in Figure 12-22.

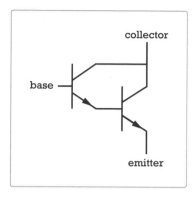 

FIGURE 12-22: *A homemade Darlington pair (schematic at left, in real life at right)*

This homebrew Darlington can replace the single 2N3904 in your circuit. Just treat its conjoined collector as though it was the collector of a single transistor. The base of the "rear" transistor in the Darlington pair then replaces the single transistor's base, while the Darlington's "front" transistor's emitter replaces the single transistor's emitter.

✳ **NOTE:** *If you really like the fuzztone sound, you might consider building the "Two-Transistor Fuzztone" on page 364, which uses a Darlington pair and adds a few diodes to get a rich British Invasion–style fuzz with increased sensitivity and sustain.*

Experiment with Resistors

To get a bit more fuzztonally experimental, consider changing the values of the two 1M ohm resistors. In the circuit we've just built, they form a *voltage divider* that *biases* the transistor's base—that is, keeps the transistor ready to hear an incoming signal at a moment's notice. My goal was to make a preamp that could reliably and faithfully amplify even very low-level signals, and I selected these resistors accordingly. But there's no reason that these two resistors have to be 1M ohm or even the same value. Try experimenting with values in the 100k ohm range or lower, eliminating the base-to-ground resistor entirely or replacing the 1M ohm resistor running between the collector and base with a potentiometer. Using a potentiometer allows you to adjust the circuit's bias on the fly, changing the amp's sensitivity, timbre, or perceived gain. Just remove the 1M ohm resistor and connect the pot's middle lug to the transistor's base and either of the outer lugs to the collector. A 1M ohm potentiometer with a linear taper is a reasonable place to start, but feel free to experiment—maybe even get weird by using a photoresistor, body contacts, or other nonstandard resistor scheme here. (The Universal LFO from Project 13? Why not?!)

Build the Preamp into an Instrument

Some instruments come with built-in preamps. This is especially the case with nicer bass guitars, heavy metal axes, and any instrument that usually produces a line-level output, such as a synthesizer, drum machine, or keyboard. If you're told that a guitar has "active electronics" or "active pickups," then it has a built-in preamp wired directly to the pickups.

You could build your Mud-n-Sizzle Preamp into an instrument, too. Be the envy of your block with a homemade cigar-box uke with active electronics! Here are a few tips:

▸ Leave off the instrument's output jack and the preamp's input jack (the preamp's output will now serve as the instrument's output, too). Run your pickup's leads directly to the preamp circuit (one pickup lead will go to the input and the other to the common ground; it doesn't matter which wire goes where).

▸ Any unshielded wire[2] that carries an audio signal should be less than 8 inches long. Play it safe and keep all your wires under 6 inches when possible. This prevents interference from seeping into the circuit. Transistor-based amps are notorious for picking up local AM stations, baby monitors, and so on.

▸ If space is at a premium, you can leave out the mud-n-sizzle tone and volume controls. These can be eliminated without further modifications to the circuit. You can also eliminate the toggle power switch by replacing the output mono jack with a stereo jack wired for "jack power" (see "Stereo Jack-Power Switch" on page 358 for details). With a stereo jack power switch, the preamp automatically powers up when you plug in your instrument—no additional hardware required. Very slick.

---

2. *Shielded wire* is insulated wire that has a woven metallic sheath (the *shield*) surrounding the signal-carrying core. Grounding the shield protects the signal from interference. The most common household application for shielded wire is the *coaxial cable* that connects to the back of your cable/satellite TV box or broadband modem.

13 THE UNIVERSAL LFO

A *low-frequency oscillator (LFO)* is a piece of equipment used to modulate the sounds created by an electronic instrument. Depending on what it's modulating, an LFO can add warbles, wobbles, vibrato, tremolo, and a variety of swelling, pulsing, or sweeping effects to your electronic instrument's repertoire.

An *oscillator* is anything that goes back and forth between two points or states, like a fan that swings from right to left, a metronome ticking back and forth, or a vibrating guitar string. The Droid Voicebox (Project 6) and the Single-Chip Space Invader Synth (Project 15) both get their voices from electronic oscillators running at relatively high frequencies, in roughly the same range as a guitar string or the column of air in a slide whistle or trumpet. An LFO, on the other hand, oscillates too slowly to produce its own musical pitch. Instead, we harness its oscillations to automate regular changes in another circuit. Think of the oscillating fan: its oscillating motor doesn't create the breeze (there's a separate motor for that), but rather controls how that breeze sweeps the room. Just as an oscillating fan can totally change the character of a room, an LFO can totally change the character of a sound.

The variable-speed Universal LFO we're going to build allows you to select from two different waveforms. The *square wave* cuts in and out, flipping between full on and full off like a light switch, while the *pseudo-sine wave* fades in and out more gradually, like a dimmer. Depending on what you connect the LFO to, these waveforms offer dramatically different effects. Additionally, the LFO's effect on each specific waveform can vary dramatically as you change the rate at which the LFO oscillates. In short, a single switch and knob offer many sonic possibilities.

We'll connect our LFO to these circuits through an *optocoupler* composed of an LED and a *photoresistor*. A photoresistor (also called a *light-dependent resistor*) is a variable resistor that changes its resistance based on how much light is shining on it; the brighter the light, the lower the resistance. With an optocoupler, we can use one circuit (our LFO) to control another—say, the Twin-T Phaser/Wah (Project 14) or an old circuit-bent kid's toy (Project 7)—while keeping them totally electrically isolated. For this reason, an optocoupler is also called an *optoisolator*. This isolated coupling eliminates the risk of weird cross talk between the circuits and of injury to either circuit or maker.

The Universal LFO can be directly integrated into projects in this book, such as the Twin-T Phaser/Wah (Project 14), or used as a stand-alone device that you can clip into projects temporarily (see Figure 13-1). The second option is tons of fun if you're into circuit bending, as described in Project 7.

Hear the LFO in action in the samples at *http://www.nostarch.com/jamband/*.

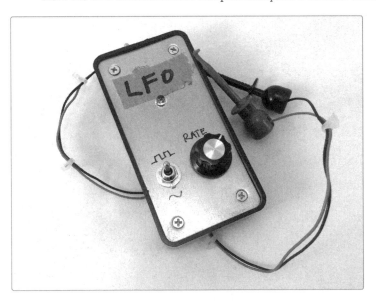

FIGURE 13-1: *The finished Universal LFO*

The following build instructions assume that you're building your LFO as a stand-alone device. Flip ahead to "Using the Universal LFO" on page 234 if you want some notion of how this circuit might be directly integrated with other projects.

Preparation

Build Time

▶ About an hour (It might take longer if you're building this into an existing device or planning a fancy enclosure.)

Tools

▶ A standard soldering kit (See page 340.)

▶ A ruler that shows 1/4-inch increments or smaller

▶ A Sharpie or other permanent marker

▶ An electric drill with bits

▶ (Optional) Other tools to work your enclosure (If you're using a wooden enclosure, you'll want sandpaper and possibly files to clean up drill holes, and maybe a hobby or utility knife.)

Supplies

▶ A 555 timer IC, such as Digi-Key part #296-1411-5-ND or Mouser part #595-NE555P

▶ An 8-pin IC socket

▶ A 1N4148 diode (Mouser part #78-1N4148 is shown here, although any generic 1N4148 or 1N4150 should work.)

▶ A 2N3904 NPN transistor, such as Mouser part #512-2N3904BU

▶ Two red LEDs

▶ A 120k ohm resistor (brown-red-yellow stripes)

▶ A 470 ohm resistor (yellow-violet-brown stripes)

▶ A 1k ohm resistor (brown-black-red stripes)

▶ A 3.3M ohm variable resistor, such as Mouser part #531-PC16SH335A1010 (Variable resistors are also called *potentiometers* or *pots*; see "The Gory Details: Audio Taper vs. Linear Taper" on page 327.)

▶ A control knob that fits your variable resistor

▶ A photoresistor (This may also be called a *light-dependent resistor*, a *cadmium sulfide (CdS) cell*, or a *photocell*. Digi-Key part #PDV-P8105-ND is an example.)

▶ A 1.5 μF electrolytic capacitor

▶ A 3300 μF electrolytic capacitor (Check out Figure 13-22 on page 239 if you're having trouble finding a 3300 μF capacitor and need options.)

▶ Two small SPST toggle switches (You can use Mouser part #108-MS550K, for example, but any similar switch will work.)

- ▶ A 9-volt battery
- ▶ A 9-volt battery clip
- ▶ 24-gauge insulated hook-up wire (Stranded wire is fine, but it's best to have both stranded and solid core in several colors. Unless otherwise specified in the build instructions, you can always assume that stranded wire is fine.)

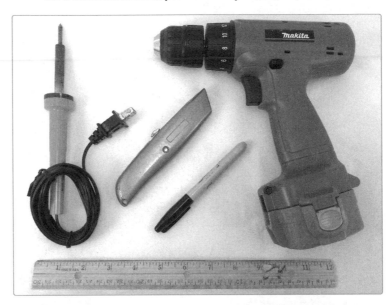

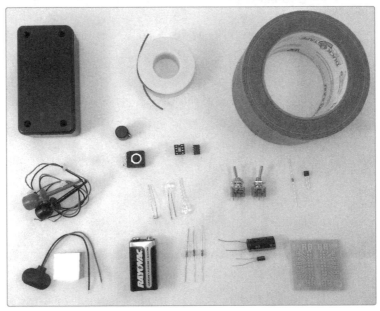

FIGURE 13-2: *Tools and supplies (not shown: sandpaper and file)*

- A general-purpose IC PC Board (These instructions show RadioShack part #276-159. Some alternatives are listed in "Use a Different PCB" on page 93; if you choose the Adafruit 1/4-Sized Perma-Proto PCB, read "Building on Another Generic PCB" on page 239 before getting started.)
- Duct tape
- Two alligator clips (Digi-Key parts #461-1013-ND and #461-1014-ND are shown in Figure 13-2.)
- A small, sturdy enclosure (See "On Enclosures" on page 214 for more information.)
- (Optional) A 9-volt battery holder clip, such as Digi-Key part #71K-ND or Mouser part #534-071

Building the Universal LFO

Before we begin: if you're new to soldering, don't start here. Look to the Droid Voicebox (Project 6) or the Mud-n-Sizzle Preamp (Project 12) for an easier first-time soldering experience. If you just need a soldering refresher, flip to "Soldering" on page 346.

There are three parts to this project: building the hardware, building the circuit, and then connecting the two. We'll first deal with the three pieces of hardware: the rate control, the waveform selector, and the power switch. The rate control sets the rate of the oscillation, the waveform selector allows you to choose between square wave and pseudo-sine wave, and the power switch—wait for it—switches on the power.

Prepare the Hardware

Step 1   Get out the two SPST switches, the battery clip, the potentiometer, the 1k ohm resistor (brown-black-red stripes), the 3300 µF electrolytic capacitor (that's the big one), and your stranded insulated wire. Cut five 4-inch lengths of insulted wire, strip both ends of each, and then tin them.

Step 2   Solder the end of one wire to lug 1 of the potentiometer and the end of another wire to lug 2 (see Figure 13-3). Lugs are numbered from left to right, as viewed with the shaft facing you and the lugs at the top. This pot will ultimately serve as the LFO's rate control. If your pot has a long shaft, which can be a pain to integrate into an enclosure later, now is a good time to trim it down. For details on this procedure, flip to "Resistors: Fixed and Variable" on page 325.

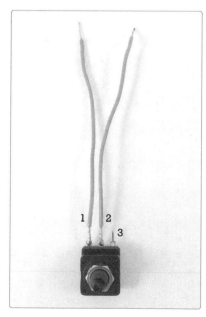

FIGURE 13-3: *The prepared rate-control potentiometer*

Step 3   Next, we'll work on the waveform selector. Take a prepared 4-inch wire and a 1k ohm resistor, and twist together the end of the wire and one leg of the 1k ohm resistor. Grab a toggle switch and solder this twisted pair to either of the lugs, as shown in Figure 13-4 (top left).

Step 4   Solder the positive leg of the big 3300 µF capacitor to the switch's other lug, as shown in Figure 13-4 (bottom left). Remember that the negative leg is marked with a stripe on the body of the capacitor; the positive leg is thus the one farther from the stripe.

Step 5   Finally, twist together the capacitor's negative leg (the one aligned with the stripe), the remaining resistor leg, and another 4-inch insulated lead, and solder the connection. Figure 13-4 (right) shows the result.

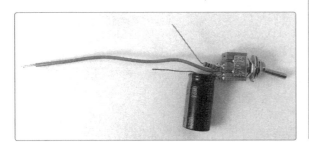
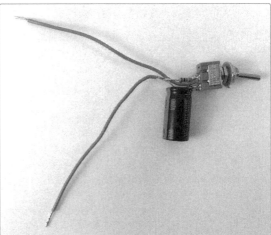

FIGURE 13-4: *Building the waveform selector*

Step 6   For the power switch, take the other toggle switch and solder the last 4-inch lead to either lug. Solder the red lead from your battery clip to the remaining lug, as shown in Figure 13-5. Good job! Set these three pieces of prepared hardware aside for now.

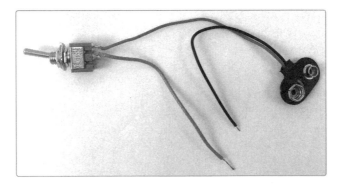

FIGURE 13-5: *The power switch assembly*

Build the Circuit

Time to look at the circuit diagram! A circuit diagram is just a map of the electrical relationship between components; it doesn't necessarily show how those components are going to be physically laid out. Figure 13-6 is a good example: the schematic looks pretty complicated, but you've actually already built about a third of it just by wiring up the hardware.

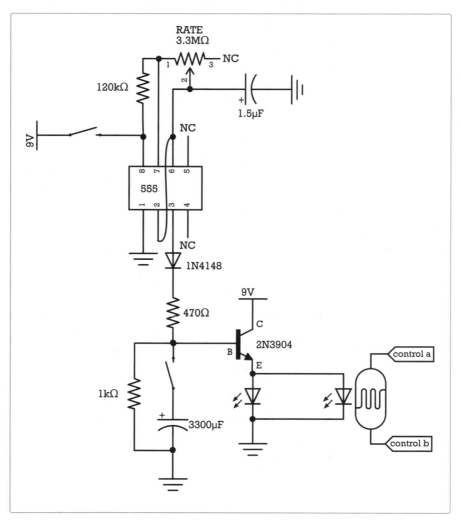

FIGURE 13-6: *The circuit diagram for the Universal LFO*

That said, building the circuit *can* be a touch tricky because the copper traces on the PCB snake around on the underside of the board somewhat counter-intuitively. For each soldering step, I'll show the component placement with a photograph of the topside of the PCB and highlight the solder points with a diagram showing the underside. To help you keep your bearings, the IC socket and transistor are marked on each diagram.

The Universal LFO is most easily understood as two circuits connected by a diode. One circuit (the top half of Figure 13-6) is built around the 555 timer IC. This circuit acts as a clock controlling the LFO rate, which you set by twiddling the potentiometer. The other circuit (the bottom half of Figure 13-6) consists of a transistor, LED, capacitor, and resistors, and it controls how the LEDs light up, whether in a square-wave or pseudo-sine-wave pattern. The LED, in turn, controls a photoresistor, which controls whatever it is you're using the LFO to modulate.

✳ **NOTE:** *If this is your first IC project, flip to the Droid Voicebox (Project 6) and read the pointers in Step 1 to familiarize yourself with 8-pin ICs. Transistors also tend to flummox folks new to electronics, so you might want to flip to "Diodes, LEDs, and Transistors" on page 334 to make sure you know which leg is which.*

Step 7 Place the IC socket and the transistor on the circuit board, as shown in Figure 13-7, sliding their legs through the holes indicated in the figure (right image). When viewed from the top, the IC socket should be as far to the left as possible and the transistor far to the right. Bend the legs to keep them on the board, but don't solder yet. You'll note that we haven't placed the 555 timer in the socket. One advantage of using an IC socket is that it allows us to entirely avoid any possibility of overheating or otherwise damaging the IC while building the circuit. Using a socket also makes it easy to replace the IC later if it somehow gets zapped. That's unlikely, but the IC is still the easiest component to damage in this project. Many folks build small projects like these without sockets, soldering the IC directly to the PCB. I'm begging you not to be one of those cowboys. Do it right the first time.

 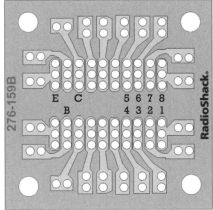

FIGURE 13-7: *Placing the IC socket and transistor. The numbers in the illustration (right) correspond to the IC socket's pins and the letters to the transistor's legs.*

Step 8 Cut a 1.5-inch length of insulated solid core wire. Stranded will do in a pinch, but tin lightly or it'll be a pain to thread through the PCB hole. This is our first jumper. Run one end into the hole directly below IC pin 2 and the other end to the hole directly above pin 6, over the top of the IC socket. Solder both ends of the jumper to their pads and then solder down pins 2 and 6 of the IC socket, as highlighted in Figure 13-8.

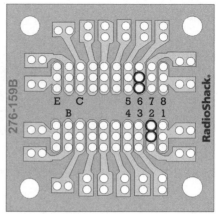

FIGURE 13-8: *The placement of the first jumper (left) and the first solder points (right)*

Step 9   Now add the 120k ohm resistor (brown-red-yellow stripes). Bend the resistor into a *V*, just as we did when building the Mud-n-Sizzle Preamp (Project 12). If you need a reminder, see Figure 12-8 on page 205. Run the resistor's legs into the first two open holes above socket pins 7 and 8, as shown in Figure 13-9. Solder both ends of the resistor and both socket pins to their respective pads.

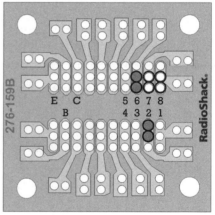

FIGURE 13-9: *Placing and soldering the 120k ohm resistor. (New solder points are white; solder points from the previous steps are gray.)*

Step 10   Add the 1.5 µF electrolytic capacitor. Place the negative leg (the one aligned with the stripe) into the hole directly below pin 1 and the positive leg in the hole beside it to the right (see Figure 13-10). Solder the positive capacitor leg (that is, the one in a column with IC pin 2) to its pad. Leave the negative capacitor leg (the one in the column with IC pin 1) to be soldered later.

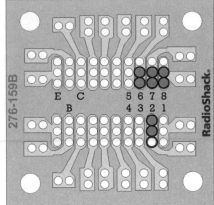

FIGURE 13-10: *Adding the 1.5 µF capacitor*

Step 11 Next, place the diode that connects the two halves of the circuit. Like electrolytic capacitors, diodes are polarized, with the black stripe indicating the negative leg. Run the positive leg into the first open hole below pin 3 and the negative leg three columns to the right in the same row. Solder the positive diode leg and socket pin 3 to their pads (see Figure 13-11).

> * **NOTE:** *You may have to look carefully for the markings on the diode, as they are notoriously illegible.*

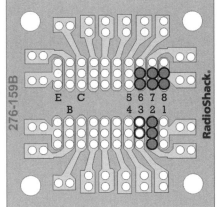

FIGURE 13-11: *Adding the diode*

Step 12 Next up is the 470 ohm resistor (yellow-violet-brown stripes). This resistor contributes to the fade-in time of the LED when the waveform is set to pseudo-sine. It runs from the negative leg of the diode to the transistor base, as shown in Figure 13-12. Bend the resistor into a *U* shape, insert one leg into the first hole immediately below the diode's negative leg, and insert the other into the second open hole below the middle transistor leg. (In the illustration on the right, the transistor's middle leg—its *base*—is represented by a *B*.) Note that placing the 470 ohm resistor this way leaves an open hole along the column connecting the transistor and the resistor. Solder the negative diode leg and the 470 ohm resistor lead to their pad now, but leave the other resistor lead to be soldered later.

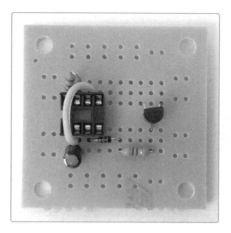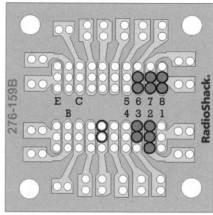

FIGURE 13-12: *Placing the 470 ohm fade-in resistor*

Step 13   Now build the optocoupler. Cut two 4-inch lengths of insulated wire, preferably stranded. Strip and tin both ends. Then solder one end of each wire to each leg of the photoresistor. Take one LED and mark its negative leg to make it easier to identify later; the negative leg is closest to the flat section on the edge of the LED lens and is usually the shorter lead. Place the LED snout to snout with the prepared photoresistor so that the head of the LED and the head of the resistor are touching, as shown in Figure 13-13 (left). Wrap the two components in a layer of duct tape. Be sure to pinch the tape down and seal it behind the resistor and LED so that no light can leak in or out. Light leaks can interfere with this sort of light-dependent control, making it annoyingly inconsistent and a pain to troubleshoot later.

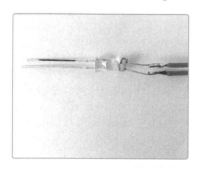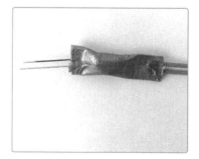

FIGURE 13-13: *Building the optocoupler. (Note the black stripe denoting the negative LED leg.)*

Step 14   Take your remaining LED. Cut two lengths of stranded wire, strip and tin them, and solder one to each leg of the LED. This is your indicator LED.

Step 15   Now connect the indicator and optocoupler LEDs to the rest of the circuit. Insert the negative LED leads into the pair of holes just below the PCB's centerline, at the right edge, and the positive leads into the pair of open holes just above the centerline. (In Figure 13-14, the lower set of leads are the negative ones, and the upper pair are the positive.) Solder all four leads to their respective solder pads now. Also, solder the transistor lead closest to the LED leads (the *emitter*, marked with an *E* in the illustration) to its solder pad.

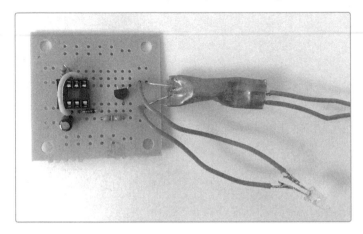 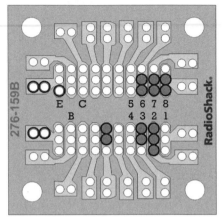

FIGURE 13-14: *The indicator and optocoupler LEDs*

Step 16   Now connect the transistor to the power. Cut a 1 1/4-inch length of insulated wire, preferably solid core. Strip both ends and tin them if necessary. Run one end to the final open hole above socket pin 8 and the other end to the second open hole up from the transistor's collector leg, marked with a *C* in Figure 13-15 (right). This jumper is the white wire running horizontally above the IC socket in Figure 13-15 (left). Flip the PCB and solder both ends of this jumper to its pads. Also solder the transistor's collector to its pad.

Step 17   Connect the LEDs to the ground with the final board-to-board jumper. Cut a 1 1/2-inch length of wire, preferably solid core. Strip both ends and tin if necessary. Insert one end of this jumper into the middle hole in the three-hole column below the centerline all the way to the right. As Figure 13-15 (right) shows, this hole does indeed connect to the lower legs of the two LEDs, even though it doesn't look like it from the top. Insert the other end of the jumper into the first hole below IC socket pin 1, as shown in Figure 13-15 (left); this jumper is the dark wire running below the IC socket and past the transistor. Flip the board over and solder both ends of this jumper to their respective pads. Also solder pin 1 to its pad and solder the negative leg of the 1.5 µF capacitor from Step 10.

✳ **NOTE:** *In Figure 13-15, the white jumper running horizontally along the top connects the transistor to the power supply. The dark jumper running horizontally below the IC and transistor connects the LEDs to ground.*

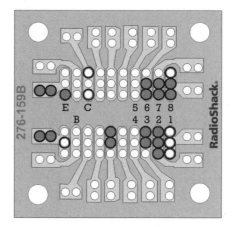

FIGURE 13-15: *Adding the last two jumpers*

Install the Hardware

Now let's attach everything you built in "Prepare the Hardware" on page 223 to the printed circuit board.

Step 18 Take the rate-control potentiometer you built in the first few steps. Run the lead from lug 1 into the topmost hole on the left edge of the board, as shown in Figure 13-16 (left). This lead connects to socket pin 7 by way of the twisting trace on the underside of the PCB. Run the lug 2 lead to the leftmost hole on the top edge of the board. Solder these in place. Next add the waveform switch. Run the lead connected to both the 1k ohm resistor and the negative leg of the big 3300 µF capacitor to the first open hole just to the left of pin 1 of the IC socket. This connects the switch to the common ground. Solder this lead in place and then run the switch's other lead to the hole directly below the transistor's middle leg (the *base* leg indicated with a *B* in Figure 13-16). Now flip the board and solder the transistor base, the switch lead, and the remaining leg of the 470 ohm resistor to this pad. *Voilà!* All of the legs and leads should now be soldered.

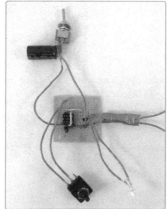
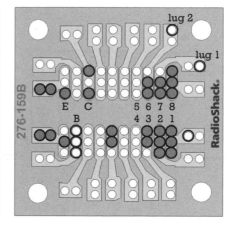

FIGURE 13-16: *Adding the rate and waveform controls*

Step 19   Now we add power! Take your power switch and run the single unconnected wire to the open hole at the left edge of the board, just above the centerline (see Figure 13-17). Run the negative (black) battery lead into the next hole down, just below the centerline. Solder both leads. Now take your IC and carefully press it into the IC socket (oriented with its dot to the left), and your circuit should be ready to go!

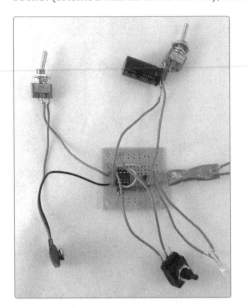

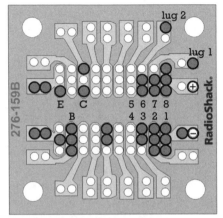

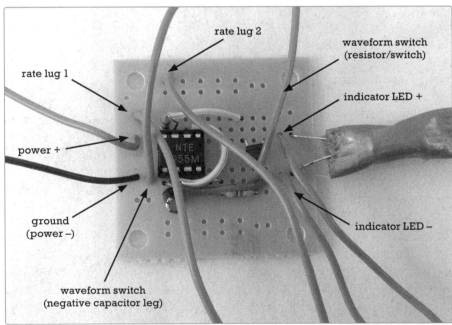

FIGURE 13-17: *The finished LFO circuit—viewed both at wide-angle (top left) and zoomed in with helpful labels (bottom). Note that the IC is now mounted in the socket.*

Troubleshooting and Packaging

Time to power up the LFO circuit and test it out! Position the rate control at its mid-point and turn on the power. The LED should blink at a rate of about one blink per second. If not, flip the power switch. If you still have no blinking, double-check to make sure you have a fresh battery, that everything is properly soldered to the PCB, and that there are no little scraps of solder connecting adjacent pads, traces, or lugs and thereby creating short circuits. If you find any short circuits, eliminate them. Loose scraps of wire or flakes of solder can usually be blown or flicked away, while most sloppy joints and solder bridges can be smoothed out by reheating them; the solder should naturally flow to the warmer joint. If a solder bridge is especially stubborn or there's just way too much solder glopped onto a joint, you may need to do a little desoldering (see "Desoldering" on page 349).

When your LED blinks, check the waveform switch. In one position, the LED should blink full on and full off at a rate of about once per second; that's square-wave mode. In the other, it should fade in and out at roughly the same rate; that's pseudo-sine wave mode.

Finally, test the rate control. Twisting it clockwise should speed the blinking up, and counterclockwise should slow it down. If the opposite happens, that's fine; everything works, but you just swapped the wires on the potentiometer, soldering to lugs 2 and 3 instead of 1 and 2. You can power down and resolder the lead, or just keep it as is; either is fine.

If you have a multimeter handy, check the operation of the optocoupler itself. Set the meter to the 2M ohm range and connect one test probe to each of the leads dangling from the photoresistor side of the optocoupler—that is, the leads sticking out of the duct tape. As the LFO swings from high to low, the LED should swing from bright to dark, and you should see the multimeter readout swing from 0 ohm (or very near to it) up to 1M ohm or more.[1]

Once everything works to your liking, mount the circuit in your enclosure. You have plenty of options here. You could use a standard plastic and metal "project box" like the one I've used, or maybe you've got a cool old wooden tea box. You could even build this LFO into your circuit-bent talking toys and keyboards—that's your prerogative. Let your freak flag fly. But however you package this project, here are a few pointers.

First, drill holes that fit your hardware. In general, that means 3/16 inch for the indicator LED, 1/4 inch for the small switches, and either 5/16 or 3/8 inch for the rate-control potentiometer. Your components may have slightly different dimensions, so double-check before you drill.

If the enclosure is large enough for the battery and PCB to rattle around, then they'll probably cause trouble. Secure the battery with a battery holder clip, such as Digi-Key part #71K-ND, or some other scheme of your own devising, such as a little double-sided tape. As for securing the PCB, most come with predrilled mounting holes you can use to screw them down, but a snip or two of double-sided foam tape will do the job just fine here, too.

---

1. There's a lot of variance between individual photoresistors, which I discuss at length in "The Gory Details: Selecting the Right Photoresistor" on page 328. If you really dig the Universal LFO, consider experimenting with different photoresistors.

If you're building a stand-alone clip-on Universal LFO, like mine, then now is the time to install the alligator clips. Attach one to each of the control leads dangling from the optocoupler.

Using the Universal LFO

The Universal LFO can be applied to basically any battery-operated electrical project, whether commercially produced or from this book. If a music or noisemaker contains a potentiometer or resistor that controls some aspect of its operation, you can almost certainly replace, augment, or hijack that control with your little LFO, often with startling results.

∗ **NOTE:** *As you explore the options I describe in this section, note that the Universal LFO may not share power well with other circuits. Plan on having the LFO be independently powered by its own battery. Then just be pleasantly surprised if it shares power nicely in any specific application.*

Integrate the LFO

When adding an LFO to a project, you have three options:

▶ **Add the LFO in parallel.**   Leave the existing resistor or pot in place and just connect one optocoupler control lead from your LFO to each of its legs. Resistance will swing between near zero and the value of the existing resistor. So, for example, if you were to add an LFO in parallel with a pitch-controlling resistor in a toy keyboard, the instrument would now oscillate between normal and high-pitched or hyperspeed. Adding the Universal LFO in parallel gives you the most flexibility. When the LFO is off, the native control behaves normally; when it's on, the LFO takes control.

▶ **Add the LFO in series.**   Desolder or snip one leg of the existing resistor or pot and connect it to one optocoupler lead from your LFO. Connect the other optocoupler lead to the pad that was previously connected to the original resistor. The resistance at this point in the circuit will now swing between close to the value of the original resistor and infinity—that is, no current passing through that portion of the circuit. With the same hypothetical instrument as in the first example, this would mean oscillating between normal operation and low-pitched or molasses-in-a-glacier-slow-mo. In this scenario, cranking the LFO to full speed will often essentially return the modulated circuit to normal operation.

▶ **Replace the existing pot or resistor altogether.**   Desolder the resistor that's already there, and then solder the two optocoupler control leads in place of the resistor leads you removed. Now the resistance at that point in the circuit should swing between near zero and near infinity—or, in terms of the hypothetical music toy, all the way from hyperspeed to glacial and back again. This scenario gives you the widest effect range; the drawback is that you can't easily return to normal operation.

These wiring schemes, illustrated in Figure 13-18, are based on the techniques explored in Project 7.

FIGURE 13-18: *Adding the optocoupler to an existing circuit: adding the LFO in parallel (left), adding the LFO in series (center), and replacing the resistor (right)*

It's super convenient to experiment with parallel LFO wiring by using a pair of alligator jumpers rather than soldering a more permanent connection (see Figure 13-19).

FIGURE 13-19: *Using alligator clips to temporarily add the LFO control to a toy megaphone*

When wiring an LFO in parallel to a potentiometer, connect one control to the middle lug of the potentiometer and then see what happens with the other control lead connected to either of the edge lugs. Running the control leads to lugs 1 and 3 isn't likely to result in much, but that shouldn't stop you from giving it a whirl.

Once the LFO is hooked up to your liking, it's time to explore! Here are a few tips:

▸ Depending on what the base sound is like and what aspect of that sound is being modulated, the square wave will often give a choppy effect at slow speeds and a shimmery effect at fast speeds. Similarly, the pseudo-sine wave can be "wobble-bass-ish" at slower speeds and warbling at higher speeds.

▸ Experiment with modulation rates. Our homebrew optocoupler "relaxes" relatively slowly, often giving different speeds very different characters. The pseudo-sine wave is *much* less pronounced at high speeds; the square wave is obvious at all rates.

▸ When wired in parallel, turning the LFO off basically returns the sound circuit to normal operation.

Create a Tremolo

Tremolo is the rapid oscillation in volume that gives a guitar that shimmery, surf rock/rockabilly sound. The guitar in Link Wray's 1958 hit "Rumble" is required listening as a standout example of tremolo. You can use your Universal LFO to create a tremolo effect in two different ways.

With many instruments, you'll be able to get a good tremolo effect in an instant by clipping the LFO directly into your instrument's output. Clip one LFO control lead to the ground lug of the output jack and the other to the tip lug of the same jack, as shown in Figure 13-20 (left). Plug into an amp as usual, and when you turn on the LFO, you'll get that sultry rockabilly shimmer. Turn off the LFO, and it's business as usual. I personally like this effect either with the LFO rate set to midrange and the pseudo-sine wave engaged (for sort of a '70s Steely Dan intro vibe) or with the LFO fairly fast and the waveform set to square (honky-tonk shimmer). But those are my tastes; start tweaking and see what you like!

For a slightly less instantaneous tremolo—but one that offers a more defined sound and works with a broader range of instruments—build an *inline tremolo*. Get a pair of jacks, solder a wire between their ground lugs, and then connect one LFO optocoupler control lead to the tip lug on each jack, as shown in Figure 13-20 (right). (If the jack anatomy has you flummoxed, there are detailed diagrams in "Quarter-Inch Phone Plugs and Jacks" on page 337.) Plug one jack into your instrument and the other into your amplifier, and then proceed as above.

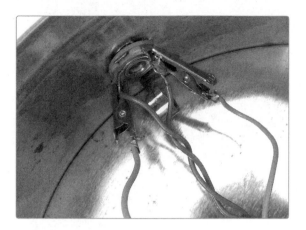 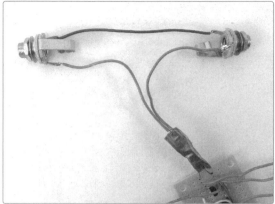

FIGURE 13-20: *A no-solder tremolo (left) and an inline tremolo (right)*

Combine the LFO with the Mud-n-Sizzle Preamp

The Universal LFO was originally cooked up for the Twin-T Phaser/Wah (Project 14), but it also pairs well with the Mud-n-Sizzle Preamp (Project 12) to create a sort of growly wobble trem. You can make this sound three different ways:

▶ Clip one LFO control lead to lug 2 of the volume pot on the preamp and the other to the ground. This gives the most traditional tremolo effect.

▶ Clip one LFO control lead to lugs 1 and 2 of the tone control pot on the preamp and the other to the ground. This creates more of an overdriven wobble effect.

▶ Clip one LFO control lead to the positive 9-volt power supply on the preamp and the other to the junction of the output cap, bias resistor, and transistor's collector leg. This produces a unique ZZ Top–ish rhythmic overdrive.

The LFO is likewise fun with the Single-Chip Space Invader Synth (Project 15) and the Bleepbox 8-Step Analog Sequencer (Project 16). In both cases, start by experimenting with connecting the LFO to pitch, mod, or volume controls—or other components. And, of course, the Universal LFO can kick your circuit-bending experiments (à la Project 7) up into the outer limits.

Tips, Tricks, and Mods

I've specified a 3.3M ohm potentiometer for the LFO's rate control because you need a fairly wide range of resistances to get a fun range of effects. But these higher-value pots can be both hard to find and a bit pricey (around $4 each). Always keep your eyes peeled for priced-to-sell surplus high-resistance potentiometers. In the meantime, you can fake your way to a high-resistance potentiometer using a handful of common 1M ohm pots. As you'll recall, the total resistance of a set of resistors wired in series is the sum of their individual resistances. While a single 1M ohm pot won't give your LFO much of a range, several wired-in series will get you where you need to be.

Take a set of 1M ohm pots and connect lug 2 of the first to lug 1 of the second (see Figure 13-21). With both cranked fully counterclockwise, you should have 2M ohm of resistance between the two free wires. You can *daisy-chain* as many pots as you like in this fashion. Want 3M ohm? Just wire lug 2 of pot A to lug 1 of pot B and then lug 2 of pot B to lug 1 of pot C. Want 5M ohm? The "lug 2 to lug 1" wiring is the same; just start with five 1M ohm pots. It may make for an awkward control, but it's cost-effective, and it opens the door to some downright glacial filter and pitch sweeps.

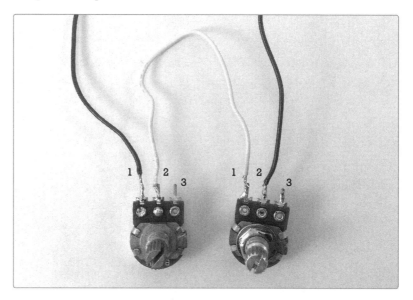

FIGURE 13-21: *Daisy-chaining pots to fake your way to higher resistance*

There are three other components that deserve your attention: the two resistors and the big capacitor connected to the waveform switch. These control the fade-in and fade-out rates of the pseudo-sine wave. To get a shorter fade-in time, replace the 470 ohm resistor that links the switch to the timer circuit with a smaller-value resistor. Try 10 ohm for a very abrupt fade in. To get a longer fade-out time, increase the value of the large 3300 µF capacitor; correspondingly, decrease the value of this cap to get a shorter fade-out.

Just as with resistors, you can fake your way to a higher capacitance by combining several smaller-value components. If you want a larger cap here (or if you don't have the 3300 µF capacitor I've specified handy), just solder a few smaller caps in parallel and add up their capacitance. For example, in Figure 13-22, I've piggybacked a pair of 1500 µF caps for a combined capacitance of 3000 µF, which is a nearly perfect substitute for the 3300 µF cap in the parts list. To try a 4500 µF cap in its place, just add one more 1500 µF cap in parallel. Give it a shot and see whether you like the results!

FIGURE 13-22: *Stacking caps to fake your way to a higher value*

The 1k ohm resistor that runs parallel to the 3300 µF capacitor also plays a role in determining the fade-out time. You can make an LFO with a variable pseudo-sine wave by replacing both the 470 ohm (fade-in) and 1k ohm (fade-out) resistors with potentiometers. I suggest trying 500 ohm and 1k ohm pots, respectively.

Building on Another Generic PCB

If you have trouble finding the single-chip PCB (RadioShack part #276-159) shown in the illustrations, don't worry! You can build this circuit on any generic pad-per-hole PCB using the schematic in Figure 13-6 and the skills you learned working on the Mud-n-Sizzle Preamp (Project 12). You can also use a permanent breadboard–style PCB and the alternate layout I've prepared, as shown in Figure 13-23.

Permanent breadboard PCBs are designed to mimic the layout of the temporary prototyping breadboards folks use to build test circuits, as described in "Building a Circuit" on page 352. This makes it a snap to transfer an experimental circuit to a permanent PCB.

My preferred permanent-breadboard PCB is the Adafruit 1/4-Sized Perma-Proto PCB, which is shown in these illustrations. This specific PCB is made and sold by Adafruit, but it's also regularly stocked by the MakerShed, Evil Mad Scientist, and many other popular suppliers. You'll find a full list of suppliers in "Guidelines for Sourcing Components" on page 323.

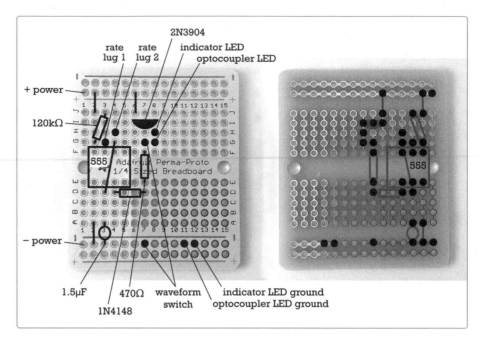

FIGURE 13-23: *An alternate layout for the Universal LFO is illustrated here. Component mounting positions are shown in a top view (on the left), and all of the solder points are highlighted as seen from the bottom of the board (right).*

Most importantly, many other manufacturers have made compatible permanent-breadboard PCBs for years. That means even if Adafruit stops making its 1/4-Sized Perma-Proto, you should still be able to use the layout shown in Figure 13-23.

Even if you have no idea what a wah-wah pedal is, you know the sound. Wah-wahs are responsible for the crying, voice-like articulation in countless rock classics— Jimi Hendrix's "Voodoo Child" is a great example—as well as the funky-see funky-do wakka-chikka-wakka rhythm guitar in tunes like Isaac Hayes's *Shaft* theme. The technical name for this effect is *spectral glide*, which means the "modification of the vowel quality of a tone." In this case, the modification emphasizes the *waaahs*, *wooowwwws*, and *aooows*. Control the spectral glide with a foot pedal, and you have a *wah-wah pedal*. Automate it with an LFO (like the Universal LFO in Project 13), and you have a *phaser effect*. Tweak it with a plain old knob, and you can do manual *filter sweeps*, just like your favorite analog synth aficionado or eurotrash club DJ. The Twin-T Phaser/Wah in this project packs all of this into one cigar box.

Traditional wah-wah pedals are tricky to build at home: they need to endure a lot of mechanical abuse, so they require a purpose-built, heavy-duty potentiometer and a specialized rocker-pedal enclosure. Our Twin-T Phaser/Wah dispenses with the mechanical pedal control and instead uses a faux pedal made from a light-sensitive photoresistor. By covering and uncovering the photoresistor with your foot, you control the amount of light hitting it, allowing you to sweep the filter back and forth across its full range, articulating the *waaah*s and *aooow*s while your hands are otherwise occupied.

There are a lot of features here, giving you plenty of options—and plenty of chances to make a wrong turn. Read this project fully in advance—especially "Using the Twin-T Phaser/Wah" on page 261 and "Tips, Tricks, and Mods" on page 262—so you can make some decisions before you get to snipping and soldering.

Hear the Twin-T Phaser/Wah in action in the samples at *http://www.nostarch .com/jamband/*.

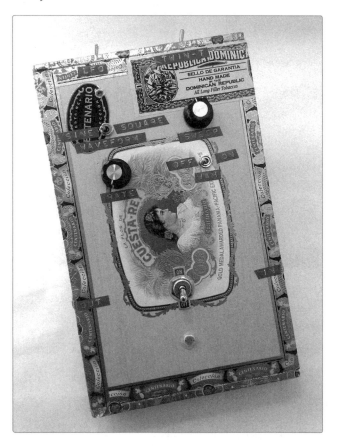

FIGURE 14-1: *The finished Twin-T Phaser/Wah*

Preparation

Build Time

▶ About an hour

Tools

▶ A standard soldering kit (See page 340.)

▶ A ruler that shows 1/4-inch increments or smaller

▶ An electric drill with bits (You'll probably need 3/16-inch, 1/4-inch, 5/16-inch, and 3/8-inch bits.)

▶ (Optional) Other tools to work your enclosure (If you're using a wooden enclosure, you'll want sandpaper and possibly files to clean up drill holes.)

▶ (Optional) Foam-backed double-sided tape or hardware to mount the circuit in its enclosure

▶ (Optional) A hacksaw for trimming down potentiometers with long shafts

Supplies

▶ A 2N3904 NPN transistor, such as Mouser part #512-2N3904BU

▶ A red LED

▶ Three 0.1 μF capacitors (marked *104*)

▶ Two 0.0047 μF capacitors (marked *472*)

▶ A 0.01 μF capacitor (marked *103*)

▶ Two 1M ohm resistors (brown-black-green stripes)

▶ A 470 ohm resistor (yellow-violet-brown stripes)

▶ Two 470k ohm resistors (yellow-violet-yellow stripes)

▶ A 33k ohm resistor (orange-orange-orange stripes)

▶ A 10 ohm resistor (brown-black-black stripes)

▶ A 10k ohm resistor (brown-black-orange stripes)

▶ A 50k ohm variable resistor (Variable resistors are also called *potentiometers* or *pots*; see "The Gory Details: Audio Taper vs. Linear Taper" on page 327.)

▶ A control knob that fits your variable resistor

▶ A photoresistor (This may also be called a *light-dependent resistor*, a *cadmium sulfide (CdS) cell*, or a *photocell*. Digi-Key part #PDV-P8105-ND is an example.)

▶ Two small SPST toggle switches (You can use Mouser part #108-MS550K, for example, but any similar switch will work.)

▶ A DPDT switch (Mouser part #612-100-DP1T8 or Digi-Key part #EG2410-ND will work, and your local hardware store likely offers plenty of other viable options.)

- ▶ A 9-volt battery
- ▶ A 9-volt battery clip
- ▶ An Adafruit Perma-Proto Quarter-sized Breadboard PCB (Adafruit product #589 or Maker Shed product #MKAD48)

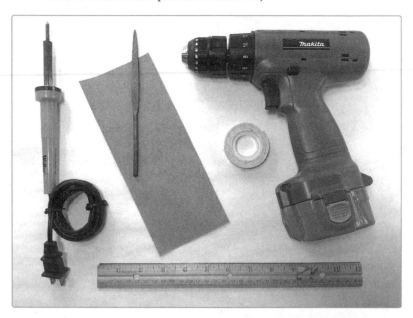

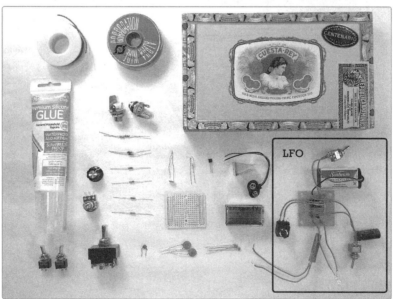

FIGURE 14-2: *Tools and supplies (not shown: hacksaw). The completed Universal LFO (Project 13) is shown in the lower-right corner.*

- 24-gauge insulated hook-up wire (Stranded wire is fine, but it's best to have both stranded and solid core in several colors.[1] Unless otherwise specified in the build instructions, you can always assume that stranded wire is fine.)

- 22- or 24-gauge bare bus wire (This is uninsulated solid core wire. Because you need only a short length of bus wire for this project, you can get away with using a scrap of wire, a leftover bit snipped from a component lead, or even a piece of a paper clip.)

- Two 1/4-inch mono phone jacks, also called *guitar jacks*

- Silicone-based household glue, sometimes called *room-temperature vulcanizing rubber* or *RTV-1*

- A sturdy enclosure (If you're going the LFO route, make sure there's room enough for two circuit boards and batteries. See "On Enclosures" on page 214 for more information.)

- (Optional) A 9-volt battery holder clip, such as Digi-Key part #71K-ND or Mouser part #534-071

- (Optional) A Universal LFO (While this is optional, it's super awesome and vital if you want the twin-T to have full phaser functionality. See Project 13 for instructions.)

Building the Twin-T Phaser/Wah

As with other projects, we'll begin by soldering the hardware. (Need a soldering refresher? Check out "Soldering" on page 346.)

Prepare the Hardware

In this section, we'll build the sweep control, the power switch, and the bypass switch.

> * **NOTE:** *If you're omitting the LFO phaser control or photoresistor wah pedal, you can skip some of the wiring in this first step (and then skip installing those pieces down the road in "Add the Phaser/Wah" on page 257). Read all of Step 1 before snipping, stripping, or soldering anything.*

Step 1 First, we'll build the sweep control. Take the 50k ohm potentiometer, or *pot*, and one of the SPST switches. Cut four 4-inch lengths of wire and two 6-inch lengths, and strip and tin the ends of each.

One pair of 4-inch wires will connect the sweep-control pot to the twin-T circuit—that pair is mandatory. The other two 4-inch wires will ultimately connect to the LFO. Both 6-inch wires and the SPST switch are for the photoresistor wah pedal. So if you're going to skip the LFO, the pedal, or both, omit their leads now.

---

1. If you're using the Adafruit Perma-Proto Quarter-sized Breadboard PCB, then you'll have no problem using stranded wire for all connections. This PCB is a double-sided, through-hole-plated board. These tend to have slightly larger holes than single-sided PCBs and thus accepts tinned stranded wire more readily than similarly laid out single-sided PCBs from other manufacturers.

For a fully loaded project (one that includes both the LFO and photoresistor wah pedal), take two of the 4-inch wires and one of the 6-inch wires and twist their ends together. Solder these three twisted wires to lug 2 of the pot. (Pot lugs are numbered from left to right when the shaft is facing you, as shown in Figure 14-3.) Repeat this with another set of two 4-inch wires and one 6-inch wire, soldering them to lug 3. If you're going to skip the wah pedal, then just solder two 4-inch wires to lug 2 and two to lug 3. If you want to skip the LFO and keep the wah, then you'll solder one 4-inch and one 6-inch wire to each of these lugs. And, if you're skipping all the extras, then you'll just have a single pair of 4-inch leads, one connected to lug 2, the other to lug 3.

If you've included the 6-inch wires for a photoresistor wah pedal control, then solder the free end of lug 3's 6-inch wire to either of the terminals on the SPST switch. The resulting fully loaded sweep-control pot is shown in Figure 14-3.

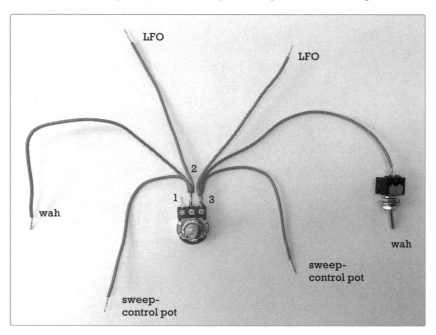

FIGURE 14-3: *The fully loaded sweep control. If you are omitting the photo-resistor wah, then this pot will just have the four 4-inch leads—no 6-inch wires and no switch. If you're omitting the LFO and the wah, then the pot will have just two 4-inch wires, one soldered to lug 2, the other to lug 3.*

Step 2   Next, we'll build the power switch with indicator LED. Take out the 9-volt battery clip, the other SPST switch, the 470 ohm resistor (yellow-violet-brown stripes), and the LED. Cut, strip, and tin a 4-inch length of insulated wire. Solder the positive (red) battery clip lead to either lug on the switch. Solder one leg of the 470 ohm resistor to the positive leg of the LED. (The positive leg is longer; the negative leg aligns with the flat spot on the side of the LED lens.) Finally, twist the remaining leg of the resistor to one end of the 4-inch lead and then solder this twisted pair to the empty switch lug. Check out the results in Figure 14-4.

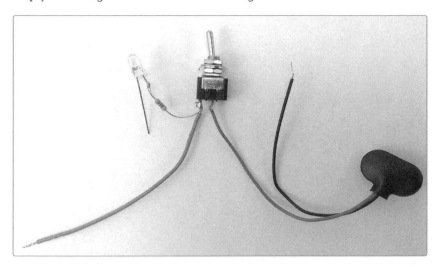

FIGURE 14-4: *The finished power switch with indicator LED*

Step 3   The last piece of hardware is the *bypass switch*, which will allow you to bring the Twin-T Phaser/Wah in and out of your *effects chain* without unplugging a bunch of stuff. We're using a standard DPDT toggle switch. Cut a short length of bare bus wire and run it diagonally from lug 1 to lug 6. Solder this in place (or screw it down if you have screw terminals like the ones shown in Figure 14-5). Then cut, strip, and tin two lengths of insulated wire. Solder one insulated lead to lug 2 and the other lead to lug 4.

✳ **NOTE:** *The left image of Figure 14-5 shows a standard DPDT toggle switch (on the left) and a heavy-duty DPDT stomp switch (on the right). The standard switch is easy to find—any hardware store will have it. The stomp switches—which are the sort used in old-school commercial guitar effects (often called "stomp boxes")— will need to be special ordered (see "Guidelines for Sourcing Components" on page 323). Both wire in the same way, as shown in the right panel of Figure 14-5.*

 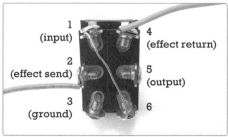

FIGURE 14-5: *Two styles of DPDT switch with two different styles of lugs (left)—both wire up the same way (right)*

Step 4   Take out your two jacks and prepare two more lengths of 4-inch wire. Solder one 4-inch wire to the tip lug of each jack; if you're unsure where the tip is, check out "Quarter-Inch Phone Plugs and Jacks" on page 337. Figure 14-6 shows the finished jacks.

Step 5   Now add the input and output jacks to the bypass switch. The jacks are identical at this point, so it doesn't matter which goes where. Connect the lead from one jack to lug 1 of the DPDT switch and the lead from the other jack to lug 5 (see Figure 14-7). Set aside all the hardware—sweep-control pot, power switch, and bypass switch—for later.

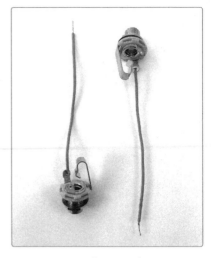

FIGURE 14-6: *Input and output jacks*

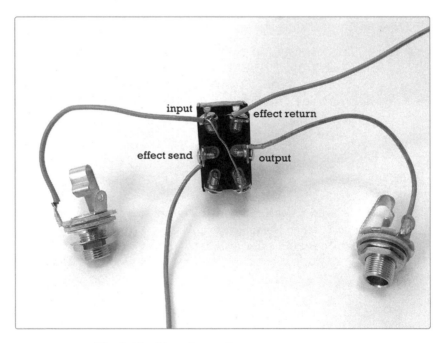

FIGURE 14-7: *The finished input/output/bypass*

Build the Circuit

It's time to look at the schematic, shown in Figure 14-8. As with the Universal LFO, this is probably best understood as two circuits linked together: the top half—centered on the transistor—is an amplifier, while the bottom half—mostly composed of resistors and capacitors—is a filter.

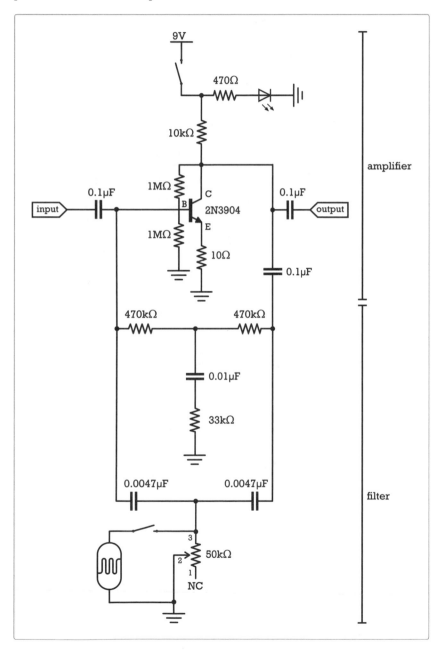

FIGURE 14-8: *The circuit diagram for the Twin-T Phaser/Wah*

The amplifier section probably looks familiar; it's basically the same preamp circuit that we used for the Mud-n-Sizzle Preamp (Project 12). The bottom half is a little more complicated. Wah effects are usually created using an adjustable *band-pass* filter, which is a filter that emphasizes a narrow band of frequencies. The popular Cry Baby wah-wah pedals rely on an inductor-based band-pass filter, and that design is widely imitated. But sourcing a suitable inductor for such a band-pass filter can be kind of tricky. Instead, we'll build a resistor/capacitor-based twin-T *band-stop*, or *notch*, filter. A notch filter attenuates—or even cuts—a narrow band of frequencies, allowing the rest to be more prominent. This is technically the *opposite* of how most wahs work, but the result sounds very similar.

Take a quick look at the filter portion of the circuit in Figure 14-8. You'll immediately see how the twin-T got its name: the heart of the filter is a set of three resistors and three capacitors, arranged into a pair of Ts. Each T bridges the input and output with its bar and connects to the ground with its stem.

Preamp Circuit

First, let's build the preamp. Since the Adafruit PCB has its columns numbered and rows lettered, I'll use those coordinates here. If you're using a different brand of PCB, make sure to check your work against the figures

Step 6   We'll start by placing the four jumpers shown in Figure 14-9. Cut, strip, and tin two 3/4-inch lengths of insulated wire. Bend them into little flat-bottomed *U*s. Run the first jumper from 4E to 4F, bridging the central divide at the 4th column. Run the second jumper from 9E to 9F, bridging the divide at the 9th column. Flip the board and solder each end of each jumper to its respective trace.

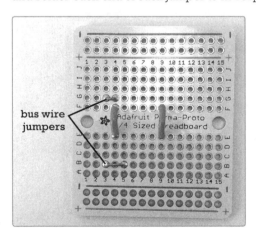 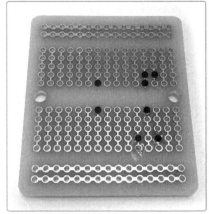

FIGURE 14-9: *The four jumpers soldered in place. Notice the bare bus wire jumpers, which are a little harder to see. The right image shows the underside of the PCB at this step. (Black dots indicate the solder points completed in Steps 6 and 7.)*

Step 7   Now snip two 3/4-inch scraps of bare bus wire. Bend one into a flat-bottomed *U*, and run it from 3B to 5B, connecting columns 3 and 5 along the row second from the bottom. Take the other snip of bus wire, fold it in half, and run it from 3G to 4G, linking columns 3 and 4 above the central divide. Flip the board and solder these. Check the results against Figure 14-9.

Step 8   Next is the transistor. These NPN transistors have a black body, flat face, and three legs. If the face is toward you and the legs down, then those legs are, from left to right, the *emitter*, *base*, and *collector*. Mount the transistor so that its emitter is in 2D, its base in 3D, and its collector in 4D, as shown in Figure 14-10. Solder these legs now.

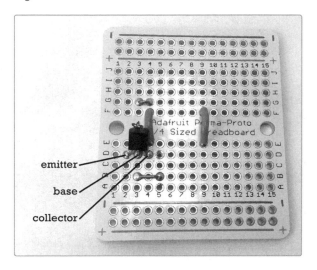

FIGURE 14-10: *The transistor is in place.*

Step 9   Now connect the 1M ohm bias and ground resistors (brown-black-green), both of which connect to the transistor's base leg. The bias resistor runs from 3E to 3F, bridging the divide. The ground resistor runs from 3A to the third pad in the row directly below the blue line (labeled with a – sign), which serves as the *common ground rail*. Figure 14-11 shows both resistors.

FIGURE 14-11: *The 1M ohm bias and ground resistors are installed.*

Step 10   Add the last two resistors to the preamp portion of the circuit. These are the ground and power resistors. Run the 10 ohm ground resistor (brown-black-black) from 2B to the second hole from the left in the common ground row directly below the bottom blue line. The 10k ohm (brown-black-orange) resistor runs from 4J to the row just above the red + line at the top—that is, the *power rail* (see Figure 14-12). Solder these now.

Step 11   Finish off the preamp with the 0.1 μF input and output capacitors. These little caps are marked *104* (although those markings can be downright infinitesimal; keep a magnifier handy). The input cap runs from 1C to 3C. The output cap runs from 4I to 6I. Solder these now and compare them to Figure 14-13.

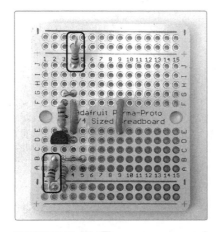

FIGURE 14-12: *The preamp ground and power resistors are installed.*

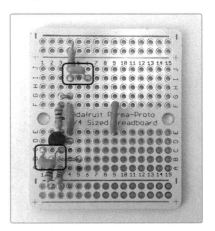

FIGURE 14-13: *Installing the input and output capacitors*

Twin-T Filter

The twin-T circuit consists of a *resistor T* and a *capacitor T*. (In Figure 14-8, the resistor T is the upper one, and the capacitor T the lower.) The resistor T is made from two 470k ohm resistors (yellow-violet-yellow stripes), a 33k ohm resistor (orange-orange-orange stripes), and a 0.01 μF capacitor (labeled *103*). The capacitor and orange-striped resistor form the stem of the resistor T, and the two 470k ohm resistors form its top bar. The capacitor T consists of the two 0.0047 μF capacitors, with the 50k ohm pot and photoresistor forming its stem.

Step 12   Run the remaining 0.1 μF capacitor from 4H to 8H, as shown in Figure 14-14. Then solder it. This cap will serve as the link between the preamp output and the twin-T filter output. The filter input connects directly to the preamp input.

Step 13   Run one 470k ohm resistor (yellow-violet-yellow stripes) horizontally from 5C to 8C and the other vertically from 8E to 8F, bridging the divide, as shown in Figure 14-15.

FIGURE 14-14: *Placing the final 0.1 µF cap, which links the filter and output*

FIGURE 14-15: *The two 470k ohm resistors in place*

Step 14   Add the 0.01 µF capacitor horizontally from 8D to 10D (see Figure 14-16). The final 33k ohm resistor (orange-orange-orange stripes) will vertically connect that 0.01 µF capacitor to the ground, running from pad 10A straight down to the ground rail, as shown in Figure 14-17. (You'll note that none of this looks even remotely like a *T* once it's laid out on the PCB.) Solder all these fellas down now.

FIGURE 14-16: *Adding the 0.01 µF capacitor to the resistor T*

FIGURE 14-17: *Finishing off the resistor T with a 33k ohm resistor linking the 0.01 µF capacitor to the ground*

Step 15   To start the capacitor T, go find the
two 0.0047 µF caps labeled *472*. Place
one horizontally from 5A to 9A and the
other horizontally from 8G to 9G, as
shown in Figure 14-18. Solder them in
place.

Step 16   Take the sweep-control pot from Step 1.
Solder one of lug 3's two remaining 4-inch
leads to 9B on the circuit board. Next,
solder one of lug 2's 4-inch leads to the
common ground at column 4 (see Figure 14-19). Reminder: if you left out the
wah or LFO leads in Step 1, you'll have
fewer leads here now. That's just fine. Do
not panic. What's important is that lug 3
is connected to pad 9B, and lug 2 is connected to the ground.

FIGURE 14-18: *Starting the capacitor T in the twin-T filter*

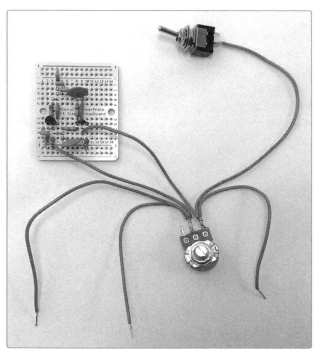

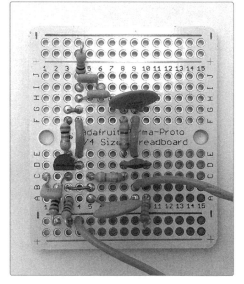

FIGURE 14-19: *The sweep control is installed (detail on the right shows more clearly where the control
hooks into the circuit).*

Install the Hardware

Now that the circuit itself is complete, we'll connect the input and output jacks and the power switch. Then, we'll be ready to test the twin-T filter.

Step 17   Solder the lead connected to lug 2 of the DPDT bypass switch to pad 1E on the PCB (that's the lead going from the switch to the left edge of the PCB in Figure 14-20). This is the circuit input. Solder the lead connected to lug 4 of the DPDT switch to 6J, which is near the top of the PCB. This is the circuit output.

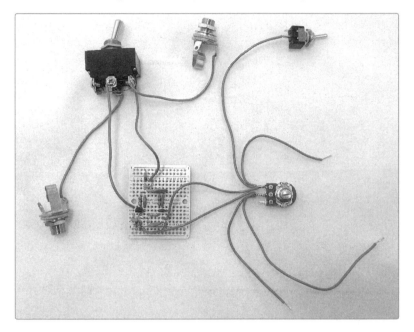

FIGURE 14-20: *The installed input, output, and bypass switch*

Step 18   Let's add power! Get out your power switch/LED assembly and connect the black wire from the battery clip to the leftmost hole on the ground. Connect the loose red wire from the switch to the power rail, just above the red line (labeled +), in the leftmost hole. Solder all of these, and your PCB should look like Figure 14-21.

Step 19   Connect the grounds. For the following connections, you can use any open hole in the common ground rail, which is just below the blue line (labeled –) at the bottom of the PCB. Start by connecting the open negative lead on the LED to any open hole on the ground rail. Then, wire each jack's open lug to any open ground hole. Finally, run a length of wire from lug 3 of the DPDT switch to the ground. Solder all of these.

* **NOTE:** *Although many folks use bare bus wire for ground connections like these, I recommend insulated wire in this case. There are so many wires floating around, it would be easy to create a short circuit with bare ground connections here. For clarity in the images, I've used black insulated wire.*

FIGURE 14-21: *Our circuit has power!*

In Figure 14-22, all of the ground connections from Step 19 are in black, as is the negative battery lead you soldered in Step 18. On the right is a close-up of the finished PCB with all of the connections labeled. Note that all the ground connections go to the common ground along the bottom of the PCB, just below the blue line marked with a minus sign.

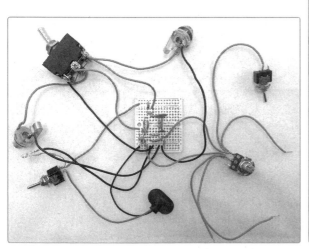

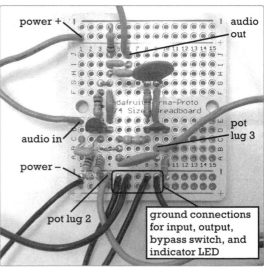

FIGURE 14-22: *The core twin-T circuit with the ground fully wired*

Test the Twin-T Circuit

Before installing the photoresistor wah control and LFO, we're going to do some quick testing.

Step 20   Snap a 9-volt battery into the battery clip and flick the DPDT bypass switch up to the lug 1/lug 4 side. If the indicator LED isn't already illuminated, flip the power switch. If the LED *still* doesn't power up, check for short circuits and solder bridges—both between tracks and solder pads on the PCB and between the lugs of switches, pots, LED leads, flying wires that run to the various pieces of hardware, and so on. I describe remedies in "General Troubleshooting" on page 355, but as a rule, if two things are touching that shouldn't touch, separate them. Also make sure your battery isn't dead.

Once you have power, plug the output into an amplifier, turn the amp on, and crank the amp to its highest setting. You should hear a hum, and when you run the Twin-T's sweep control pot clockwise and counterclockwise, you'll give that hum a sweeping whoosh. If you don't get any whoosh, flip the DPDT switch again.

Still nothing? Search for more short circuits and solder bridges, especially in the twin-T filter section. If you're hearing AM radio, try connecting the common ground to a piece of metal. You can do so using an alligator clip for now: just attach one end of the clip to one of the grounded legs of one of the resistors or to the ground lug on one of the jacks, and attach the other end to a steel saucepan, metal desk, aluminum pie pan, or something similar.

When you have the twin-T circuit working, plug in an unpowered instrument, such as a standard electric guitar or a Robo-Tiki Steel-Stringed Ukulele (Project 10) sporting the hotter 42-gauge wire pickup. Turn the volume on your amp down to a comfortable level. (Remember that the built-in preamp will give the signal a pretty hearty boost.)

Pluck a string, run the pot, and soak up the funk. You should hear a distinct spectral sweep as you twist the knob. Once you've noodled around a little, power everything down and remove the battery. If you're skipping the photoresistor wah pedal and LFO control, then you're done, and you can jump to "Final Testing and Enclosure" on page 259. Otherwise, keep going with "Add the Phaser/Wah" next.

Add the Phaser/Wah

The Universal LFO isn't a compulsory part of this build, but I'd strongly advise including it, as automating the filter sweep is what it takes to make a true phaser effect. If, however, you're content with manual filter sweeps, then you can skip to Step 23.

Step 21   Flip to Project 13 and get cracking on the Universal LFO. Build the complete project, including the power supply. Yes, that means you'll have separate power supplies for the LFO and the Twin-T. Although keeping the power supplies separate is a little unwieldy to package, it's definitely easier to build and troubleshoot, so it's worth the extra effort.

Step 22   Once you're done building and testing your Universal LFO, solder one of each of its control leads to the open 4-inch insulated wires on the Twin-T Phaser/Wah's 50k ohm sweep pot. It doesn't matter which lead goes where, as long as each 4-inch lead gets one control lead (see Figure 14-23).

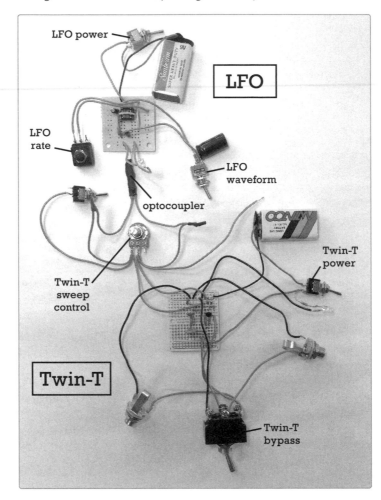

FIGURE 14-23: *The finished guts of the Twin-T Phaser/Wah with LFO. The LFO is at the top, the Twin-T at the bottom, and the two are linked through the sweep-control pot in the middle.*

Step 23   Now for the photoresistor-based faux-pedal wah controller. Cut, strip, and tin a 6-inch length of insulated wire. Solder one end to one of the legs of your photoresistor. Solder the other end of the lead to the open lug on the *wah control switch*, which is the SPST switch connected to lug 3 of the 50k ohm pot. Then, solder the free leg of the photoresistor to the open 6-inch lead from lug 2 of the 50k ohm pot.

   Ta-da! You're done building! In the next section, we'll test the whole thing.

Final Testing and Enclosure

Make sure everything has fresh batteries and powers up on its own. Hook an appropriate instrument to your Phaser/Wah, hook the Phaser/Wah to your amp, and power everything up. With the LFO off and the photoresistor covered with something opaque—such as a ceramic coffee mug—the Phaser/Wah should function as it did in "Test the Twin-T Circuit" on page 257.

Now turn off the sweep-control pot by running it fully clockwise, and then activate the LFO. The filter effect should come in and out in sync with the blink of your LED. If it doesn't, that might mean you installed the sweep pot backward. That's fine. Just twist it fully counterclockwise instead of clockwise to let the LFO take over.

Set to square wave, the effect should cut in and out rather sharply, an effect most useful at relatively high speeds. Set to pseudo-sine wave, the filter should sweep in as the LED glows and then dims, an effect that's most enchanting at slower speeds.

Finally, turn off the LFO and uncover the photoresistor. You should now be able to control the filter by covering and uncovering the photoresistor, exposing it to more or less light, for an overall effect similar to a traditional wah-wah pedal.

When your controls check out, mount everything in your enclosure. I prefer a nice large, flat cigar box to give me room for all my bells and whistles. For advice on choosing and modifying enclosures, see "On Enclosures" on page 214.

Most of the mounting process is pretty straightforward: keep the power and control switches and pots a fair distance from the big DPDT bypass switch so you have room to operate. Also, placing the input and output jacks on the sides of the box, as opposed to the lid, usually makes it easier to keep the cords out of your way.

The one tricky bit is mounting the photoresistor for the wah pedal. The photoresistor should be mounted on the outside surface of the enclosure, which is going to necessitate a little desoldering and resoldering.

To install the photoresistor, start by drilling a hole somewhere on your enclosure that'll be convenient for toe-tap operation. (Figure 14-1 shows that I installed mine on the top of my enclosure, away from the other controls; it's centered near the bottom edge of the box lid.) The hole needs to be slightly smaller than the head of the photoresistor. In most cases, drilling with a 3/16-inch or 1/4-inch bit is just right, although smaller photoresistors might call for an 1/8-inch hole. For truly huge photoresistors, you might be best off drilling two small holes, one for each lead.

If you're using a metal-topped enclosure, insulate the leads on your photoresistor; a couple snips of shrink tube or a fold of duct or electrical tape should do. Now smear some silicone-based glue around the hole on the exterior of the case, thread the photoresistor leads through the hole, and press the resistor into place so that it faces outward.

Let it dry for at least 30 minutes before resoldering the leads, and give it a full 24 hours to dry before putting the wah into action. (Smearing semi-dry silicone-based glue all over the light-sensitive surface of the resistor will hamper its operation.) Figure 14-24 shows the installed photoresistor wah pedal controller from the inside and outside.

FIGURE 14-24: *An interior view of the photoresistor wah control installed in the lid of the enclosure (left). The switch activates the photoresistor wah. The photoresistor itself is soldered to the two visible leads. The exterior close-up (right) shows the photoresistor glued directly to the outside of the box lid.*

When you're finished drilling holes, stuff the enclosure (see Figure 14-25). Make a point of securing the batteries and PCBs, since they are highly likely to bang around and cause hassles otherwise.

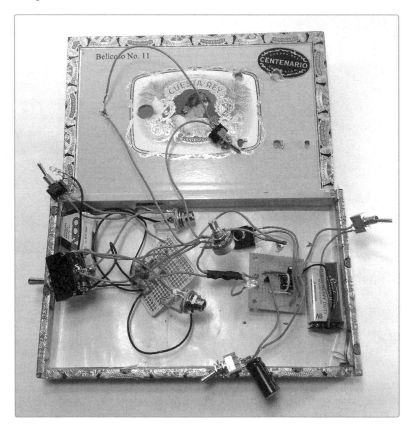

FIGURE 14-25: *Stuffing the enclosure for the Twin-T Phaser/Wah*

Using the Twin-T Phaser/Wah

Like other effects, such as the Mud-n-Sizzle Preamp (Project 12), the Twin-T Phaser/Wah is pretty self-instructive: plug an instrument into the input, plug an amp into the output, and begin exploring the sounds you can coax out of your instruments and the filter. Let's quickly run through the three modes in isolation.

▶ **Wah-wah mode** Turn on the wah switch and then cover and uncover the phototransistor with your toe to produce a wah-wah pedal effect of the sort you might hear in a Hendrix or Hayes tune. How you play your instrument has a big impact on how these come across. For classic wakka-chikka effects, work the wah-wah pedal while muting the guitar strings. For "Voodoo Child"–style yowls, play longer single notes with a little bend to them.

▶ **Manual filter mode** Filter sweeps are super common in dance and club music. It's that whooshing effect, where the music seems to lose and then gain "presence" without changing volume. To do filter sweeps using the Twin-T Phaser/Wah, just switch off the LFO and photoresistor and then work the sweep-control knob back and forth manually (in other words, you create a filter sweep by sweeping the filter's notch frequency).

With the LFO and wah off, you can also dial in a particular frequency and then leave the filter alone, emphasizing a sweet spot in your tone. This set-and-forget technique is a really common off-label use of wah pedals: Hendrix did it when he played the "Star-Spangled Banner" at Woodstock, as did Mark Knopfler on Dire Strait's "Money for Nothing."

▶ **Phaser mode** As for LFO-controlled phaser effects, you'll easily pick these out in pop recordings from the 1970s[2] and modern electronica, but they're also widely used to more subtle effect. Listen for phasers in the Sex Pistols's "Anarchy in the UK," which layers several guitars with phasers panned hard right and left. To use your phaser, switch off the photoresistor wah pedal and turn the sweep control fully clockwise. Then, turn on the LFO and select a rate and waveform that work for you.

Of course, you can also mix-and-match these modes. Get the LFO going and then occasionally override it on the fly with the sweep control or wah. Or dial in a sweet spot on the sweep control and then vary from that using the photoresistor wah. Or set a sweet spot and then occasionally activate the LFO to transform the mood of your jam. In addition to playing with different settings, experiment with placing the Twin-T Phaser/Wah in different positions in your effects chain, as it can behave very differently as its input changes.

One note for general use: our photoresistor wah pedal, being light dependent, is less precise than a traditional mechanical wah pedal and will need a steady light source in order to behave consistently—or, alternatively, it can be augmented with an unsteady light source in order to behave erratically, if that's your thing.

---

2. The rhythm guitar in The Rolling Stones' "Shattered," the opening bass riff in Thin Lizzy's "Dancing in the Moonlight," and many Steely Dan tunes—including "Peg" on the album *Aja*—are nice examples of a phaser in action.

Tips, Tricks, and Mods

This project is the most complicated so far in the book, which means there's a lot of room for modification and improvisation.

Troubleshooting Mods

For starters, here are some simple tricks to improve your sound. If you're picking up radio broadcasts, baby monitors, or other RF interference, try running a ground connection from the common ground on the PCB to the metal enclosure. If you didn't use a metal enclosure, you can add a *ground plane* by lining the bottom of your enclosure with a piece of thin aluminum cut from a disposable pie tin or catering tray—or you can even use a few layers of aluminum foil. Either solder your ground connection to that ground plane or bolt a ground wire to it. Place a layer of construction paper or self-adhesive contact paper between the ground plane and PCB to prevent short circuits.

If you find that hotter instruments or inputs are roaringly loud and seem little impacted by the Twin-T's sweep, try cutting the power to the twin-T portion of the circuit. This kills the preamp but largely keeps the passive filter working. Another option is to break the ground between the instrument and effect. This option is especially effective with electronic instruments, like synths. To make this more convenient, consider adding a *ground interrupt* switch on the Twin-T Phaser/Wah's input jack, as shown in Figure 14-26. To do this, install a regular SPST switch on the ground wire connecting the twin-T input jack to the circuit. When you open the switch, the ground on the instrument no longer connects to the twin-T circuit, significantly cutting the amplification while maintaining the effect.

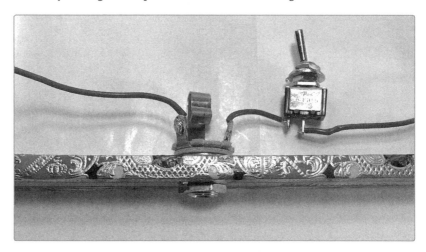

FIGURE 14-26: *A simple ground interrupt switch*

Tweak Components

In terms of purely fun mods, the most obvious place to start is with component values. The output cap can be increased up to about 10 µF. The higher the value, the more bass-heavy your output will sound, which can give a very different color to your tone. If you're using electrolytic caps (anything over 1 µF or so will likely be electrolytic), connect the positive leg of the electrolytic cap to the circuit itself and the negative leg to the output jack.

You may also want to monkey with the matched capacitors and resistors in the twin-T filter section of the circuit. Each of the caps in the crossbar of the capacitor T should have about half the capacitance of the capacitor in the stem of the resistor T. In my design, for example, the capacitor T's two crossbar caps are 0.0047 µF and the resistor T's single stem cap is 0.01 µF. The matched resistors in the resistor T's crossbar can go as low as 100k ohm. Using lower-value resistors will shift the effective range of the wah down into lower frequencies. The current design, with 470k ohm resistors in the resistor T's crossbar, mostly affects higher frequencies.

As with the Mud-n-Sizzle Preamp (Project 12), you can boost this effect into something much more aggressive and distorted by replacing the single transistor in the preamp with a Darlington transistor pair. For details, flip to "Boost the Fuzz with a Darlington Pair" on page 216.

You could save space by omitting the bypass switch. This is especially handy if you decide to build the Twin-T Phaser/Wah into another instrument or effect. To skip the bypass switch, wire the input jack's tip directly to the circuit's input (at pad 1E on the AdaFruit PCB) and the output jack's tip directly to the circuit output (pad 6J). Both jack's ground lugs continue to go to the common ground rail. If this sounds confusing, just flip back to Figure 14-22; the detailed picture on the right shows where the audio in and audio out connect to the PCB.

Mount the Twin-T Inside a "Broken" Pedal Enclosure

The real obstacle to building a commercial-style wah pedal isn't the circuit. The twin-T notch filter is a pretty easy build, and if you'd prefer a band-pass filter (the sort used in the very popular Cry Baby wah pedal), there are plenty of straight-forward inductor-based wah circuits out there and many suggestions on how to scrounge up a suitable 500 millihenry inductor. Just Google *inductor wah circuit* and work from there.

The real trick is that darn custom-built, heavy-duty pedal enclosure and the beefy wah potentiometer. Fortunately, there are plenty of "broken" wah pedals out there in secondhand music stores, pawnshops, garage sales, message boards, and so on. Most of the time, these are broken because the circuit itself got trashed; leaky batteries, beers, and decades of abuse are frequent culprits. The pedal enclosure and wah pot are usually fine, ready for your circuit to move in, hermit-crab style.

To install your Twin-T Phaser/Wah in a pro enclosure, start by pulling out the dead wah's guts: clip the leads to the dead wah's potentiometer and discard everything else. Test the dead wah's potentiometer with your multimeter. All that really matters is that the pot consistently changes resistance as you rock the pedal. Now snip off your Twin-T's photoresistor and solder those two leads to the middle lug and one outer lug on the dead wah's pot. Bolt this beautiful monster together, and you're done.

It's a pretty straightforward half-hour's work, although you might need to get a little creative in order to accommodate all your pots and switches. Need ideas on how to accomplish this? Flip back to "Package Your Project" on page 110 and consider an add-on enclosure for this Frankensteinian phaserwahstrosity.

Now you've got something truly wicked-awesome: a fully custom, heavy-duty, heavy-metal wah-wah pedal with full filter-sweep and phaser capabilities. Rock on!

15

THE SINGLE-CHIP SPACE INVADER SYNTH

The Single-Chip Space Invader Synth sounds a lot like the old-school console and arcade games that inspired its name because it relies on many of the same sound-generation techniques. Early video games couldn't play prerecorded sounds—not even brief, low-resolution samples. All of their sound effects had to be created by mixing tones generated by simple circuits like those found in this project. Did the results sound realistic? No! Every explosion is a simple burst of static, and the entire *Space Invaders* soundtrack is just four notes steadily looping at an ever-increasing tempo. These sounds are primitive, but like many primitive expressions, they're also primal. The games themselves have faded—when was the last time *you* paid 25 cents to bull's-eye regularly descending alien crafts?—but artificially lo-fi *chiptune* music is thriving. And you can craft yourself a shiny fistful of that primal electronica.

The Single-Chip Space Invader Synth has two nonidentical sound-generating circuits, or *oscillators*. (In this case, they're *square-wave oscillators*, which means that the oscillators generate square-shaped waves, rather than triangles, sawtooths, or smoothly curved sine waves.) One oscillator is responsible for producing the pitch of each tone, while the other is used to *modulate* that primary pitch. The two square-wave oscillators are diode mixed, resulting in a very basic kind of *FM synthesis* that readily lends itself to old-school chiptune effects, like revving light-bikes and laser blasts.

The Single-Chip Space Invader Synth has two playing modes:

▶ You can individually trigger notes with the red button—much as you individually trigger notes by plucking a guitar string or striking a piano key.

▶ You can also flip the hold switch to continuously produce sound while you run the knobs, making the synth a bit like a theremin[1] with a much richer tonal palette and much less frustrating controls.

As an added bonus, this 9-volt brute is capable of driving a set of headphones or its own little speaker, so you can easily switch from rocking out loud-and-proud to practicing in private without ever getting off the sofa.

FIGURE 15-1: *The finished Single-Chip Space Invader Synth*

---

1. The theremin is that instrument that looks like a box with two metal antennas sticking out. You play it by waving your hands around the antennas. If you don't know what the heck I'm babbling about, search for videos online; you'll definitely recognize the sound from a billion blurry, low-budget black-and-white sci-fi and horror films.

The Single-Chip Space Invader Synth purposefully relies on cheap commodity parts used in slightly unconventional ways. The result is that every Single-Chip Space Invader Synth is unique, with its voice and control response influenced by the idiosyncrasies of your components, case, hardware, lengths of wire, and much more.

Hear the Single-Chip Space Invader Synth in action in the samples at *http://www.nostarch.com/jamband/*.

Preparation

Build Time

▶ About 1 hour

Tools

▶ A standard soldering kit (See page 340.)

▶ A ruler that shows 1/4-inch increments or smaller

▶ An electric drill with bits (You'll probably need 3/16-inch, 5/16-inch, and 3/8-inch bits.)

▶ An amp and instrument cable for testing the synth

▶ (Optional) Other tools to work your enclosure (If you're using a wooden enclosure, you'll want sandpaper and possibly files to clean up drill holes.)

▶ (Optional) Foam-backed double-sided tape or hardware to mount the circuit in its enclosure

▶ (Optional) A hacksaw for trimming down potentiometers with long shafts

Supplies

▶ A CD4093 integrated circuit, such as Digi-Key part #296-2068-5-ND

▶ A 14-pin IC socket

▶ A red LED

▶ Two 1N4148 silicon diodes (Mouser part #78-1N4148 is shown in Figure 15-2, although any generic 1N4148 or 1N4150 is fine.)

▶ Two 0.1 µF capacitors (marked *104*)

▶ A 220 ohm resistor (red-red-brown stripes)

▶ A 470 ohm resistor (yellow-violet-brown stripes)

▶ A 10k ohm audio potentiometer (This is a variable resistor with an audio taper; see "The Gory Details: Audio Taper vs. Linear Taper" on page 327.)

▶ A 100k ohm potentiometer (A linear taper is best; if the potentiometer isn't otherwise labeled, you can assume it has a linear taper.)

▶ A 500k ohm potentiometer with a linear taper

▶ Three control knobs that fit your potentiometers

- ▶ Two small SPST toggle switches (You can use Mouser part #108-MS550K, for example, but any similar switch will work.)
- ▶ An SPST pushbutton switch, such as Mouser part #103-1012-EVX or a non-illuminated doorbell button

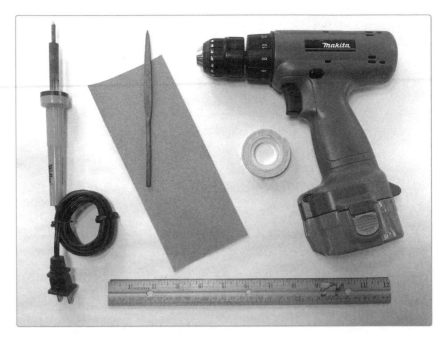

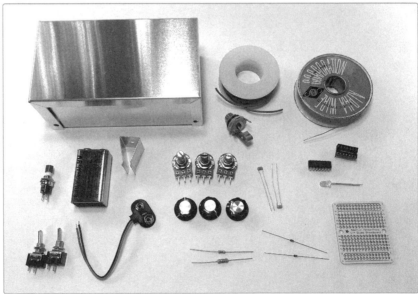

FIGURE 15-2: *Tools and supplies (not shown: hacksaw, instrument cable, and amp)*

- A 9-volt battery

- A 9-volt battery clip

- 24-gauge insulated hook-up wire (Stranded wire is best.)

- 22- or 24-gauge bare bus wire (This is uninsulated solid core wire.)

- A 1/4-inch mono phone jack, also called a *guitar jack*

- An Adafruit Perma-Proto Quarter-sized Breadboard PCB (Adafruit product #589 or Maker Shed product #MKAD48)[2]

- A sturdy enclosure (Cigar boxes are fun—folks are delighted by the inherent contrast of the synth-in-a-cigar-box—while a simple all-metal enclosure lends an apt sort of space-age retrofuturism. See "On Enclosures" on page 214 for more options.)

- (Optional) A 9-volt battery holder clip, such as Digi-Key part #71K-ND or Mouser part #534-071

Building the Single-Chip Space Invader Synth

As with other electronics projects, we'll begin with the hardware to allow you to freshen up your soldering skills on the least delicate components. Then, we'll move on to building the circuit itself before connecting it all and installing the functioning project in an enclosure.

Prepare the Hardware

Step 1   Start with the power supply and indicator LED. Take out the 9-volt battery clip, one of your SPST switches, the 470 ohm resistor (yellow-violet-brown stripes), and the LED. Strip and tin the positive (red) lead on the battery clip and then solder it to either of the two lugs on the switch. (If you're new to soldering, flip to "Soldering" on page 346 now for instructions.) Next, solder the 470 ohm resistor to the positive leg of the LED; the positive leg is longer, while the negative is closer to the flat section of the LED lens. Finally, cut, strip, and tin a 4-inch length of wire. Twist together one end of the wire and the free leg of the 470 ohm resistor, and solder this twisted pair to the empty lug of the switch. Figure 15-3 shows the results.

----

2. I've sung the praises of this prefab printed circuit board elsewhere in the book, and you can see front and back pics in Figure 13-23 on page 240. If this specific printed circuit board is hard to find, fear not: several other manufacturers make similar boards. Just Google *permanent breadboard* or *solderable breadboard*. One caveat: the AdaFruit Quarter-sized Perma-Proto is significantly smaller than most alternatives and is the just-right size for our Single-Chip Space Invader Synth. So if you use a different circuit board, you may need to cut it down or plan for a larger enclosure.

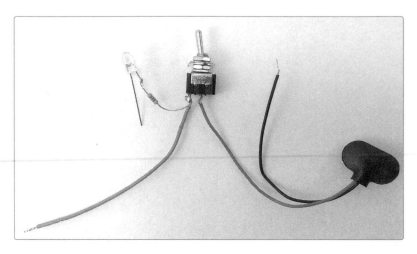

FIGURE 15-3: *The prepared power switch and supply with indicator LED*

Step 2   Next up are the three potentiometers, or *pots*. Start by preparing their leads: cut six 4-inch lengths of insulated wire, strip them, and tin the ends. The 10k ohm pot and 500 ohm pot are wired identically. In each case, one wire is soldered to lug 2 and one wire to lug 3. Remember that lugs are numbered from left to right when viewed from the front, with the pot shaft pointing toward you and the lugs on top. In Figure 15-4, the 10k ohm and 500k ohm pots are the left and right pots, respectively. For the 100k ohm pot, solder one wire to lug 1 and one wire to lug 2, as shown in Figure 15-4, where the 100k ohm pot is in the middle of the lineup. Set the 100k ohm and 500k ohm pots aside and keep the 10k ohm pot handy for Step 4.

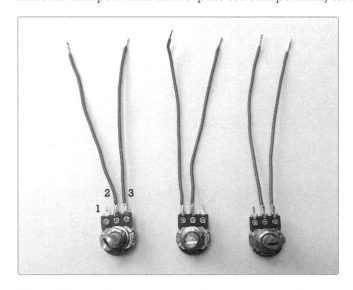

FIGURE 15-4: *The prepared pots, from left to right: 10k ohm, 100k ohm, 500k ohm*

Step 3   Now we'll start working on the output controls, which offer two options for playing notes: a pushbutton trigger for playing individual notes and a toggle hold switch that you can flip for continuous output. Cut four 4-inch insulated wires, and strip and tin the ends. Dig out your remaining SPST toggle switch and your single SPST pushbutton switch; solder one wire to each lug on each switch (see Figure 15-5).

Step 4   Next, we'll connect the trigger and hold switches, 10k ohm volume pot, and output jack. First, take the 10k ohm pot and solder the wire from lug 2 to the tip lug of your 1/4-inch output jack. This is the lug that connects to the jack's metal tongue. (If you need a more detailed diagram of instrument jack and plug anatomy, flip to "Quarter-Inch Phone Plugs and Jacks" on page 337.) Take the two switches, twist together the ends of one wire from each switch, and solder this pair to lug 1 of the 10k ohm pot, as shown in Figure 15-6. Set aside the hardware until Step 12.

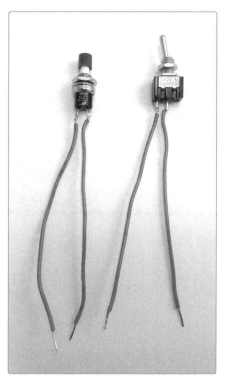

FIGURE 15-5: *The prepared trigger (left) and hold (right) switches*

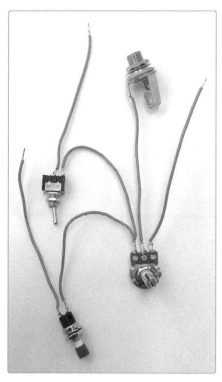

FIGURE 15-6: *The finished output assembly*

Build the Circuit

Now that the hardware is done, it's time to take a look at the circuit diagram, shown in Figure 15-7.

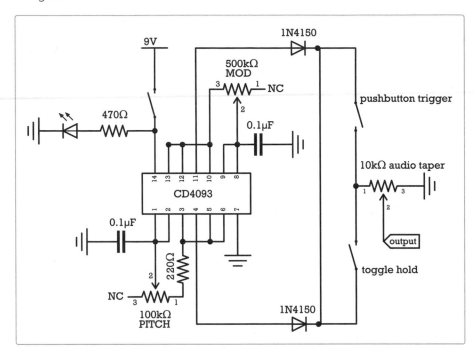

FIGURE 15-7: *The circuit diagram for the Single-Chip Space Invader Synth*

One neat thing about this circuit is that we get both oscillators from a single IC. In fact, the chip could provide as many as four oscillators: it has four identical logic sections, and each can serve as a usable oscillator by itself. We'll use two logic sections per oscillator[3] and mix those outputs using the diodes you see along the top and bottom of the circuit diagram.

Most of the time, audio signals are mixed using resistors. For example, that's what you're doing when you use a soundboard to mix the signals from several instruments and mics so that they can all come out of one set of speakers. On the Twang & Roar Kalimba (Project 11), we wired up a single pot to work as a simple mixer so that we could blend the signals from two different kinds of pickups. Resistor mixing generally only allows you to add signals together. If the resistors are variable, you can also control the ratio of each signal.

But diode mixing of two very similar signals—like the pair of square waves being produced by the CD4093 in this configuration—is different. It allows the signals to either add or subtract from each other, creating complex new waveforms. Each oscillator by itself can produce only a string of evenly spaced, identical

---

3. This results in the most stable output across the widest range of frequencies. Options for building stripped-down single and multi-oscillator circuits using this IC are illustrated in "CD4093 Oscillator Circuits" on page 360.

square waves. But if you hook the Space Invader Synth's final diode-mixed output to an oscilloscope and watch the readout while adjusting the synth's pitch and mod controls, you'll see regular square waves, irregular square waves, regular groupings of square waves, or even more complex shapes, like repeating patterns of stairsteps or ziggurats.

If this is your first IC project, please take a moment to flip back to Step 1 of the Droid Voicebox on page 76 to familiarize yourself with the layout and orientation of an IC (which often confuses newcomers). You'll note that we're using an IC socket in this project. Lots of folks just solder their ICs directly to the circuit board. Please don't do this. The IC is sensitive to both heat and static electricity. Soldering to an empty socket and then inserting the IC in the end will avoid headaches and save you a lot of time if the IC fails later. Swapping out a socketed IC takes a few seconds; desoldering 14 IC pins is fiddly business that takes longer than rebuilding the circuit from scratch.

The AdaFruit PCB listed in the supplies has its columns and rows conveniently numbered and lettered. I'll use these coordinates to indicate where components should be placed. If you've used another brand of breadboard-style circuit board, just follow along with the illustrations.

Step 5 Place the IC socket so that it straddles the horizontal divide along the middle of the PCB, with one end of the socket at holes 3F and 3E and the other at 9F and 9E (see Figure 15-8). Flip the PCB over and solder socket pins 1, 2, 4, 5, 6, 7, 8, 9, 10, 11, 12, 13—that is, every pin other than pins 3 and 14—to their respective pads. IC pins are numbered counterclockwise from the lower left, as shown in Figure 15-8.

Step 6 Place your nine jumpers. Cut eight 3/4-inch lengths of bare bus wire. Bend four of these into skinny *V*s and mount them from pad 3D to pad 4D, 7A to 8A, 8G to 9G, and 4G to 5G. Solder these now. Next, bend four lengths of bus wire into flat-bottomed *U*s and mount them from 5C to 7C, 5H to 7H, 11F to 11E, and 9A directly to the lower ground rail. (The ground rails are the rows of holes at the top and bottom of the PCB, directly below the blue lines marked with minus signs.) Solder these jumpers, being mindful not

FIGURE 15-8: *Positioning the IC socket. Sockets are symmetrical, but if yours has a notch at one end, as mine does, it's traditional to place that to the left so that it aligns with the orientation markings on the IC.*

to push the jumpers down flush with the surface of the PCB, where they might cause short circuits with holes they're supposed to be jumping over. Figure 15-9 shows all these connections.

Step 7   For the final jumper, cut a 2 1/4-inch length of bare bus wire and run it along column 14, from the top ground rail to the bottom ground rail, as shown in Figure 15-9.

Step 8   Add the two 0.1 µF capacitors, or *caps*. Run one cap from 4A to the lower ground rail and the other from 9J to the upper ground rail (see Figure 15-10). It doesn't matter which leg goes in which hole as these caps are nonpolarized. Solder the caps now.

FIGURE 15-9: *All the jumpers are in place.*

FIGURE 15-10: *Mounting the 0.1 µF capacitors*

Step 9   Connect the diodes. Unlike the caps, the diodes *are* polarized, so each has a positive and negative leg. The negative leg is indicated by a dark stripe on the body of the diode. For both diodes, that stripe will go to the right, as shown in Figure 15-11. Run one diode from 6I to 11I and the other from 6B to 11B. As with other components spanning multiple columns, be mindful not to press the legs down flush with the surface of the PCB because this could create short circuits in the intervening columns. Solder the diodes now.

Step 10   Add the 220 ohm resistor (red-red-brown stripes). Run this resistor from 5D to 10D. Solder both ends of the resistor as well as IC socket pin 3, as shown in Figure 15-12.

FIGURE 15-11: *Adding the diodes. Note that in both cases, the negative-leg stripe is to the right.*

FIGURE 15-12: *The 220 ohm resistor*

Install the Hardware

Remember all that hardware from the first few steps? It's time to connect it to the PCB. In this section, we'll install the pitch control pot, the modulation pot, the trigger button, the hold switch, and the power.

Step 11 First, install the pitch control, which is the 100k ohm pot. Run the lug 2 lead to solder pad 3A and the lug 1 lead to 10A. Solder these leads now. Next, install the modulation control, which is the 500k ohm pot. Run the lug 2 lead to 9H and the lug 3 lead to 7J. Solder these now. Figure 15-13 shows the results.

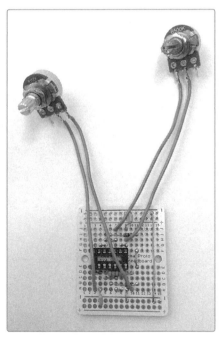

FIGURE 15-13: *The pitch-control and mod-control pots are installed.*

Step 12   Grab the output assembly you completed in Step 4 on page 271. Run the remaining lead on the toggle switch to 11D and the lead on the pushbutton switch to 11G. Solder both now. Then, take the lug 3 lead on the 10k ohm pot, which is the volume control, and connect it to the rightmost hole on the PCB's lower ground rail (see Figure 15-14).

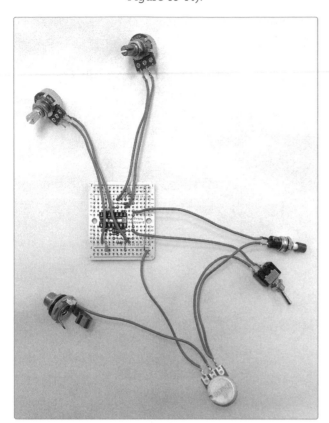
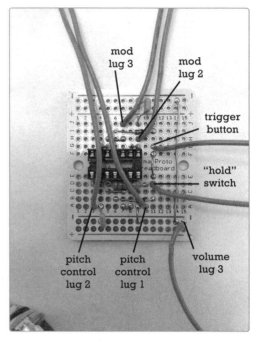

FIGURE 15-14: *The circuit with completed output (left) and a labeled detail of the PCB (right)*

Step 13   Now to add power! Run the negative (black) battery wire to the lower ground rail, connecting it to the rightmost empty hole, as shown in Figure 15-15. Then, run the insulated wire connected to the power switch—that is, the positive power supply—to pad 3G, which will connect it to pin 14 of the IC. Solder both connections now.

Step 14   Finally, add the last two ground connections. Cut a pair of 4-inch insulated wires and strip and tin both ends of each. Solder one wire to the open leg of the LED, which is the one that's aligned with the flat-edged LED lens and which was originally the shorter leg. Run the other end of this lead to the rightmost empty hole in the lower ground rail, next to the negative power lead you just installed. Connect the remaining 4-inch lead to the unwired lug on the output jack. (This is the ground lug that connects to the metal sleeve inside the jack.) Run the other end of this lead to the rightmost hole of the PCB's upper ground rail, in the top right corner of the PCB. These ground connections are illustrated with black wires in the labeled diagram in Figure 15-15. Solder both ground wires to the PCB.

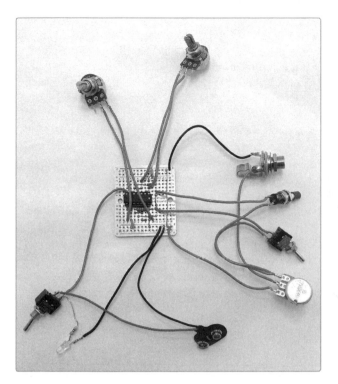
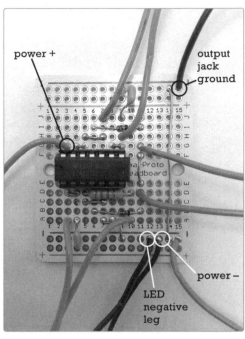

FIGURE 15-15: *The finished guts of the Single-Chip Space Invader Synth (left) and a labeled close-up of the new ground connections added in Step 14 (right). Note that the IC is now mounted in its socket.*

Step 15   Now, press the IC into its socket—the notch goes to the left, just like the notch on the socket—and you're ready to fire her up!

Troubleshooting

Time for troubleshooting! Connect a fresh 9-volt battery; if the LED doesn't light up, flip the power switch. If you still don't get a response from the LED, look for short circuits; stray snips of wire or little scraps of solder touching adjacent pads, traces, or lugs are often the culprit. Check the jumper connections, too, to make sure the wires bridging several columns are not brushing against unintended solder pads.

   If you find any short circuits, eliminate them. Loose scraps of wire or flakes of solder can usually be blown or flicked away, while most sloppy solder joints can be smoothed out by reheating them. The solder naturally flows to the warmer joint. If a solder bridge is especially stubborn or there's just too much solder gunked onto a joint, you may need to desolder (see "Desoldering" on page 349) and redo the connection.

Once your synth powers up, turn the 10k ohm volume pot fully counterclockwise, hook the synth to your amp, set the amp volume to its midpoint, and flip it on. Depending on the position of your toggle hold switch and the sensitivity of your amp, you may already hear the synth buzzing. If you don't hear anything, hold down the pushbutton trigger and turn up the volume on the synth itself. If you get to full volume and still hear nothing, turn the volume back down, flip the hold switch, and repeat. If you now have sound, then there was something wrong with your pushbutton switch. Look for shorts on its terminals, and if everything appears fine, replace the pushbutton. Still no sound, regardless of the hold switch's position? Look for more short circuits. The hardware—especially the output jack itself—is frequently the guilty party at this stage.

Once you *have* sound, test the two output switches. The pushbutton trigger should give short bursts of sound that last as long as you hold the button, while the toggle hold switch should give a continuous sound. If the toggle is off and you aren't pressing the pushbutton, then you should hear nothing. The toggle overrides the pushbutton, so if the toggle is on, the pushbutton should have no effect.

Next, flip the toggle so you have a continuous tone. Turn the 500k ohm pot—the modulation controller—fully counterclockwise, and then twist the 100k ohm pot—the pitch controller—back and forth. The pitch of the tone should clearly go higher as you turn clockwise and lower as you turn counterclockwise. Set the pitch controller to the middle of its range and turn the mod pot back and forth. The timbre should vary wildly as you do so: at the extreme clockwise position, it should take on a rhythmic beating, revving up as you turn the knob counterclockwise, until it reaches zip-zappy extremes.

If either pot behaves in the *opposite* way to what I've described, it just means that you used lugs 1 and 2 instead of lugs 3 and 2 (or vice versa) when building the controller. This common mistake is really no biggie. You can swap the leads on the pots to match my build or leave them as they are and operate "goofy foot." It's totally a matter of personal preference and aesthetics.

Enclosure

Once you have the synth working, box it up. I've chosen an aluminum project box enclosure (as shown in Figure 15-1 and Figure 15-16) because it shields against interference and looks suitably "space age." (I've also already used plenty of cigar boxes in this book.) There are many ways to play the Space Invader Synth, so consider several packaging options and see which you like best. You might decide that the tastiest sounds are easier to achieve with the synth sitting on a tabletop. In this case, use a heavier case that won't get pulled off the table when someone brushes against your instrument cables. If, however, you have more fun holding the synth and tweaking knobs like a mad hatter, then you'll probably want to use a lightweight, easy-to-hold box.

FIGURE 15-16: *The Single-Chip Space Invader Synth mounted in its aluminum case*

With any enclosure, remember the following:

► Choose a box you can machine with the tools you have. That is, if you only own a rechargeable hand drill, don't try to package your synth in an old steel ammo case.

► Secure the batteries to the box with clips and the PCB with screws or double-sided tape if you think they'll otherwise rattle around.

► If you have a metal enclosure, either insulate the PCB, bare wires, and solder points, or line the inside of the case with something nonconductive, like electrical tape, duct tape, or contact paper.

✳ **NOTE:** *For more detailed instructions on insulating or shielding your enclosure, see the Twin-T Phaser/Wah's "Tips, Tricks, and Mods" on page 262.*

When drilling holes for your hardware, you'll probably want a 3/8-inch drill bit for the output jack, a 5/16-inch bit for the pushbutton switches and pots, a 1/4-inch bit for the toggle switches, and a 3/16-inch bit for the LEDs. Your components may have slightly different dimensions, however, so make sure to double-check before you drill.

Playing the Single-Chip Space Invader Synth

I'm going to come clean: this instrument is purposefully obtuse. The point of the Space Invader Synth is to make it easier to create your own jams than it is to once again plunk out "Twinkle, Twinkle, Little Star." Yes, someone with a decent ear can definitely pick out tunes on this synth, but even total newbies will immediately construct their own riffs, rhythms, and yowling solos.

Trigger Mode

An easy way to start is to flip the hold switch off, dial in a volume and modulation you like, and then restrict yourself to working only the pushbutton trigger and pitch control. Aim to produce a repeating pattern of three or four steady pitches, with the occasional slide or slur (achieved by holding down the trigger button while you twist the pitch control knob). If you can operate a combination padlock, then you can work that pitch knob with enough precision to hit the same notes repeatedly and lay down a zappy little groove.

Theremin Mode

If you want something more cosmic and synthy, play in theremin mode: flip the toggle hold switch and then control the pitch with one hand and the volume with the other. This allows you to glide smoothly from note to note,[4] articulating and differentiating individual notes by cutting the volume. If you carefully work the volume knob as you move from pitch to pitch, you can get that wavering, swooping operatic sound emblematic of the theremin—think of the original *Star Trek* television series theme[5] or the swooping synth licks in the chorus of The Beach Boys' "Good Vibrations." This last example was played on the Electro-Theremin,[6] a simple single-oscillator sine-wave synth that is operationally very similar to our Space Invader Synth.

Experimental Free Style

Try tweaking the pitch and modulation knobs simultaneously, turning them either in the same or alternate directions at various rates. Do this in both theremin and trigger operating modes. You'll quickly discover a rich world of old-school video game sounds, including zip-zap space blasters, racing lightbikes, speeding saucers, and deeply disgruntled androids.

One quick tip as you explore: a simple volume pot, like the one we've used here, doesn't attenuate all frequencies evenly. This is something you can address with a little soldering (check out "Extra-Presence Volume Control" on page 366), but I like it as is. Because we're mixing two independent square waves, this uneven

---

4. Theremin enthusiasts call this a *glissando*, while *portamento* is the name usually used for this effect with synthesizers. A guitarist or other string player would call it a *slide*.

5. Performed by operatic soprano Loulie Jean Norman, who was imitating a theremin

6. Invented by trombonist Paul Tanner

attenuation is especially noticeable, uncovering stark new timbres, textures, and aural delights as you alter the volume, even if you leave the pitch and mod controls untouched.

If you want a more conventional playing experience, check out the resistor ladder mod described in "Add a Keyboard" on page 284.

Tips, Tricks, and Mods

Despite the low part count, there's a lot going on in this little Single-Chip Space Invader Synth—which means lots of opportunities to mod!

Add Speakers

This synth produces a nice, strong line-level output, which is sufficient to drive headphones or a small speaker, albeit somewhat quietly in many cases. Have an old set of "cans" (i.e., big ole headphones that cover the entire ear) with a 1/4-inch jack? Plug 'em in! You can also connect any pair of headphones you own using a 1/4-inch-to-1/8-inch adapter plug. (Such an adapter looks like a 1/4-inch plug with a built-in modern headphone jack; you can find one at most electronics shops or in the audio/video section of any big-box store.) Now you can jam without annoying everyone you live with.

* **NOTE:** *Do you hear your synth through only one side of your headphones? That's totally normal, as headphones are built for stereo sound and your synth is strictly mono.*

As for adding external speakers, both new and secondhand speakers in the 3 ohm to 8 ohm range should work when wired directly to the output. In general, a speaker with a larger diameter will give you better bass response, and a lower-resistance speaker will be louder.

If you have a large enclosure, you could include a built-in speaker. This could replace or complement the output jack; flip back to Project 7 and check out "Modify the Output" on page 100 for ideas. Alternatively, you can easily build an add-on monitor speaker. Just solder your speaker's two leads to the tip (positive) and sleeve (ground) connections on a standard 1/4-inch plug. For example, the little *dandelion* speaker shown in Figure 15-17 is a 1-inch diameter 8 ohm speaker with its leads soldered to an old guitar-cable plug. Secure a piece of coat hanger inside the plug, glue it to the back of the speaker's magnet, and you have a sturdy little add-on accessory. You can see it mounted on the Space Invader Synth in Figure 15-17 (right).

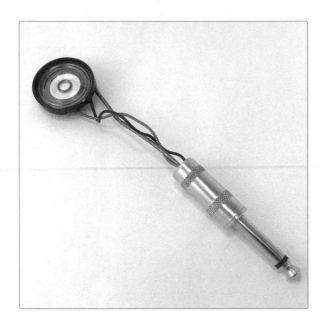
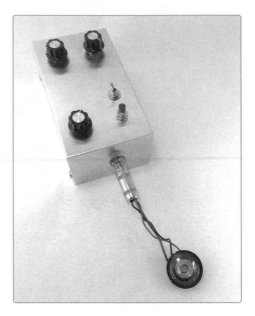

FIGURE 15-17: *A simple dandelion speaker for your Single-Chip Space Invader Synth*

Work with Pedals and Effects

The Single-Chip Space Invader Synth plays well with many commercial and home-brew effects. On the commercial end, try pumping your Space Invader Synth through a digital delay pedal,[7] chorus pedals, phasers, and some distortions. You can buy super dinged-up—but perfectly functional—stomp boxes at secondhand guitar shops and pawn shops, where they often go cheap. If you live in a college town, as I do, Craigslist can also be a fairly reliable source for some fantastic deals.

As for homebrew effects, the Space Invader Synth works well with the low-pass filter built in the Mud-n-Sizzle Preamp (Project 12), the Blinkie Tremolo (full build instructions are in my first book, *Snip, Burn, Solder, Shred*), and the Universal LFO (Project 13). The tremolo effect works best if you keep the synth's output volume fairly low, but you can boost it back up by placing the Mud-n-Sizzle Preamp after the trem unit.

The Space Invader Synth can get along well with the Twin-T Phaser/Wah (Project 14) if you use the "ground connection break" trick described in its "Tips, Tricks, and Mods" on page 262. Unfortunately, completely breaking the ground connection tends to kill most of the low-end from the Space Invader Synth's output. Inserting a 1M ohm resistor in the ground connection between the two circuits, instead of breaking the ground completely, results in a decently balanced output signal that's still responsive to the twin-T filter.

---

7. I especially love the Boss DD-3.

Tone Mods

The CD 4093–based oscillator in the Single-Chip Space Invader Synth can be savagely loud, capable of overdriving basically any amplifier. The way we've wired the LED indicator lamp significantly tames that tone for an old-school video-gamey bleep-bloop vibe. For a much more aggressive sound, simply omit the indicator LED and its resistor. For a sound that's roaringly aggressive while still fairly manageable, remove the LED/resistor pair from the power line and instead connect the LED's positive leg to pin 11 of the IC and its negative leg to the ground (leaving out the resistor). In this configuration, the LED will dirty up your tone some and also blink at the rate of the modulation oscillator.

To smooth out the synth's reedy buzz-jangle, you can add a simple adjustable low-pass filter, such as the one in "Low-Pass Filter" on page 365. (This is fundamentally the same filter as the one we built into the Mud-n-Sizzle Preamp.) Once you build the filter, it's really easy to install: just connect lug 1 of the filter's control pot to the output lug on the synth's jack (so that both the filter and the PCB are wired to the jack). You can solder the filter's ground connection to any point on the PCB's ground rail.

Pitch Mods

The 220 ohm resistor mounted on the PCB sets the highest pitch the oscillator can generate. Wanna be able to hit higher notes? Decrease the value of this resistor. Wanna be able to hit lower notes? Increase the value of this resistor. Just keep in mind that changing this resistor *shifts* the oscillator's range; it doesn't *expand* it. If you swap in a lower-value resistor, you'll be able to hit higher notes at the expense of losing some of the low notes, and vice versa. This particular oscillator can produce tons of sounds that are too high to be heard by human ears or too low frequency to be perceived as continuous tones. If you want to mostly keep your synth's voice in the audible tone range, stick with a resistor around 200–300 ohm.

If you want to *increase* the range of the pitch controller, use a larger variable resistor than the 100k ohm pot I've specified. Because the highest note the oscillator can play corresponds to zero ohm (a resistance any standard pot can be set to), increasing the value of this pot effectively adds more low-frequency pitches to your controller.

If you're thinking about swapping in a higher-value pot, there are a couple things to keep in mind. First, expanding the range means cramming more notes onto the same size dial. Yes, you have more notes to choose from, but zeroing in on any specific note while playing becomes increasingly difficult. If you make it hard to repeatedly hit the same note, then it's going to be hard to play a recognizable riff. Second, because the synth outputs square waves, your pitch can go only so low before the continuous tone falls apart and becomes a series of discrete beats, which are the flat-topped peaks of those square waves. This starts to happen with pots around 500k ohm.

The 100k ohm pitch pot is a good compromise between range and conventional playability. With the 100k ohm pitch pot, you end up with about two octaves centered on middle C—which is fairly useful in traditional music, while still offering an enjoyably broad sonic landscape. It's still fairly plausible that an average musician or dedicated newbie will be able to coax out a passable version of "When the Saints Go Marching In" from such a synth, if that's what floats their boat.

＊ **NOTE:** *You can make the same changes—adding a resistor in series with the pot or swapping pot values—to the modulation pot and thereby alter its range or cap its maximum frequency.*

Finally, the 0.1 µF caps set the lowest pitch each oscillator can hit; the higher the capacitance, the lower the lowest pitch you can play. But again, bear in mind that you can go only so low before your tones become thumps. If you choose to go with the higher-value polarized electrolytic caps, be sure to connect the positive leg to the chip and the negative leg—the one aligned with the stripe on the cap body—to the ground.

Control Mods

There are several circuit mods you can try that change how you play the synth and open up further sonic possibilities. Here are three examples:

▶ Add a pushbutton mute—like the one in the Scratchbox (Project 5)—to the modulation oscillator. Muting the modulation oscillator allows you to immediately convert your complex, multitone timbre to a pure square-wave tone. To add such a mute, add a normally closed pushbutton switch— abbreviated NC[8]—just before the diode in the upper half of the circuit diagram in Figure 15-7. (This will involve desoldering the diode's positive lead from pad I6 and inserting the switch between the diode and PCB).

▶ Build in a Universal LFO (Project 13) to augment the modulation control, the volume control, or even the pitch control itself. See "Using the Universal LFO" on page 234 for guidance.

▶ Replace the 100k ohm pitch control pot with either a photoresistor or a set of body contacts, like the ones used in the Droid Voicebox (Project 6). This drastically changes the way you interact with pitch control and is great for sound sculptures. Both of these mods also dramatically increase the sonic range of your synth, injecting a good deal of randomness into performance and play.

Add a Keyboard

Each pitch your Single-Chip Space Invader Synth produces corresponds to a specific resistance dialed in on the pitch-controller pot. For example, on my prototype synth, middle C is 52k ohm, C♯ is 49k ohm, D is 46k ohm, D one octave higher is 19k ohm, and so on.

---

8. A *normally closed* switch might also be called *normally on*, especially in a hardware store, where terminology is less technical. But don't buy one abbreviated as *NO*, which means *normally open* or *normally off*, like a doorbell switch.

* **NOTE:** *If all these Ds and Cs are throwing you off, flip to Appendix C for a crash course in music theory.*

It stands to reason that you should be able to remove the 100k ohm pitch pot from your Space Invader Synth and replace it with a set of resistors—each corresponding to a specific note—wired to a set of pushbutton switches (as illustrated in Figure 15-18).

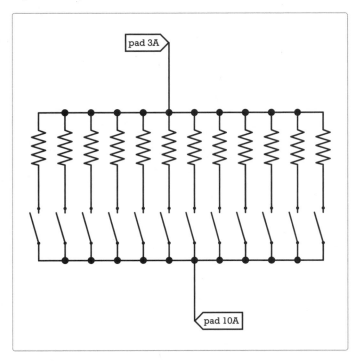

FIGURE 15-18: *An abstract (and impractical) circuit diagram for a resistance-based synth keyboard*

Theoretically, each switch should behave much like a piano key. For example, when you activate the middle switch, that single resistor—let's say it's a 52k ohm resistor—will connect pad 10A to pad 3A, and thus the synth would play middle C.

But there are several problems with this scheme. First and foremost, 52k ohm resistors are as rare as hen's teeth. Resistors are generally available only in certain common values, and few of your notes will map exactly to those values. In two octaves—24 notes, in this case, centered on middle C—I found that only about one-fourth of the notes my synth played corresponded to commercially available resistors.

But even if resistors were made in every conceivable value, I still couldn't just give a chart of resistances and the notes they produce, as I did for the tine lengths and notes of the Twang & Roar Kalimba (Figure 11-21 on page 190). This is because my list of resistances probably wouldn't work for you. Individual component tolerances, the lengths of your wires, the quality of your supplies and soldering, the charge on your 9-volt battery, the temperature of the circuit, local weather, and other similar seemingly trivial factors all stack up to impact the pitch your synth plays. Any list of values that work for my synth likely won't sound great on yours.

That said, there are still lots of ways to build a keyboard for a synth such as this one—it's just that many of these approaches are either pricey or maddeningly futzy. If you're feeling ambitious, I've included the least insane keyboard scheme in Figure 15-19, which shows the guts of a simple one-octave *string resistor ladder keyboard* prototype I whipped up for my Space Invader Synth.

FIGURE 15-19: *A single-octave string resistor ladder for my Single-Chip Space Invader Synth*

This simple network of resistors makes it possible to hardwire a series of very specific resistor values and, thus, to (fairly) reliably play a fixed set of notes, such as two full *chromatic scales* (explained in Appendix C), which is pretty darn handy if you have a thing for popular Western music produced in the last several centuries. String resistor ladders like these are exceedingly cheap and easy to build, which is why cheap toy keyboards generally rely on them. They're also very flexible, allowing for a variety of interface choices for your keyboard.

To begin experimenting with a two-octave string resistor ladder keyboard for your Single-Chip Space Invader Synth, gather an 18k ohm resistor (brown-gray-orange stripes), a small 10k ohm linear resistor (even a trim pot will do), a fistful of 1k ohm resistors, a few 2k ohm and 3k ohm resistors, a chromatic tuner, and your soldering tools.

Step 1 Begin with your 18k ohm resistor (brown-gray-orange stripes) and solder a string of about 30 1k ohm resistors to it. As you finish each solder joint, don't trim the excess lead from both resistors. Instead, bend out one leg at each junction, as shown in Figure 15-19. These "rungs" will become important later.

* **NOTE:** *This might sound like a lot of soldering, but you can easily do this entire process—from soldering the initial ladder to finding notes, adding any needed resistors, and finishing off the ladder—in under an hour.*

Step 2 Tape this resistor ladder to a piece of cardboard; this will keep it steady and give you a place to keep track of the results as you search for notes. Position the ladder with the 18k ohm resistor to the right.

Step 3 Turn the pitch control pot on your synth fully counterclockwise, connect the 18k ohm end of this string of resistors to lug 2 of your control pot, and connect a stripped and tinned length of insulated wire to lug 1. Henceforward, we'll call this insulated wire your *stylus*.

Step 4   Power up your amp and synth, and then touch the end of your stylus to the rung at the junction between the 18k ohm resistor and the first 1k ohm resistor. You should hear a high pitch. Check your chromatic tuner; it'll likely tell you that this is a fairly out-of-tune D_5 or E_5, but it could be a higher note. If it's a recognized, in-tune note, write the note down on the cardboard above that junction. Then, check the next junction. Work your way up the ladder in this fashion, one rung at a time, checking each and noting where the in-tune notes are. You'll probably find usable notes at every few junctions; they'll tend to space out as you move up the ladder.

Step 5   If you don't find all the notes you need, add more resistors to the end of your ladder. Try 2k ohm and 3k ohm now because you'll be in the lower registers. Test each of these new junctions, jotting down where your in-tune notes are. Try to find about 24 consecutive notes for a highly playable keyboard.

Step 6   Once you've found your notes, go back to each junction where there's no useable note and trim away the rung as I've done in Figure 15-19. Now each of your remaining rungs should correspond to a recognized note in the chromatic scale.

Step 7   Finish your string ladder keyboard by adding a small linear pot, such as a 10k ohm, as shown in Figure 15-19. The overall pitch of your keyboard will inevitably drift with the weather, battery power, component age, jostling, and so on. This pot will allow you to retune the keyboard as a whole or to do cool detuning warbly effects as you play. At this point, you can wire pushbutton switches to each rung (see Figure 15-20).

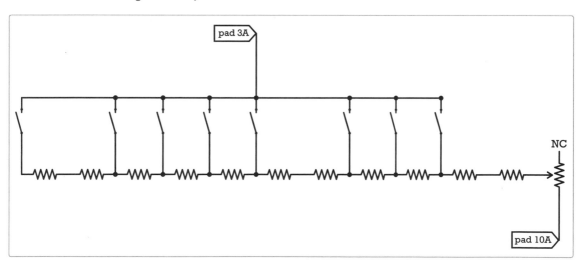

FIGURE 15-20: *An abstract circuit diagram of an abridged string resistor ladder keyboard*

These buttons work like piano keys, with your highest note at the far right and your lowest at the far left. Personally, I prefer to forgo the buttons and just keep using the stylus to play the ladder.

One neat way to package the ladder is to mount a series of metal screws in the top of a wooden or plastic enclosure, laying them out like piano keys, and to connect each of these virtual piano keys to its corresponding rung inside the case. Then, you can tap each screw head with your stylus to play the desired note. This description might seem clumsy, but that's exactly how commercial Stylophones, like the one shown in Figure 15-21, work. Be the envy of your block with your very own completely custom DIYlophone!

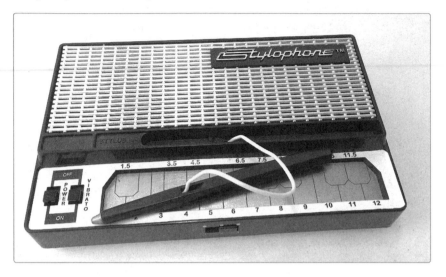

FIGURE 15-21: *A commercial Stylophone uses a resistor ladder and stylus for the keyboard control.*

Resources

If you've dug building this synth, then you should get your hands on a copy of Nicolas Collins's book, *Handmade Electronic Music: The Art of Hardware Hacking* (Routledge, 2006). It includes *tons* of great, open-ended projects for building simple synths like this.

I also recommend you look up Tim Escobedo's Synthstick from his legendary, sadly defunct, *FolkUrban* DIY musical instrument site, archived here: *http://www .oocities.org/tpe123/folkurban/*. That build taught me to hack a single oscillator from a CD 4093 Quad NAND Schmitt Trigger IC, and I've never regretted it for a moment. If you're interested in Tim's other electronica hijinks, you can find archived PDFs in the "Resources" section of my site: *http://www.davideriknelson.com/sbsb/ resources/*.

This project combines a simple, if idiosyncratic, 8-step analog sequencer with a modified version of the two-oscillator Single-Chip Space Invader Synth (Project 15). A *sequencer* is any device that controls and automates a sequence of events. Although originally applied strictly to industrial contraptions, the term has now been entirely swallowed by electronic music, where sequencers are responsible for funky bass lines, lush synth pads, and rock-steady robot drumbeats.

The *analog* in this project's name refers to both the pair of analog oscillators that provide the synth's voice and the sequencer controls themselves, all of which give you access to the full sweep of states their components can create, rather than limiting you to a set of discrete, predetermined settings. The *8-step* means that the sequencer repeats a steady pattern of eight individually controlled pitches, enough for a single sprightly bar of music or a pair of alternating 4-step bars. This might not seem like much by modern standards, where 16 steps is the norm for a hardware sequencer and a computer can hold a functionally infinite number of steps. But it's perfect for a performance sequencer meant to be tweaked on the fly by a live musician working a groove.

* **NOTE:** *You'll see more musical terminology in this project than you have in the others. If you find yourself confused about bars, 16th notes, rests, and so on, flip to Appendix C.*

Hear the Bleepbox in action in the samples at *http://www.nostarch.com/jamband/*.

FIGURE 16-1: *The finished Bleepbox 8-Step Analog Sequencer*

Preparation

Build Time

▸ About 3 hours

Tools

▸ A standard soldering kit (See page 340.)

▸ A ruler that shows 1/4-inch increments or smaller

▸ A Sharpie or other permanent marker

▸ An electric drill with bits (You'll probably need 3/16-inch, 5/16-inch, and 3/8-inch bits.)

- ▶ Other tools to work your enclosure (If you're using a wooden enclosure, you'll want sandpaper and possibly files to clean up drill holes.)

- ▶ A multimeter or continuity tester (See "Using a Multimeter" on page 349 or "Super-Basic Continuity Tester" on page 357.)

- ▶ An amp and instrument cable for testing the synth

- ▶ (Optional) Foam-backed double-sided tape or hardware to mount the circuit in its enclosure

- ▶ (Optional) A hacksaw for trimming down potentiometers with long shafts

✳ NOTE: *For clarity, I've split the supplies list into two parts: one for the 8-Step Sequencer and the other for the modified Space Invader Synth that the sequencer controls.*

Supplies for the 8-Step Sequencer

- ▶ A CD4017 integrated circuit, such as Digi-Key part #296-2037-5-ND

- ▶ A 16-pin IC socket

- ▶ A 555 timer IC, such as Digi-Key part #296-1411-5-ND or Mouser part #595-NE555P

- ▶ An 8-pin IC socket

- ▶ Two 1N4148 silicon diodes (Mouser part #78-1N4148 is shown in Figure 16-2, although any generic 1N4148 or 1N4150 is fine.)

- ▶ 10 red LEDs

- ▶ Nine 220 ohm resistors (red-red-brown stripes)

- ▶ Two 10k ohm resistors (brown-black-orange stripes)

- ▶ A 470 ohm resistor (yellow-violet-brown stripes)

- ▶ Nine 1M ohm variable resistors (Variable resistors are also called *potentiometers* or *pots*; Mouser part #313-1000F-1M is shown in Figure 16-2, but you can use any 1M ohm pot with a linear taper. See "The Gory Details: Audio Taper vs. Linear Taper" on page 327 for details.)

- ▶ A 100k ohm variable resistor with a linear taper

- ▶ 10 control knobs that fit your variable resistors

- ▶ A 1.5 µF electrolytic capacitor

- ▶ Nine small SPST toggle switches (You can use Mouser part #108-MS550K, for example, but any similar switch will work.)

- ▶ A small SPDT toggle switch with two *on* positions (These are often called *on-on* or *changeover switches*. Something like Mouser part #108-1MS1T2B3M2QE-EVX will work.)

- ▶ A small center-off SPDT toggle switch (This has three positions: one side turns on one circuit, the other turns on another circuit, and the center position turns off both circuits. These are often called *on-off-on switches*. Something like Mouser part #612-100-C1111 will work.)

- ► Two SPST pushbutton switches, such as Mouser part #103-1012-EVX or non-illuminated doorbell buttons
- ► A 9-volt battery
- ► A 9-volt battery clip
- ► 24-gauge insulated hook-up wire (Stranded wire is fine. It's best to have several colors.)
- ► 22- or 24-gauge bare bus wire (This is uninsulated solid core wire.)
- ► An Adafruit Perma-Proto Quarter-sized Breadboard PCB (Adafruit product #589 or Maker Shed product #MKAD48)[1]
- ► Duct tape and electrical tape, or heat-shrink tubing
- ► A sturdy enclosure (Cigar boxes are good here, as are plastic lunchboxes. See "On Enclosures" on page 214 for more options.)
- ► (Optional) A 9-volt battery holder clip, such as Digi-Key part #71K-ND or Mouser part #534-071

Supplies for the Modified Single-Chip Space Invader Synth

✳ **NOTE:** *It's perfectly possible to modify a prebuilt Space Invader Synth to work here; check Step 23 on page 310 for details on the changes that need to be made.*

- ► A CD4093 integrated circuit, such as Digi-Key part #296-2068-5-ND
- ► A 14-pin IC socket
- ► Two 1N4148 silicon diodes (Mouser part #78-1N4148 is shown in Figure 16-2, although any generic 1N4148 or 1N4150 is fine.)
- ► A photoresistor (This might also be called a *light-dependent resistor*, a *cadmium sulfide (CdS) cell*, or a *photocell*. Flip to "The Gory Details: Selecting the Right Photoresistor" on page 328 for some important tips on selecting a photoresistor for this project.)
- ► A 220 ohm resistor (red-red-brown stripes)
- ► A 470 ohm resistor (yellow-violet-brown stripes)
- ► A 10k ohm audio potentiometer (This is a variable resistor with an audio taper; see "The Gory Details: Audio Taper vs. Linear Taper" on page 327.)
- ► A 500k ohm variable resistor with a linear taper
- ► Two control knobs that fit your variable resistors
- ► Two 0.1 µF capacitors (marked *104*)
- ► Two small SPST toggle switches (You can use Mouser part #108-MS550K, for example, but any similar switch will work.)

---

1. I've sung the praises of this *PCB (prefab printed circuit board)* elsewhere in the book—you can see front and back pics in Figure 13-23 on page 240. If this specific PCB is hard to find, fear not: several other manufacturers make similar boards. Just google *permanent breadboard* or *solderable breadboard*. Be forewarned that the Adafruit Perma-Proto Quarter-sized Breadboard PCB is smaller than most alternatives, so you'll likely have to cut other boards down to fit this project.

- ▶ A 1/4-inch mono phone jack, also called a *guitar jack*
- ▶ An Adafruit Perma-Proto Quarter-sized Breadboard PCB (Adafruit product #589 or Maker Shed product #MKAD48)

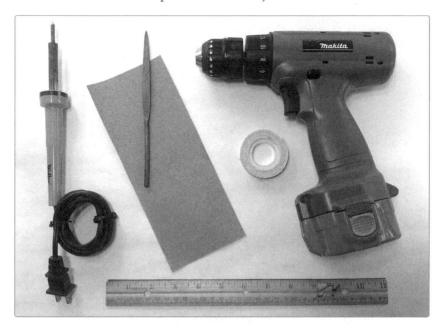

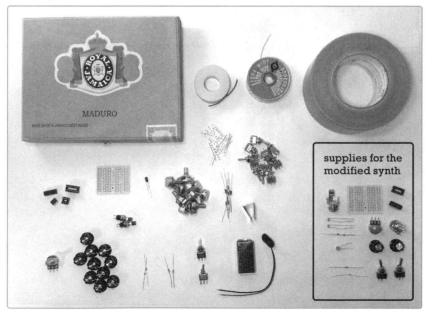

FIGURE 16-2: *Tools and supplies; the parts for the modified Single-Chip Space Invader Synth are clustered in the lower-right corner of the bottom image (not shown: amp, instrument cable, hacksaw, marker, and multimeter)*

The Bleepbox Control Breakdown

The Bleepbox 8-Step Analog Sequencer is definitely the most complex project in this book, so let's take a tour of the controls shown in Figure 16-3 before we begin.

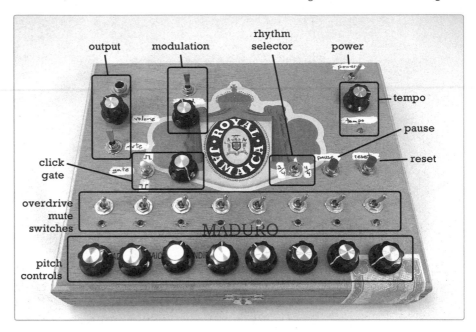

FIGURE 16-3: *The Bleepbox control panel*

Output Controls

In the upper left, we have the output section, with an *output jack*, *output volume control*, and *output mute*. These are pretty self-explanatory.

Modulation Controls

To the right of the output controls are the *modulation controls*, which give the synth its signature phat chiptune timbre. When you turn the mod oscillator off with the *mod switch*, you hear a pure square wave. With modulation turned on, you hear the result of the mod oscillator and the pitch oscillator interacting with each other, giving you a wide range of effects and textures. The *mod rate knob* controls the speed of the mod oscillator.

Click-Gate Controls

Below the modulation controls are the controls for the Bleepbox's automated click-gating effect, which is distinctive to this design and a bit of an odd duck. When you flip the *click-gate switch* either up or down, it creates a steady rhythm within your pattern by inserting two evenly spaced clicks in each step, one at the beginning and one in the middle. It's similar to the click track you hear in recording software or a very stripped-down version of a drummer keeping the beat on the closed hi-hat. This click rhythm isn't muted by the output mute.

When the click-gate switch is in *square-up* mode, as noted by the label, it accentuates the first half of each step and nearly mutes the second half. In *square-down* mode, it mutes the first half of each step while accentuating the second half for a funky "riding behind the beat" feel. In the middle position, the click-gate effect is off. The *depth* of this effect, or how emphatic it is, can be adjusted with the *depth knob* to the right of the switch.

Step Controls: Pitch and Overdrive Mute

The rows of eight switches, eight LEDs, and eight knobs along the bottom edge form the *step controller* section. Each switch-LED-knob cluster controls a single step in the sequence. Each step's *indicator LED* indicates when the step is active and gives a relative sense of that step's pitch (brighter light = higher pitch). The *pitch control knob* sets the pitch for that step. The toggle switch is an *overdrive mute*, another odd-duck control: it mutes the corresponding step by running the pitch oscillator too fast to be audible. Because our sequencer slides between pitches rather than jumping from note to note (*trés analogique!*), muting in this fashion leads to the voice sliding zappingly back into the mix, rather than simply popping back on.

Tempo, Power, and Pause/Reset Controls

The *power switch* is in the upper-right corner of the sequencer control panel. Beneath that is the *tempo control*, which sets the rate of the sequence, ranging from a sludgy 50 beats per minute (bpm) up to self-oscillation. *Self-oscillation* means that the pattern runs so fast it becomes its own complexly layered pitch, a sort of meta-voice of the synth. In this case, the self-oscillation sounds a lot like an angry robo-wasp shifting her space-race speederbike during the final lap of a post-apocalyptic motocross race. The pause and reset buttons sit below this tempo control. *Pause* holds the pattern at a specific step, humming the chosen pitch. *Reset* restarts the pattern.

Rhythm Control

Finally, the switch marked *3/4* and *4/4* is the *rhythm selector*. It allows you to choose between an *8-step pattern*, which is recognizable as the 4/4 rhythm you hear in almost any pop song, and a *6-step pattern*, which naturally falls into a 3/4 rhythm, like a waltz. If these terms are unfamiliar, just flip to Appendix C for a quick overview.

Building the Sequencer

The Bleepbox looks a bit daunting at first glance. So many functions! So many components! So many flashing lights and flying wires! But this build is no more difficult than any other in this book. It's really just a combination of concepts and circuits you've already built. When it comes down to it, the hardest part is drilling all the holes for the pots, switches, and LEDs.

Take a look at the circuit diagram in Figure 16-4, which shows the sequencer portion of the Bleepbox circuit.

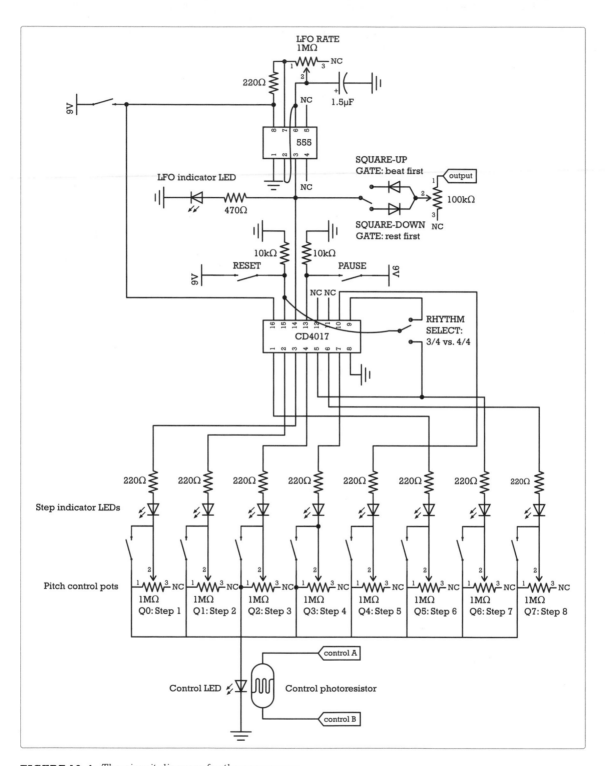

FIGURE 16-4: *The circuit diagram for the sequencer*

The LFO section at the top is the same 555 timer-based circuit that drives the Universal LFO (Project 13). This IC feeds a steady beat to the input of the CD4017 chip. The CD4017 is also called a *decade counter*: it watches for a signal on its input (pin 14), and every time it detects a beat, it advances its count by turning off the current output and turning on the next output. It starts with its first output—labeled *Q0*—turned on. When it detects a beat coming from the 555, it turns off Q0 and turns on the second output, labeled *Q1*. At the next beat, it turns off Q1 and turns on Q2, and so on all the way to Q9, for a total of 10 possible steps before looping back—hence the *decade* in the IC's name, although we've wired it up to loop back after its eighth step.[2] Each step in the count drives an indicator LED wired to a switch and pot that allow you to control the brightness of the indicator LED.

The LED/switch/pot assemblies from each step are then tied together to drive a final control LED that matches the brightness of the currently lit step indicator LED. This final control LED is strapped to a photoresistor: the brightness of the control LED thus determines the resistance of the photoresistor, which in turn controls the pitch of a slightly modified version of the Single-Chip Space Invader Synth. This "LFO → decade counter → indicator LED → pot → control LED" scheme allows us to fully automate a sequence, with control over the pitch of each step and the overall tempo of the pattern.

✳ **NOTE:** *For clarity, Figure 16-4 doesn't show the schematic for the modified Space Invader Synth. This schematic is in Figure 16-21, and we'll give it a look when we're ready to build the synth itself.*

Prepare the Hardware

As usual, let's begin the actual build with the hardware. If you've never, ever soldered before, this probably isn't the place to start; try building the Droid Voicebox (Project 6) first. That's a great first-time soldering project, and it will also familiarize you with ICs. If you just need a soldering refresher, flip to "Soldering" on page 346.

First, we'll prepare four switches: power, rhythm selector, pause, and reset. We'll then assemble the eight switch-pot-LED pitch controllers and finish up by building the click-gate control.

Step 1 Warm up your soldering iron and grab the insulated wire, the 9-volt clip lead, one SPST toggle switch, the SPDT toggle switch, and both pushbutton switches. Cut seven 6-inch wires and one 4-inch wire, and strip and tin both ends of each wire. If need be, strip the ends of the 9-volt clip lead wires (many ship with the leads already stripped and tinned). Now, solder one 6-inch wire to each lug of each pushbutton switch and solder one 6-inch wire to each of the three lugs on the SPDT switch. Then, solder the 4-inch wire to one of the SPST toggle switch terminals—it doesn't matter which—and finish up by soldering the positive (red) lead from the 9-volt battery clip to the other SPST toggle switch terminal. You now have your power, reset, pause, and rhythm selector switches, as shown in Figure 16-5.

----

2. In Figure 16-4, I've labeled only the outputs we're using. If you're curious about which pin is which for all 10 outputs, google *CD4017 pinout* and look at the image search results.

Step 2   Prep the tempo control pot next. Cut, strip, and tin two 6-inch lengths of insulated wire. Take one of your 1M ohm pots and solder one wire to lug 1 and the other to lug 2 (see Figure 16-6).

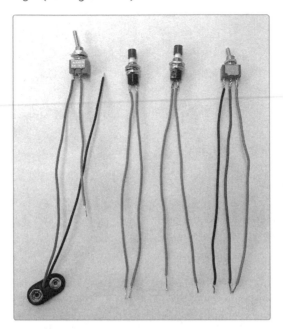

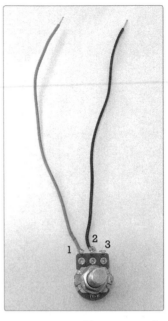

FIGURE 16-5: *From left to right: power, reset, pause, and rhythm selector switches*

FIGURE 16-6: *The prepared tempo control pot*

Step 3   Prep the eight step controllers. Use your multimeter—or, barring that, a simple continuity tester circuit, like the "Super-Basic Continuity Tester" on page 357—to make sure all of your eight remaining SPST toggle switches are turned off. In the off position, the switch leans toward one lug; for example, in Figure 16-7, my switch is turned off and leaning toward the left lug. Let's call this left lug—the one the switch is leaning toward when turned off—*Lug A*. On each switch, mark Lug A with a dot so you can easily orient the switches when building each step controller. (You're going to be really annoyed later if you discover that you installed some switches backward.) The dot-free lug is now *Lug B*. Use your marker to number your eight 1M ohm pots 1 through 8, and you're ready to start soldering.

Step 4   Build your first pitch controller. Gather the 1M ohm pot you labeled *1*, an LED, a 220 ohm resistor, and an SPST toggle switch. Cut two 6-inch lengths of insulated wire and two 2-inch lengths of insulated wire. Strip and tin both ends of each wire.

✳ **NOTE:** *In Figure 16-7, I've used four different colored wires (from left to right: red, black, green, and white). This minimizes the likelihood of wiring connections incorrectly. I strongly recommend using a similar color-coding scheme.*

Step 5   Solder one 2-inch wire to each toggle switch lug and solder one 6-inch wire to one leg of the 220 ohm resistor (that's the white wire in Figure 16-7). Solder the other leg of this resistor to the positive leg of the LED, which is the longer one, *opposite* the flat area on the side of the LED body. Twist together the remaining (negative) LED leg and the 2-inch wire attached to switch Lug B. Solder this pair to pot lug 2, as shown in Figure 16-7. Then, twist the 2-inch wire attached to switch Lug A with the remaining 6-inch wire. Finally, solder this pair to lug 1 on the pot. *Ta-da!* Your first step controller is complete. Repeat seven times for a total of eight assemblies.

Step 6   Once you've finished all eight pitch controllers, gather the leads connected to lug 1 of each pot and twist the ends together into a big bundle, using pliers if necessary. Solder that bundle to the longer, positive leg of the final LED—which will serve as the optocoupler's control LED. Then, cut, strip, and tin another 6-inch length of insulated wire and solder it to the negative leg of this LED (the shorter leg nearest the flat area on the LED body). Consider reinforcing the eight-wire LED connection with some electrical tape or shrink tube, as I have, to prevent short circuits. Figure 16-8 shows the finished 8-step controller assembly.

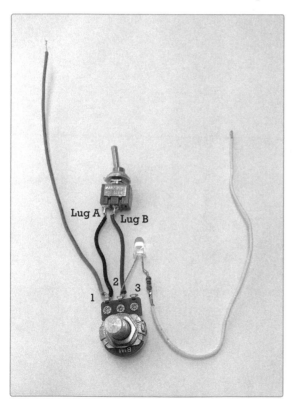

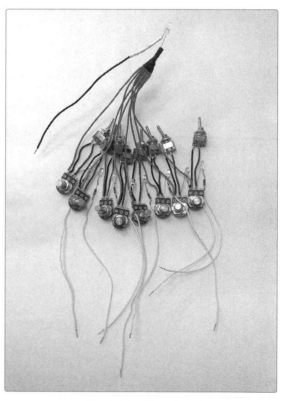

FIGURE 16-7: *A single pitch controller. Contrasting wire makes it easier to keep the connections straight.*

FIGURE 16-8: *The finished bundle of pitch controllers connected to the control LED*

Step 7   Next, build the click-gate control. Gather your on-off-on SPDT switch, two diodes, and the 100k ohm pot. Cut, strip, and tin two 6-inch lengths of insulated wire. Solder one to lug 1 on the 100k ohm pot. Solder the other wire to the center terminal on the SPDT switch. Pick one of the switch's two edge terminals—it doesn't matter which—dub this *Terminal A*, and mark it with a dot of marker or some correction fluid.

Step 8   Find the black stripe on each diode, as shown in Figure 16-9 (left). The negative diode leg is the one closest to that stripe. Keeping this in mind, solder the positive leg of one diode to Terminal A on your on-off-on SPDT switch. Solder the negative leg of the other diode to the remaining switch terminal. Twist together the free legs on the two diodes and solder them to the middle lug of the 100k ohm pot.

When the switch is leaning toward Terminal A, the click gate is in square-up mode and will accentuate the beginning of each step. When the switch is leaning the other way, the click gate is in square-down mode, accentuating the latter half of each step. When the switch is in the middle, the click gate is disengaged. Set this assembly aside with all the other hardware and get ready to stuff your circuit board.

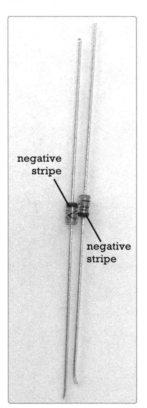
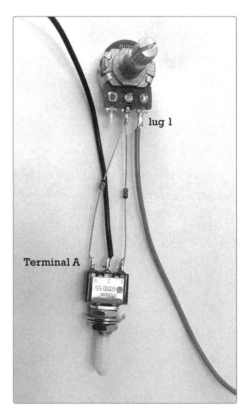

FIGURE 16-9: *The completed click-gate control assembly*

Build the Circuit

After all the measuring, stripping, twisting, soldering, and repeating you just did, building the PCB will feel like a breezy afternoon stroll. As in earlier projects, we'll refer to solder pads using the number-letter coordinates etched on the topside of the PCB. For example, the hole farthest to the left directly above the central divide is called *1F*.

Step 9   First, mount the IC sockets so that both straddle the PCB's center divide, as shown in Figure 16-10. The 8-pin socket runs from 2F to 5F above the divide and 2E to 5E below. Leave one column open and then place the 16-pin socket from 7F to 14F and 7E to 14E. If your sockets have notches like those in Figure 16-10, place the notches to the left; that's just good form. Bend out the top left and bottom right legs on both sockets to hold them in place on the PCB. Then flip the board and solder all the pins to their solder pads. Take care not to create solder bridges between any adjacent pins.

Step 10   Now, add the jumpers shown in Figure 16-11. There are seven: two insulated fly-overs and five bare bus connections. Start with the flyovers. Cut two 1 1/2-inch lengths of insulated wire. Use solid core insulated wire if you have it; otherwise, use stranded wire. Strip both ends of each wire, tin them if your wire is stranded, and then run one wire from 3D to 4G over the 8-pin socket, connecting its pin 2 to its pin 6. Run the other wire from 4D to 9G, connecting pin 3 on the 8-pin socket to pin 14 on the 16-pin socket. Solder all four ends of these two wires now.

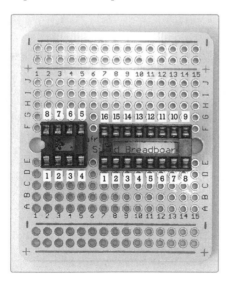

FIGURE 16-10: *The IC sockets are in place. Pins are numbered counterclockwise from the lower-left pin of each socket.*

FIGURE 16-11: *The seven jumpers on your Bleepbox circuit board*

Step 11   Snip four 3/4-inch lengths of bare bus wire. Bend each into a flat bottomed *U*. Run wires from 2A and 14A to the lower ground rail at the bottom of the PCB. This is the row of holes just below the blue line with the negative signs. Run two more jumpers from 2J and 7J to the power rail at the top of the PCB; that's the row of holes just above the red line with the plus signs. Solder all four of these wires now.

Step 12   For the final jumper, cut a 2 1/4-inch length of bare bus wire and run it down column 15 at the right edge of the PCB from the upper ground (the topmost row of holes) to the lower ground. Solder this wire, and your jumpers are done.

Step 13   Place the three board-mounted resistors, which are two 10k ohm resistors (brown-black-orange stripes) and a 220 ohm resistor (red-red-brown stripes). Bend the 220 ohm resistor into a tight *V* and mount it from 2G to 3G to connect pins 7 and 8 on the 555 timer. Bend the two 10k ohm resistors into flat-bottomed *U*s. Run one from 8J to the upper ground rail (the row of holes at the top of the board) and the other from 10J to the upper ground rail. Solder all three resistors, as shown in Figure 16-12.

Step 14   Now for the 1.5 µF electrolytic cap, which connects pin 5 on the 8-pin socket to the ground. Electrolytic caps are polarized, so start by identifying the negative leg, which is aligned with the stripe on the cap's body. Run this negative leg to the top ground rail and the positive leg to pad 4J (see Figure 16-13). Solder both legs.

FIGURE 16-12: *Placement of the three board-mounted fixed resistors*

FIGURE 16-13: *Adding the 1.5 µF electrolytic cap*

Step 15   Next, add the indicator LED. Cut, strip, and tin two 6-inch lengths of insulated wire. Dig out your final LED and the 470 ohm resistor (yellow-violet-brown stripes). Solder the resistor to the positive LED lead. Then, solder one insulated wire to the open resistor lead and the other wire to the negative LED lead. Run the positive lead, which is attached to the resistor, to pad 4A and the negative LED lead to the lower ground rail at the hole immediately below 4A (see Figure 16-14).

✳ **NOTE:** *If your enclosure has a clear or translucent top, you can get a pretty neat effect if you skip the insulated leads here and instead solder the indicator LED— with its resistor—directly to the board. This light blinks on and off with each clock cycle of the 555 timer. Mounted underneath a translucent top, it will make your sequencer appear to have a throbbing red heartbeat. Trés chic!*

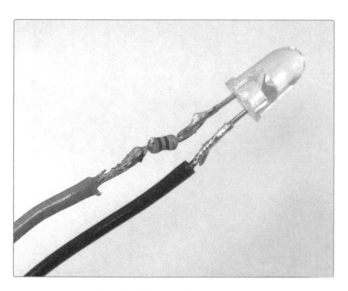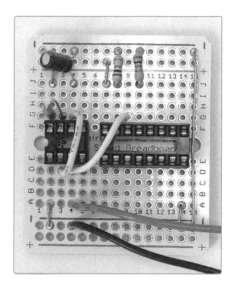

FIGURE 16-14: *Detail of the indicator LED with resistor (left). Close-up of the PCB at this stage (right). Note that the two wires near the bottom of the PCB go to the LED.*

Install the Hardware

Now it's time to install all the hardware from the first few steps. We'll start with the pause and reset pushbuttons you built in Step 1. These switches send voltage to pin 13 or 15 of the CD4017, respectively, in order to force the chip either to hold at its current step, thereby pausing the sequence, or to jump back to its first step, thereby resetting the sequence.

Step 16   Take one pushbutton switch—it doesn't matter which—and connect either of its leads to pad 10I on the PCB. Run the switch's other lead to the power rail at column 11, as shown in Figure 16-15 and in detail in Figure 16-16 (right). Solder both wires. Take the other switch and connect one lead to pad 8I and the other lead to the power rail at column 9. Solder both of these.

Step 17   Now, install the tempo control. Take the only 1M ohm pot left—the one you added leads to in Step 2. Connect the lug 1 lead to pad 3I and the lug 2 lead to 4I. Solder both leads and compare your progress to Figure 16-16.

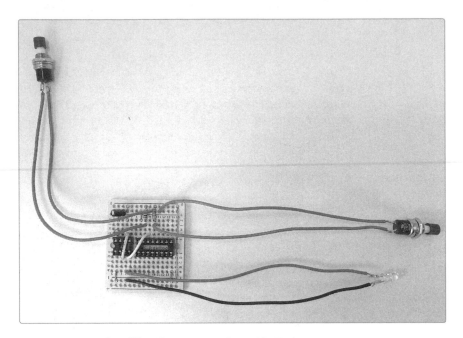

FIGURE 16-15: *Installing the pause and reset buttons*

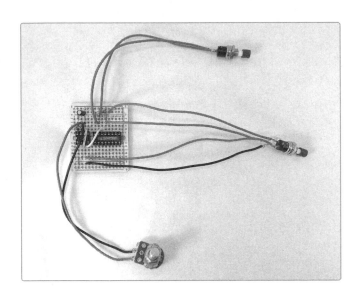
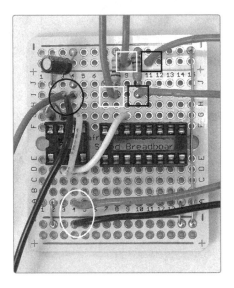

FIGURE 16-16: *The tempo control is installed. Detail on the right shows the tempo control pot connections in a black circle, the indicator LED leads in a white circle, the pause button wires in black squares, and the reset button connections in white squares.*

Step 18   Install the rhythm selector switch, which is the on-on SPDT switch with three leads. Run the center-lug lead to pad 8H, one side lead—it doesn't matter which—to 14G, and the final lead to 11D (see Figure 16-17). Solder it all now.

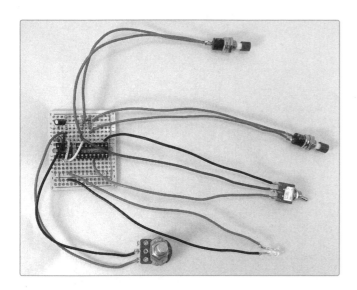

FIGURE 16-17: *The rhythm selector is in place. In the detail on the right, the center-lug connection is squared and the other two lugs circled.*

Step 19   Install the click-gate control, which is the center-off SPDT switch with the diodes and 100k ohm pot wired to it. Run the lead from the center lug of the switch to pad 4B and solder it. You'll solder the other wire on this assembly—the insulated lead dangling from lug 1 of the 100k ohm pot—to the output jack after we build the modified Space Invader Synth.

Step 20   Finally, install your 8-step controller bundle—that knot of resistors, LEDs, switches, and wires we built in Steps 3 through 6. There should be eight free wires, each connected to one of the numbered pots via a resistor and LED (see the white wires in Figure 16-18). We're going to connect these to the PCB in order and solder each immediately so that it doesn't escape. The lead attached to pot 1 goes to pad 9A, pot 2 to pad 8A, pot 3 to 10A, pot 4 to 13A, pot 5 to 13J, pot 6 to 7A, pot 7 to 11A, and pot 8 to 12A. Figure 16-18 identifies which wire goes where and shows a picture of the PCB at this stage. To finish the pitch-control bundle, run the single open wire on the shared pitch-control LED—which is that LED's ground—to the lower ground rail at the foot of column 5.

Step 21   Wire the PCB for power. Adding power is super easy: run the negative (black) wire on your 9-volt snap-on connector to the upper ground at the upper-left corner of the board (see Figure 16-19). Run the open wire on the power switch to the power rail in the hole immediately below this.

Step 22   Place your ICs into the sockets. The little 555 timer goes in the 8-pin socket, while the larger CD4017 goes in the 16-pin socket. Push your flyover jumpers out of the way or sneak the ICs beneath them. Both ICs should be positioned with their orientation marking—be it a dot, a divot, or a stripe—to the left, as shown in Figure 16-19.

IC output	Pot #	PCB solder point
Q0	Pot 1	9A
Q1	Pot 2	8A
Q2	Pot 3	10A
Q3	Pot 4	13A
Q4	Pot 5	13J
Q5	Pot 6	7A
Q6	Pot 7	11A
Q7	Pot 8	12A

FIGURE 16-18: *A wiring chart for the step controllers and a picture of the PCB thus far. The wires connecting to each step controller are white. The black-wired ground connection for the shared control LED is circled, and the step controller wires are boxed.*

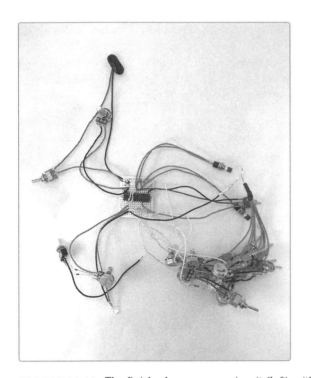

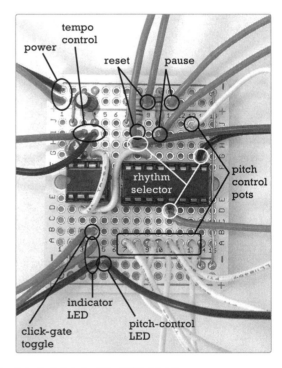

FIGURE 16-19: *The finished sequencer circuit (left) with a labeled detail of the PCB (right)*

Troubleshooting the Sequencer Circuit

Attach a fresh 9-volt battery and power up your sequencer. Turn the tempo control—the 1M ohm pot wired directly to the board, with no switches or extra components—to its midpoint. Set all the other pots fully counterclockwise and turn all the switches in the pitch-control bundle on so they lean *away* from Lug A, which you marked in Step 3.

Is the indicator LED blinking? This is the LED that connects directly to the 555 timer at pad 4A. (Other LEDs might also blink, but we'll get to them in a bit.) If the indicator LED isn't blinking, flip the power switch and check again. If you still get no response, check that your battery is fresh and look for shorts on your PCB and hardware, especially around the 555 timer, the power switch, and the tempo control pot. Eliminate any short circuits: separate uninsulated wires that are brushing against each other, flick away loose scraps of wire or flakes of solder, and reheat sloppy joints and solder points to smooth them. (The solder should naturally flow to the warmer joint.) If a solder bridge is especially stubborn or there's just too much solder glumped to a joint, you may need to desolder and redo the connection.

> * **NOTE:** *If you're unsure how to desolder, see "Desoldering" on page 349.*

Once the indicator LED blinks, run the tempo control clockwise and counterclockwise. The blinking should speed up as you go clockwise until it becomes a blur or the LED is constantly illuminated, and it should slow as you go counterclockwise to about 50 bpm. If it does the opposite, it just means you wired the pot backward in Step 2. You can either swap the leads by transferring the side lead to the other side or leave them alone and use the knob backward.

Once the tempo control and its indicator LED check out, set the tempo control back to its midrange and turn your attention to the step indicators, which are the eight LEDs wired to the pitch control bundle. They should illuminate brightly and steadily one at a time. The blink rate of the step indicator LEDs is twice as long as the tempo rate LED. That is, each step's LED should stay illuminated for a full on-off cycle of the tempo rate LED. It will then blink off as the next step blinks on. The common control LED, connected to all eight red step control leads, should stay lit more or less steadily.

> * **NOTE:** *The blink of each step indicator LED—and the pitch control LED they share—may have a two-tone flavor, whereby it's slightly brighter while the tempo rate LED is on and slightly dimmer when the tempo rate LED is off. That's normal and perfectly acceptable. In fact, it will ultimately lend a little extra color to your synth's voice.*

If any of the steps aren't lighting up, look again for shorts. A solder bridge or stray fleck of metal on the PCB might be the problem, but the hardware is a much more likely culprit. These LEDs have a tendency to brush up against each other when not secured, and that creates short circuits. Consider temporarily mounting the switches and LEDs on a piece of cardboard or tying them to a chopstick to keep the wires unsnarled and save your sanity while troubleshooting (see Figure 16-20).

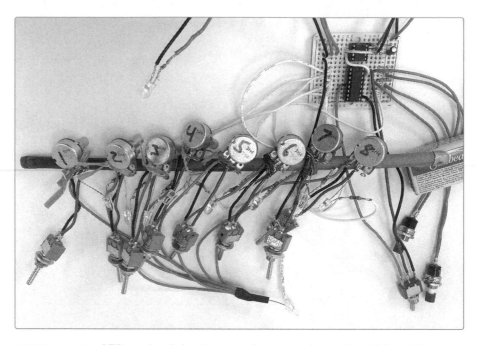

FIGURE 16-20: *LEDs and switches temporarily mounted on a chopstick to aid troubleshooting. I used kitchen twist ties to hold the wires and pots to the chopstick because they're insulated and easy to attach one-handed—and my kitchen junk drawer is crammed with them.*

Similarly, if two or more step indicator LEDs blink simultaneously, that might be due to a short on the PCB, but it's more likely due to the legs of neighboring LEDs touching. Consider temporarily mounting the LEDs on something to prevent leads from brushing against each other. If all but two of the steps are lighting up, the rhythm selector is probably just set to 3/4 instead of 4/4; try flipping the stand-alone on-on SPDT switch. Now each pot's indicator LED should light up in order, from 1 to 8. If they light out of order, you just swapped a wire in Step 20. No harm done: renumber the pots so that they match the order the LEDs illuminate and move along.

Next, check the rhythm selector switch. In one position, all eight LEDs should light up before the sequence loops back to the first step. In the other position, only LEDs 1 through 6 should light before the sequence starts over.

If the rhythm selector checks out, test the overdrive mute switches and pitch pots. Flick the mute switches off so the toggles all lean *toward* Lug A. The step indicator LEDs should now all be either dark or very dim. Test each pot, checking to be sure that you can set the brightness of the LED for each step; the light should get brighter as you turn clockwise (turning the tempo way down will probably make this easier). Each pot controls the individual brightness of its indicator LED *and* the brightness of the control LED that they all share. The control LED should shift in brightness to match the brightness of the currently illuminated indicator LED, rather than blinking on and off. Is that happening? If not, then short circuits, especially between those dang LEDs, are probably the cause. And if the LEDs and pots are performing as described? Perfect! Move on to the final test: pause and reset.

The pause button is the one with a lead going to solder pad 10I. Push and hold this button. The current step indicator LED should stay lit, even as the tempo rate LED keeps blinking. Release the button to resume the sequence. Now, press and release the reset button wired to 8I on the PCB. Every time you hit this, the sequence should jump back to LED #1.

When your sequencer circuit passes inspection, you're ready to put together the synth.

Building the Modified Single-Chip Space Invader Synth

Right now you've got a sequencer that doesn't control anything other than a row of blinking lights. That's kinda fun, but something that controls a row of blinking lights *and* makes a rad noise will be much, much more fun. A modified version of the Single-Chip Space Invader Synth (Project 15) will be the Bleepbox's voice. Let's build this now, using the modified parts list from the beginning of this project, the circuit diagram shown in Figure 16-21, and the instructions that follow.

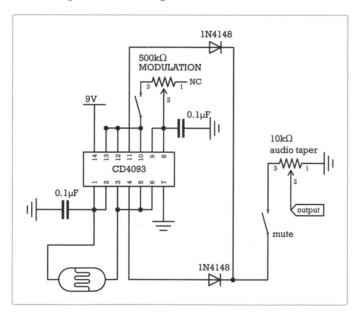

FIGURE 16-21: *The circuit diagram for the modified Single-Chip Space Invader Synth*

Step 23   Flip to Project 15 and, starting with Step 1, build the synth with the following modifications:

a.   Omit the power supply—the switch, 9-volt battery clip, and LED power indicator. Don't build it in Step 1 or install it in Step 13. Instead, cut, strip, and tin two 6-inch lengths of insulated wire. For clarity, I've used red and black wires. Run one of these wires (I've used red) to pad 3 on the synth PCB. Run the other wire (I've used black) to the synth's lower ground rail, at the foot of column 13 (see Figure 16-22). We'll connect these to the Bleepbox sequencer PCB's power and ground rails in a few steps.

b.   Omit the pitch control pot. Skip the 100k ohm pot in Step 2 and ignore all references to it thereafter; it's dead to us.

c.   Add an SPST toggle switch to the mod pot. In Step 2, cut the lead attached to lug 3 of the 500k ohm pot in half, strip and tin both ends, and solder one end to each lug on the SPST toggle switch.

d.   Omit the trigger button. Don't build it in Step 3 and ignore its mention from then on.

e.   Replace the 100k ohm pitch control pot in Step 11 with a photoresistor-based pitch controller. Cut, strip, and tin two 6-inch insulated wires. Solder one wire to each leg of the photoresistor. Solder one of these leads to pad 3A on the Space Invader Synth's PCB and the other to pad 10A.

Your Bleepbox-ready Space Invader Synth is good to go! Figure 16-22 shows the results.

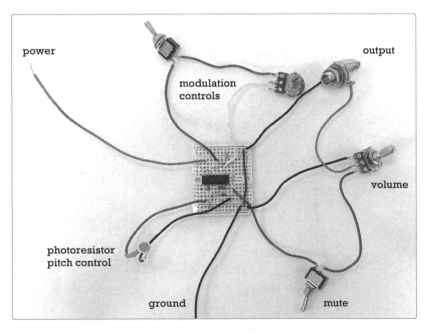

FIGURE 16-22: *The modified Space Invader Synth circuit, ready to be grafted to its new master*

Connecting the Synth

The synth voice will be both powered and controlled by the sequencer circuit. Let's wire them up now.

Step 24   Connect the free end of the insulated wire connected to pad 3G on the Space Invader Synth PCB to the power rail on the sequencer PCB at column 13 (see Figure 16-23). Connect the black ground wire on the synth PCB's lower ground rail at column 13 to the sequencer's upper ground rail at column 13, above the power connection you just soldered. Solder these now. Finally, solder the remaining lead from the click-gate controller to the tip on the Space Invader Synth's output jack—that's the positive lug, not the ground.

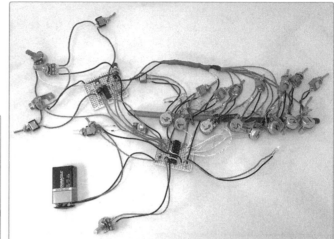

FIGURE 16-23: *A close-up of the optocoupler (left). I sealed mine with black heat-shrink tube instead of tape; both work fine. The finished guts of the entire Bleepbox 8-Step Analog Sequencer (right). I added an extra reinforcing wrap of tape to the optocoupler prior to installing the guts in their enclosure.*

Step 25   To finish the Bleepbox, we need to bind together the common control LED that's wired to all eight step controllers with the photoresistor on the synth, forming an optocoupler. We built this same style of optocoupler for the Universal LFO (Project 13); the three steps are illustrated in Figure 13-13 on page 229. Place the common control LED and the photoresistor snout to snout, as shown in Figure 13-13, and wrap them in duct tape, making sure to seal them completely.

Your Bleepbox is complete and ready to test!

Testing the Sequencer and Synth Together

Set the volume pot, the tempo pot, and all of the pitch pots to halfway. Turn the click-gate depth pot fully clockwise and turn the mod rate pot fully counterclockwise. All of the overdrive mute switches should be off—that is, leaning toward Lug

A and, therefore, not engaged—and the output shouldn't be muted. Set the rhythm selector to 4/4 and the click-gate switch to its middle position so neither gate option is engaged. Connect your 9-volt battery, plug in your amp, and then power up the amp and the Bleepbox.

The tempo rate LED should blink at about 60 bpm, and you should hear a relatively steady tone, varying only slightly from step to step. You should also see each step LED light up in sequence. Don't worry if they're fairly dim; that's normal.

Flick all of the overdrive mute switches on. Now there should be no droning tone, and each LED should light in sequence at maximum brightness. Try the tempo control pot; it should still make both the tempo rate LED and the individual step LEDs blink faster as you go clockwise and then slower as you go counterclockwise. At full clockwise, all of the LEDs will probably appear to be continuously lit. You may also hear a sudden, quiet squeal when you get the tempo control pot to its extreme clockwise position; that's normal. At full counterclockwise, the blink rate may be downright glacial, especially if your 9-volt is wearing down from all this testing.

Return the tempo control pot to halfway and then flick the rhythm selector switch from 4/4 to 3/4. Instead of each of the eight steps lighting in sequence, the Bleepbox should now light only the first six steps before starting over at Step 1. Flip the selector back to 4/4.

Let's test the click-gate function. Flick the switch so it's leaning toward Terminal A, which you marked earlier. You should now hear two clicks per step, corresponding to the light and dark period of the tempo rate LED. This is square-up mode.

When you flick to the other position on the click-gate switch, you should hear two louder clicks in a more distinct alternating "one-two" pattern. This is the square-down mode. Turning the click-gate pot counterclockwise should make these clicks quieter, but you should still hear them even when the pot is fully counterclockwise. Return the click-gate switch to its center position, and you should be back to virtual silence, with maybe just a very distant clicking.

Check the pause and reset buttons. Hit pause to make the sequencer hold on the current step; hit reset to make it hop back to the first step. The tempo rate LED should keep beating steadily throughout this. You might find that these buttons behave a touch erratically, especially if you press them quickly or right on top of the tempo rate LED blinking on or off. This is to be expected.

* **NOTE:** *Erratic button behavior is caused by* contact bounce. *A simple switch contains two springy copper contacts, and when you close the switch, the contacts don't snap together cleanly. Instead, they smack against each other, creating a flurry of electronic pulses before they settle in place. In most situations—like when flicking a light switch—you don't notice contact bounce because the chatter is fast and tiny while the overall voltage of the system is large. But in logic circuits like this, contact bounce can confuse the chip. If you want to get rid of this, you can build a* debounce circuit; *just google* bounceless switch *or* debounce circuit *for examples. I didn't debounce for this project because I like the pseudo-randomness.*

Now, test the modulation oscillator controlled by the mod rate knob and mod switch. Twist the knob clockwise, and you should hear a high-pitched, buzzy square-wave tone that gets lower as you continue clockwise. A slight shift in the tonal color as the tempo rate LED clicks on and off is normal. If you flick the toggle switch, the mod square wave should be muted. Once tested, flick the toggle switch back and twist its knob fully counterclockwise.

Finally, disengage all of the overdrive mutes. You might want to turn the output volume down here. Use the pause button to pause on a few different steps and then change the pitch using that step's pitch-control pot. Turning the pot clockwise should make the pitch higher, while turning counterclockwise should make it lower.

Release the pause button. Turn the mod oscillator's knob clockwise, and you should hear the mod oscillator modifying each step's pitch. Flick the mod switch off, and you should just hear the square-wave pitch of each step. Turn the mod back on, and you're back to a richly textured dual-oscillator voice. Flick the output mute switch, and all should go quiet.

Because this project is so complex, expect some hiccups. If the sequencer was working as of Step 22, then any gremlins you come across now are probably related to the new Single-Chip Space Invader Synth board or to the places that the two boards connect. It's also possible that you inadvertently created a short circuit in the sequencer circuit while wiring it all up.

If you do find problems, they'll almost invariably be simple. Check for dead batteries, bum switches, or short circuits, which are usually LEDs brushing against each other or the body of a pot. Persevere, and you shall overcome.

Enclosure

When you know the Bleepbox guts work, you're ready to package them up. Selecting and customizing an enclosure is totally a matter of personal preference, but in a build with so many wires and so much hardware, it can be bit of a bear. Here are a few tips as well as some process shots of my enclosure:

▶ Ask yourself what looks good, what will be most playable, and what will work best with the circuit. Figure 16-24 shows a cigar box that's about 9 × 7 inches and just over 1/2 inch deep. In my opinion, these are ideal dimensions: the enclosure is portable, sits well on a table or lap, and has plenty of room to house everything without risking shorts and pinched wires. Plus, you can place the controls close enough together that it's easy to play live.

▶ Sketch your button layout and then mock it up using some spare parts or paper cutouts. When you're happy, sketch the layout on the inside of the enclosure before you start drilling (see Figure 16-24). If you drill from the inside, your holes will be cleaner, and the exterior top of the Bleepbox will be unmarred. But there's a downside. When you sketch your layout on the underside of the panel like this, you're forced to pull a Ginger Rogers: you've gotta do everything Fred Astaire does, but backward and in high heels. Make a sketch in advance, double-check everything, and constantly ask yourself, "Am I about to accidentally drill this whole layout as a mirror image or upside down?"

- Space your hardware out. You want parts grouped conveniently, but not so close that they create short circuits or are impossible to install. I placed the step-control hardware 1 inch apart horizontally, with about 1 inch between the pot and LED and 1/2 inch between the LED and switch. Most of the hardware is similarly laid out in inch multiples in rough columns and rows. It's a tidy look, and I recommend it.

- Use the right-sized bit. For the hardware in the parts list, use a 3/8-inch drill bit for the output jack, a 5/16-inch bit for the pushbutton switches and pots, a 1/4-inch bit for the toggle switches, and a 3/16-inch bit for the LEDs. These dimensions should give nice, snug fits without a lot of tedious reaming and filing.

- Be careful with switch orientation! It will get really confusing if some of your overdrive mutes engage when flicked to the left and others engage when flicked to the right. It's also much more intuitive to have the click gate in square-up mode when flipped up and in square-down mode when flipped down. This is all a matter of personal preference, but like twisting clockwise to speed up or increase pitch, it just *feels* right.

- Consider your labeling options. In Figure 16-1, I fully embraced the junkyard aesthetic: I used extra fine point marker on swipes of correction fluid. On the other end of the spectrum, you could use a slick metal enclosure and print the text on self-adhesive labels. Want that 1980s analog synth/drum machine vibe? Search online for the Orbitron and TR-909 fonts; these are free look-alikes inspired by the typefaces used by Roland, Kawai, Korg, Alesis, and their ilk. Whatever you decide, make your life easier and label those knobs and switches.

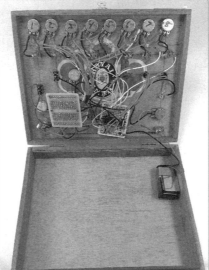
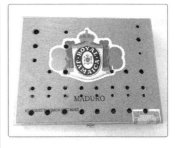

FIGURE 16-24: *A few process shots of the enclosure build*

Playing the Bleepbox

Let's program a sequence together. This'll give us an opportunity to work through all the features of the Bleepbox. Once you've heard them in action, opportunities for improvisation will begin to present themselves.

Your First Sequence

For starters, prepare the synth. Make sure it has a fresh 9-volt battery and is powered down. Then, program it as follows.

Control	Setting	Control	Setting
Volume	75% or higher	**Step 1**	9 o'clock
Output mute	Off (unmuted)	**Step 2**	12 o'clock
Modulation	On, 50%	**Step 3**	9 o'clock
Tempo	75%	**Step 4**	12 o'clock
Rhythm selector	4/4	**Step 5–8**	3 o'clock
Click-gate	Middle position (neither gate option engaged), with pot fully clockwise	**All overdrive mutes**	Off

Set your amp to about 25 percent, plug in the synth, power up the amp, and then power up the synth. The sequencer should play the following programmed sequence: an alternating high and low pitch, followed by a more-or-less continuous lower pitch—something like *be-bop-be-bop-ba-ah-ah-ah*. You'll note that the sequence plays in a *portamento* style, sliding between notes instead of crisply jumping from one to the next. This is a consequence of the photoresistor-based analog pitch controller. The effect gets really neat when it moves from a fairly low tone, such as step 8, to a high one, such as step 1. You'll hear a fast, pronounced slide into the new note.

The overdrive mute takes full advantage of this portamento playing style. Engage the overdrive mutes in steps 5 through 8. When your sequencer hits step 5, the pitch should drop out, muting the last four steps of the sequence, but when step 8 wraps back to step 1, you should get a big squealing slide to the step 1 note.

Experiment with the Click-Gate Control and Mute

To deemphasize this portamento effect and emphasize the rhythmic aspects of the automated sequence, engage the click-gate control in square-up mode. You should hear a click twice per step, corresponding with the off and on blinks of the tempo rate LED.[3] Note that the first half of each step is significantly louder than the second half. If you turn down the click-gate depth knob, you'll decrease the depth of the effect, making the difference in emphasis between the first and second half of each step less obvious. The click should remain present but will get quieter as you turn down the effect depth.

....

3. And thus coming in on the 16th notes, if you think of each step as an 8th note.

Switch to square-down mode, crank the click-gate knob to full, and listen for two changes: the click should be less apparent, and each step should start with a rest, bringing the note in on the second half of each step. This greatly deemphasizes the slide between notes and gives your groove a funky backbeat feel. The square-up gate effect can be fairly subtle, while the square-down gate tends to be more emphatic, with a wider variety of variation as you change the depth.

You'll notice that you still hear the 16th-note clicks even when steps are overdrive muted, with the square-up clicks more emphatic than the square-down clicks. When you engage the output mute, it will silence the square-up clicks, but the square-down clicks will remain audible. This is a really useful trick, as it makes it possible to drop a break into your set—muting all the notes—while keeping the clicks running, maintaining the beat. It's sort of like having a bass-line synth with a very rudimentary built-in drum machine.

✱ **NOTE:** *When the square-down gate is engaged and its depth is set to 100 percent, the synth's voice may* leak *through the output mute. Turn the click-gate depth back to 90 percent or lower to completely mute the synth voice even as the 16th notes click away.*

Cut and Loop with Pause and Reset

Flip the click gate back to square-up mode and unmute the output and all the steps. This sequence program has two distinct parts and is designed to be easily cut and looped using the reset and pause buttons. Hit reset just as step 5 is about to come on. You'll send the sequence back to step 1, looping the *be-bop-be-bop* section of the sequence. Repeatedly doing this will build tension, which you can then release by letting the sequence run all the way through uninterrupted. You can also build tension with the pause button: pause a specific step, let it hold for a few beats, and then release it. Once you're comfortable cutting, looping, and building tension, try stuttering the pause and reset buttons, either individually or together, to create random glitch effects and irreproducible improvisational breakdowns.

Explore the Modulation and Per-Step Pitch Knobs

When you have a feel for your sequencer controls, try tweaking the mod rate knob as the sequence plays or even muting the mod entirely. Explore sounds and textures by adjusting the mod slowly or quickly; zero in on ranges or settings where interesting multitone timbres arise. Tweak a few step pitch knobs. If you do this while a step is active, you can warble or slide the note. If you tweak while the step is inactive, the sequence will seem to slowly evolve. And don't neglect the power of muting all or some of the steps and then unmuting them as the sequence comes back around. Music is all about establishing patterns, violating them, and then restoring them.[4]

---

4. In fact, I'd argue that all art is about this simple "three-act" process of presenting an orderly system, tangling it up, and then giving your audience the tools to restore some order to that tangle—but this argument probably stretches a little beyond the scope of one footnote in a DIY book.

Change the Rhythm

Finally, switch the rhythm selector to 3/4. You'll find that the sequence now loops at step 6, instead of going all the way to step 8. This is most immediately recognizable as waltz time—although at the moment, it doesn't sound that waltzy. Slow the sequence down a touch, set the click gate to square-up mode at 100 percent, and engage the overdrive mutes on the even-numbered steps. Count it out as *one-ee-and-ah-two-ee-and-ah-three-ee-and-ah* and *voilá*! You have "The Space Cowboy Waltz," as arranged for phasers set to stun.

One more nifty trick: your Bleepbox makes a great metronome! Turn off the mod oscillator, turn all eight step controller knobs fully counterclockwise, engage all eight overdrive mutes, mute the output, set the rhythm selector to 4/4, flip the click gate to square-down mode, and set its depth to 100 percent. Turn on the Bleepbox, and you'll hear a steady pulse of 16th notes. You can break this string of 16th notes into recognizable bars by disengaging the output mute and turning off the overdrive mutes on specific steps. For example, try turning on steps 1 and 5 to get two tones separated by eight clicks, or try turning on steps 1, 3, 5, and 7 for four tones separated by clicks. Adjust the tempo to your liking and play along with the beat. You can even cut the audible tones and clicks entirely (by setting the click gate to its middle position and engaging the output mute) and just use the blinking lights to help you keep time.

Tips, Tricks, and Mods

We've loaded the Bleepbox 8-Step Analog Sequencer with every damn feature we could squeeze out of these chips, so most of the mods offered here are omissions.

Strip Away Features

If the click-gate controls don't do anything for you, you can eliminate them with no ill effects; just leave out the paired diodes, SPDT on-off-on switch, and 100k ohm pot. You can similarly skip the pause button, reset button, or rhythm selector without dire consequences. If you omit the rhythm selector, you'll want to run one extra jumper from 8H to either 11D (for a 6-step sequencer) or 14G (for an 8-step sequencer).

Build a 4-Step Bleepbox

You could likewise reduce this to a simpler and cheaper 4-step sequencer by omitting the rhythm selector and half the pitch-control assemblies. Just wire pin 10 on the CD 4017 chip directly to the pin 15 reset to force the sequence to loop after four steps. Figure 16-25 shows the circuit diagram for this minimal 4-Step Bleepbox.

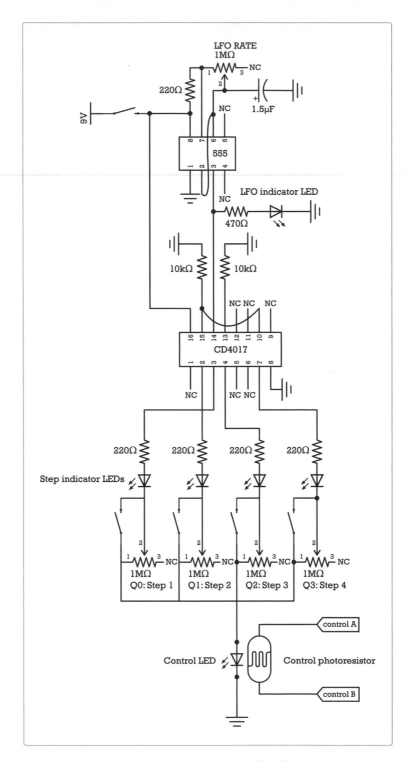

FIGURE 16-25: *The circuit diagram for a 4-Step Bleepbox*

Add Speakers

Because the Bleepbox uses the same sound circuit as the Single-Chip Space Invader Synth, it also outputs a hot enough signal to drive a set of headphones. Just plug 'em in, as shown in Figure 16-26 (left). Or you can outfit the Bleepbox with its own built-in or detachable monitor speaker, as discussed in the Space Invader Synth's "Tips, Tricks, and Mods" on page 281. The same speaker selection rules apply. Figure 16-26 (right) shows a Bleepbox sporting the dandelion speaker described in Project 15.

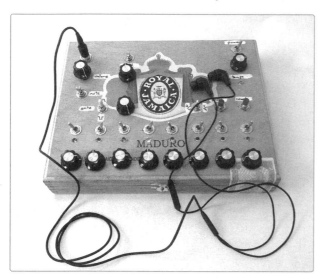 

FIGURE 16-26: *A Bleepbox kitted out with headphones (left) and a small dandelion speaker (right)*

Add More Sequence Lengths

If you really like creating tight 4-step loops, consider swapping the SPDT rhythm selector switch for an SP3T switch, also sometimes written as *1P3T*. This style of switch can select among *three* circuits instead of just two. Such switches are relatively rare and often expensive, but try googling *SP3T rotary* or *SP3T toggle* to see if you get lucky. These switches also occasionally turn up in the usually very low-priced surplus and junk bins at electronics shops. When adding a three-position switch, wire the extra terminal to pin 7 on the CD4017, which will enable 4-step patterns. Label this new setting *4*, the old *3/4* as *6*, and the old *4/4* as *8*. Also, relabel the rhythm selector something like *sequence length*.

Circuit Bending

One final mod is to add an extra pot to bend the overall pitch of the synth. Wire the pot in series with the photoresistor side of the optocoupler that connects the sequencer control to the modified Space Invader Synth board (see Figure 16-27).

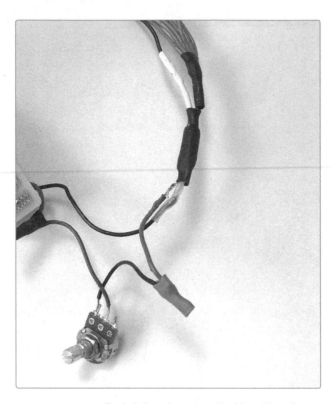

FIGURE 16-27: *A pitch-bend pot installed in a Bleepbox*

Clip one of the wires connecting the photoresistor to the synth—it doesn't matter which. Strip and tin both free ends. Then solder one end to lug 1 of your pot and the other to lug 2. Twisting this new *pitch-bend* pot counterclockwise will lower the pitch of the entire sequence as it plays. This makes it possible to tweak and warble an entire sequence at once without disrupting the relationship between the notes. Wired in this fashion, the pitch-bend pot affects only the main pitch oscillator, not the mod oscillator. One thing to keep in mind: bending the pitch down like this usually cancels out the overdrive mutes because it drags them down into the audible range.

Finally—and above all else—your finished Bleepbox is a sound circuit; it can be circuit bent just like any commercial music machine or noise-toy. Poke it and prod it. Add body contacts, extra pots, or a Universal LFO wherever and however you deem fit. Take your cues from Circuit Bending for Beginners (Project 7) and start exploring and exploiting.

Resources

If you've gotten this far, you're ready to build just about anything thrown your way in books like these:

- *Handmade Electronic Music: The Art of Hardware Hacking* by Nicolas Collins (Routledge, 2009)

- *Electronic Projects for Musicians* (Music Sales America, 1992) and *Do-It-Yourself Projects for Guitarists* (Backbeat Books, 1995) by Craig Anderton

- *Getting Started in Electronics* by Forrest M. Mims III (Master Publishing, 2003), which includes schematics for "100 Electronic Circuits" in the back

- *Timer, Op Amp & Optoelectronic Circuits & Projects* by Forrest M. Mims III (Master Publishing, 2004)

- Anything in the pages of *MAKE* magazine

- Most simple circuits you find online. If you like this book's projects, google *Tim Escobedo circuits* or *Atari Punk Console*[5] and go from there.

You're also ready to begin exploring the world of synth and musical effect kits. Check out these sites:

- *PAiA* (*http://paia.com/*) offers lots of kits—including full analog monosynths and modular synthesizer systems as well as their venerable Theremax Theremin. More importantly, they sell reasonably priced kits for many of the best projects from Anderton's *Electronic Projects for Musicians*. Don't miss some of the awesomely priced little kits in the website's "Building Block Kits" section.

- *Music from Outer Space* (*http://www.musicfromouterspace.com/*) has lots of neat kits; I've been sorely tempted by their Noise Toaster and Echo Rockit kits.

- The music section of the *Maker Shed* (*http://www.makershed.com/collections/music/*) also carries a variety of electronic music kits and is adding more all the time.

Congratulations! You've most definitely earned your old-school electronics geek merit badge. Wear it proudly, let your freak flag fly—and go make your own sound now!

---

5. Fun fact: the popular Atari Punk Console project is actually based on one of Mims's circuits featured in *Timer, Op Amp & Optoelectronic Circuits & Projects*.

ELECTRONIC COMPONENTS, TOOLS, AND SKILLS

Welcome to the appendices! This first appendix will introduce you to all the standard electronic components, the tools you need use them, and the skills it takes to harness their awesome powers.

Guidelines for Sourcing Components

It's almost always cheaper to buy electrical components and tools online, but if you live near a shop serving electrical tinkerers or ham radio enthusiasts, go there first. These shops generally offer excellent deals and recommendations, especially on bigger investments, like a good multimeter or benchtop soldering iron station. Their stock often includes a mind-boggling array of awesome rarities salvaged from estate sales and auctions. As a bonus, these shops are invariably staffed by wise old folks with a bottomless reservoir of tips, tricks, and quick fixes. Sourcing supplies from your local ham shop is a great way to get an idiosyncratic education in shade-tree EECS, plus the occasional story about the early days of DARPA or off-color anecdote about World War II–era South Pacific remote listening posts.

Resistors, transistors, diodes, switches, and the like generally don't vary a lot between manufacturers; as a rule, I buy the cheapest version of whatever fits the parameters.

Jacks are the one exception I make to this rule. Normal operation puts a great deal of strain on jacks, especially the springy metal tongue that presses against the plug's tip. In a cheap jack, this tongue deforms over time, leading to a loose, noisy connection. Buy a high-quality jack, like those made by Switchcraft, and save yourself headaches down the road. Switchcraft models L11, L12A, L12B, and 112AX will be suitable for all of the projects in this book.

Sources for buying online:

▶ **Digi-Key** (*http://www.digikey.com/*) or **Mouser Electronics** (*http://www .mouser.com/*)   I do most of my general component shopping at these sites. Mouser is often slightly cheaper, but Digi-Key offers the cheapest shipping options as of this writing.

▶ **Jameco Electronics** (*http://www.jameco.com/*)   I've never ordered from Jameco, but it comes fondly recommended by friends and colleagues and has an easily navigable website.

▶ **Micro Center Computers & Electronics** (*http://www.microcenter.com/*) This is a good choice if you find the above sites too daunting. The search function is more straightforward, and you can click the Site Map link to navigate to the Hobby Electronics section. Also consider **All Electronics** (*http:// www.allelectronics.com/*).

▶ **Antique Electronic Supply** (*https://www.tubesandmore.com/*)   For music-related supplies and kits, including vintage parts and hard-to-find music-specific components, try this site. It's my preferred source for pickup winding wire; search for *42 AWG 750* to find reasonably priced small spools.

▶ **Small Bear Electronics** (*http://smallbearelec.com/*)   An excellent source for classic knobs, enclosures, stomp switches, and more. Small Bear also has some great DIY articles in its Deep Cave section (*http://diy.smallbearelec.com/*).

Finally, always keep an eye out for weirdo components that might inspire future projects. Check out **American Science & Surplus** (*http://www.sciplus.com/*) as well as garage sales, resale shops, and local freecycle groups.

Components Primer

Many components, especially capacitors and some resistors, have wattage and voltage ratings printed on them. All you need to know is that any 1/4- or 1/2-watt component is fine for these projects, as is anything rated more than 9 volts. Most will be 16 or 35 volts. You should always use components rated higher than the amount of voltage you plan on applying to them.

* **WARNING:** *Because the batteries used in the projects in this book are small and supply relatively low currents, every project featured here is safe. Don't try to rework any of these to run on more than a 9-volt battery, and please don't use the information in this book as permission to start messing with wall current!*

Resistors: Fixed and Variable

Resistors are electrical components that conduct current but impede its forward flow. Resistance is measured in *ohms*, abbreviated with the symbol Ω, with values usually in the hundreds or thousands.

Fixed resistors are little lozenges with wires sticking out of either end (see Figure A-1). They are labeled with three colored bands grouped as a set, then a space, and then a fourth band that's almost always silver or gold.

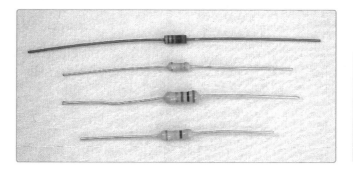

fixed resistor

FIGURE A-1: *Several fixed resistors and the resistor symbol (right) used in circuit diagrams*

If you hold the resistor so that the three-band group is to the left, the first three colored bands give the resistor's *value*, or its resistance in ohms, and the last band tells you its *tolerance*, or how close to that value it really is. Resistors are cheap, so you should always purchase those with a gold fourth band, which corresponds to a tolerance of ±5 percent. That tolerance means the resistor's true value is no more than 5 percent above or below the labeled value; for example, if you use your multimeter to measure the actual resistance of a "100 ohm" resistor with a gold band, you'll find that it measures somewhere between 95 ohms and 105 ohms. Silver is ±10 percent, meaning the resistor's true value could deviate a lot more from the labeled value.

When you're reading a resistor value, the first two bands provide the numerical value, and the third tells you how many 0s to tack onto the end, as shown in Table A-1.

For example, a resistor with brown-black-orange bands has a value of 1 (brown) and 0 (black) followed by three 0s (orange), and is thus 10,000 ohms, or 10k ohms. If you need to find a 470 ohm resistor in your bin, look for the one marked with a yellow band (4), a purple band (7), and a brown band (add one 0).

* **NOTE:** *If this is confusing—and it usually is at first—google* resistor value decoder. *The Internet abounds with Java-based resistor decoders.*

TABLE A-1: Resistor Value Decoder

Color	Value
Black	0
Brown	1
Red	2
Orange	3
Yellow	4
Green	5
Blue	6
Violet	7
Gray	8
White	9

The order of the colors in the chart is the same as the order of the colors in the rainbow (ROY G BV). In other words, the system has a built-in mnemonic: just remember "black-brown-ROY G BV-gray-white," and you'll remember the values from 0 through 9.

Variable resistors can be adjusted to offer different amounts of resistance—usually between 0 ohms and their marked value (see Figure A-2).

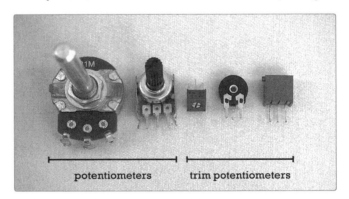
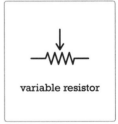

FIGURE A-2: *Variable resistors: two potentiometers and three trim potentiometers (left) and the variable resistor symbol (right)*

The larger variable resistors, which are usually accessible from the outside of the finished project's enclosure, are meant to be adjusted often and are called *potentiometers*, or *pots*. Very small variable resistors that are meant to be adjusted only during the initial building stage are usually called *trim pots*, or *trimmers*, though they're sometimes referred to as *set-and-forget* variable resistors. A variable resistor can have either an *audio* (also known as *logarithmic*) *taper* or a *linear taper*. Unless told otherwise, use a linear taper pot in most applications and an audio taper pot for volume controls.

The Gory Details: Audio Taper vs. Linear Taper

A strip of resistive material inside your pot connects lugs 1 and 3. Lug 2 connects to the *wiper*, a separate conductor that you move by turning the pot's shaft. Current can flow from lug 2 to lug 1 or 3, but it has to pass through that conductive strip first. The resistance across lugs 1 and 3 matches the marked value of the pot, but the resistance between the middle lug and either outer lug depends on the position of the wiper: as you move the wiper, the resistance between either outer lug and the middle lug varies between 0 and the marked value of the pot. That resistive material running between lugs 1 and 3 is *tapered* so that it offers consistently varying resistance as the wiper moves from one end to the other. Most pots have a *linear taper*, so when the wiper is at its halfway point, the resistance is half the marked value of the pot; when it's at its 3/4 point, the resistance is 75 percent; and so on.

This is great for most electrical signals, but lousy for controlling the apparent volume of audio sources because we perceive audio amplitude logarithmically: for a sound to seem twice as loud, it must be exponentially more powerful. To address this problem, some pots have a taper that approximates a logarithmic progression; these are said to have an *audio taper* or—less often— a *logarithmic taper*.

The pot's value is usually printed on its face, with an *A* prefix for audio tapers and either a *B* or no prefix for linear tapers. In this book, a pot's lugs are numbered from left to right when the pot is viewed from the front with the lugs on top, like a crown. This means they're numbered from right to left when viewed from the back, which is the view you'll often have when you're soldering or troubleshooting (see Figure A-3).

FIGURE A-3: *Variable resistors, viewed from the front (left) and back (right), with lugs labeled*

The Gory Details: Selecting the Right Photoresistor

Photoresistors are variable resistors that change resistance according to the level of light that falls on them. Most commonly available photoresistors offer near-zero resistance when exposed to bright light and then shoot up to around 1.2M ohms when plunged into full darkness. This allows for all sorts of neat effects—machines that "magically" start as you approach, synths that change their tune as the sun sets, automatic nightlights—making them a favorite of DIYers. But despite their hobby popularity, photoresistors are still mildly infuriating from an engineering standpoint: every photoresistor is unique and often a touch finicky. They're wickedly inconsistent from manufacturer to manufacturer—even batch to batch—and relatively slow to change states (especially by modern microprocessor standards, where lag times are measured in milliseconds). Most vexing for modern applications, a photoresistor doesn't jump from 0 to 1M ohms, no matter how quickly the nightlight seems to shut off when you turn on the lights. That photoresistor always moves through every intervening value between 0 and 1. We live in a digital world, but these little guys are stubbornly analog.

You can directly observe this analog transition with your multimeter. Set the meter to the 2M ohm range, clip one probe to each photoresistor lead, shine a bright light on the photoresistor, and watch the resistance drop. Now, turn off the light and quickly cover the photoresistor so it's in full darkness. The resistance will climb over the course of a second or so.

This totally analog response is awful for digital applications, which generally want something to be crisply on or off, without a whole lot of wiggle in between. But it's tons of fun for music. For example, the relatively slow relaxation of a photoresistor gives us smooth legato slides between notes on a simple synth and those zappy Bleepbox glissandos as a sequence moves from an overdrive-muted step to an active one. That said, many photoresistors are too slow even for our loosey-goosey art-house applications. For example, if the photoresistor in your Bleepbox relaxes too slowly, it doesn't have time to move between high and low pitches before the sequence advances, drastically limiting your effective range. An especially slow photoresistor can similarly limit the responsiveness (and thus usefulness) of a Universal LFO.

For the projects in this book, you'll likely be happiest with a photoresistor that goes to nearly 0 ohms in full light, exceeds 1M ohms in full dark, and slides between those extremes in a second or less. Most common photoresistors should work, but you might want to check a few with your multimeter in order to get a sense of the range of your batch. Larger photoresistors tend to have slower response and less range, while smaller ones have a broader range and quicker response. Photoresistors that measure 3 mm to 5 mm along their widest axis usually go all the way to 0 ohms and as high as 1.2M ohms or more with a pretty sprightly response time. They work fine for these projects, although I prefer photoresistors around 6 mm wide. These tend to be slightly slower, enabling, for example, a slightly more pronounced Bleepbox glissando. Anything larger is too sludgy to be fun. If you can, buy a bag of assorted photoresistors and experiment to find one that pleases your aural palette. Search online for *photoresistor assortment*, and you'll find plenty of sellers. Expect to pay about $9 per 100 photoresistors.

Cutting Down Potentiometer Shafts and Snapping Off Anchors

Occasionally, you'll find a pot shaft too long for your final project enclosure. To deal with this, just measure the depth of the knob cover you want to use—usually around 1/4 inch—mark this on the pot's shaft, clamp the shaft in a vice, and cut it down with a hacksaw. If you're worried about getting grit in the pot—which isn't a huge problem, as most modern pots are fully enclosed—wrap the body of the pot in masking tape before sawing.

Pots usually have small anchors protruding from their bodies. You can either drill an extra hole in your enclosure to accommodate the anchor or snap it off with a pair of pliers (see Figure A-4).

FIGURE A-4: *Snapping off the anchor. Quick quiz: what's the maximum resistance this pot offers, and does it have a linear or audio taper?[1]*

✳ **NOTE:** *Resistors—fixed, variable, and otherwise—are basically indestructible; you don't need to treat them with TLC as you start soldering.*

---

1. This is a 1M ohm variable resistor with a linear taper. The *1M* printed in white tells you the pot's value, and the *B* means it has a linear taper.

The Gory Details:
Voltage, Current, Resistance, and Ohm's Law

In 1827, Georg Ohm described the relationship among voltage, current, and resistance in an electrical circuit. This was codified in *Ohm's law*, often formulated like so:

$$I = \frac{V}{R}$$

This equation can be stated as "current equals voltage divided by resistance." What does that mean? Well, imagine filling your bathtub. Your building's water supply is the *voltage* (V), measured in *volts*. The faucet's valve is a variable resistor (R), measured in *ohms*. The water flowing out of the tap is the *current* (I), measured in *amperes* and often shortened to *amps*.[*]

Assume the building's water supply is stable. When the valve is wide open, it offers very little *resistance*, so the flow is high—which is to say there's a strong *current*. Increase the resistance by closing the valve, and the current flowing from the tap decreases to a trickle. Close the valve far enough—in other words, increase the resistance as much as you can—and no current flows at all.

Plug some numbers into the equation, and you'll see that this description holds in the formula: if V is fixed and R goes down, I goes up. If you decrease the voltage (the building's water supply) while leaving the resistance the same (don't touch the tap), then you know less water is going to flow out of the faucet. In other words, if R is fixed, then as V decreases, so does I.

Here's where the tried-and-true "electricity is like water" analogy breaks down: a spike in water pressure isn't a big deal; you might get sprayed in the face or slop water on the floor. But if you overload even simple electrical circuits like the ones in this book, they'll overheat. An LED overloaded by a 9-volt battery will fail, melting and sparking in the process; it can easily damage adjacent circuitry or burn you. This is why it's important to have an appropriate level of resistance, which is the thing hobbyists like us most often use Ohm's law to figure out.

....
* Because an ampere is a very large amount of current—most houses run on 10 amp supplies, and less than 1 amp can stop the heart—you'll usually see current measured in *milliamperes (mA)*, which are 1/1000 of an amp.

Finding the Right Resistance

Using Ohm's law, we can calculate the best resistor value for a given circuit. Consider the simple indicator LED circuit used in many of the projects in this book (see Figure A-5).

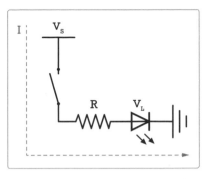

FIGURE A-5: *A simple LED circuit*

To model this circuit, you need to rearrange Ohm's law[2] so that we can focus on resistance instead of voltage:

$$\frac{V}{I} = R$$

Next, break the voltage into two components:

$$\frac{(V_S - V_L)}{I} = R$$

V_S is the battery voltage, and V_L is the forward voltage the LED requires to light up. For red LEDs, this is usually about 2 volts, but check your LED's data sheet to be sure. The current in this circuit is determined by how many milliamps the LED draws, usually around 20 mA or 0.02 amps. (Again, check the data sheet for your LEDs to be sure.) Plug these in on the left side of the equation, and your ideal resistor value will pop out on the right:

$$\frac{9 \text{ volts} - 2 \text{ volts}}{0.02 \text{ A}} = 350 \text{ ohms}$$

You want to buffer the LED with a 350 ohm resistor or larger. While 350 ohm resistors are rare, 470 ohm resistors are ultra-common, which is why I've buffered my 9-volt-driven power-indicator LEDs with 470 ohm resistors throughout this book.

What if you'd rather use a blue or white LED as the indicator for a project? Those usually have a forward voltage near 4 volts. Plug in the numbers, and you find the following:

$$\frac{9 \text{ volts} - 4 \text{ volts}}{0.02 \text{ A}} = 250 \text{ ohms}$$

It would be safe to swap our red LED for blue or white because our 470 ohm resistor is higher than 250 ohms. But the downside is that you have almost twice as much resistance as you need, which might dim that dazzling blue LED more than you like. A value closer to 250 ohms would give a brighter blue LED. While 250 ohm resistors aren't commonly manufactured, 240 ohm and 270 ohm resistors are. Go with

---

2. If this algebraic hocus-pocus is mystifying, google *Ohm's Law transformation* to check out some charts showing the various algebraic permutations of this formula.

270 ohms. Although 240 ohms is within a ±5 percent tolerance of 250 ohms—and so, statistically, many "240 ohm" resistors *are really* 250 ohm resistors—always allow for a margin of error: fresh 9-volt batteries can measure up to 9.8 volts; I've seen rechargeables go over 10 volts. And you could wind up with one of the "240 ohm" resistors that's really a 228 ohm resistor. Play it safe, and use a 270 ohm (red-violet-brown stripes) or higher resistor to buffer your white or blue indicator LED.

What if you need a specific resistor value you don't have? For example, you thought you'd want to use blue LEDs—and so stocked up on 270 ohm resistors—but then changed your mind and decided you'd rather use reds. Never fear: if you need a specific resistor value that isn't in your jackstraw-box-o'-tangled-resistors, you can do a little *resistor math* (see "Resistor Math: Series vs. Parallel" on page 92) to fake your way to any value you need.

Capacitors

Capacitors, or *caps*, are composed of two wires separated by a thin insulating layer called a *dielectric* (see Figure A-6). When voltage is applied to a cap, it stores energy like a water tower: pump the water in, and it stays put until you open the valve; give that water a path out of the tank, and it forcefully pours back out. *Capacitance*, or a component's ability to hold this charge, is measured in *farads*, abbreviated *F*. A farad is a huge amount of capacitance, so most components are labeled in terms of *picofarads* (*pF*), which are smaller than *nanofarads* (*nF*), which are smaller than *microfarads* (*μF*). Most hobby-caliber circuit diagrams use micro-farads. The smallest fixed caps—like the ones in this book—are often marked in a frustratingly opaque code, so see Table A-2 for a cheat sheet. For clarity's sake, this book always works in microfarads.

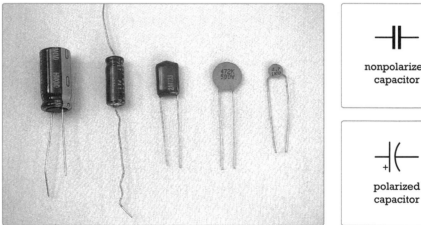

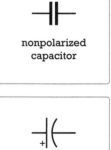

nonpolarized
capacitor

polarized
capacitor

FIGURE A-6: *Some capacitors—electrolytic, or polarized, caps (two on the left); Mylar-film cap (middle); ceramic-disc caps (two on the right)—and their symbols*

TABLE A-2: Commonly Used Ceramic/Mylar
Capacitor Markings and Their Values

Marking	Value (in µF)
102	0.001
202	0.002
472	0.0047
103	0.01
104	0.1
105	1
205	2

Most small caps are little beige ceramic discs, sometimes called *ceramic capacitors*. Occasionally, they're Chiclet-shaped plastic lozenges called *Mylar capacitors* because the dielectric is a thin Mylar film. The larger caps are barrel-shaped *electrolytic capacitors*. Ceramic and Mylar caps are *nonpolarized*, meaning that it doesn't matter which way which leg goes, while electrolytic caps are *polarized* and must be oriented properly in circuits. The negative leg of an electrolytic cap is always marked with a thick stripe, minus signs, or both. Added bonus: electrolytic caps are large enough to be labeled with their value, rather than relying on the terrible code. Most caps under 1 µF are ceramic or Mylar, and most over 2 µF are electrolytic.

Just as you can do "resistor math" to create a specific resistor that you don't have handy, you can do "capacitor math" to get a custom capacitance. If you want to add capacitances—that is, for example, use two 10 µF caps to give you 20 µF of capacitance—then wire the two caps in parallel, as demonstrated in Figure 13-22 on page 239.

Alternatively, if you need a lower capacitance, you can use several caps in series. When you wire caps in series, add the inverse of their capacitances to get the inverse of the total capacitance, which sounds confusing but is exactly the same as the formula you use to calculate total resistance when wiring resistors in parallel (flip to "Resistor Math: Series vs. Parallel" on page 92 for all the excruciating details). In the case of our two 10 µF caps, wire them in series and you end up with the following:

$$\frac{1}{10} + \frac{1}{10} = \frac{2}{10}$$

The resulting fraction reduces to 1/5, and you invert that to get your final capacitance: 5 µF. This is handy if you have a circuit that calls for a 5 µF cap but you have only 10 µF caps (once you start buying in bulk, this sort of thing happens more often than you'd expect).

Like resistors, caps are hardy components. Although it's theoretically possible to apply so much current or heat to an electrolytic cap that it pops, I've never known anyone who did so accidentally.

Diodes, LEDs, and Transistors

Diodes and *light-emitting diodes (LEDs)* allow voltage of only a certain level or greater to pass. That is, the voltage must exceed the diode's *forward voltage*. Diodes and LEDs will also let current travel in only one direction (see Figure A-7). In other words, they have *polarity*, and like all polarized components, they must be mounted in the proper direction to work. The stripe on a physical diode is closest to the diode's negative leg, or its *cathode*, and it corresponds to the crossbar at the tip of the triangle on the diode symbol. On an LED, the negative leg is the shorter leg that's on the side closest to the flat area on the LED lens's edge.

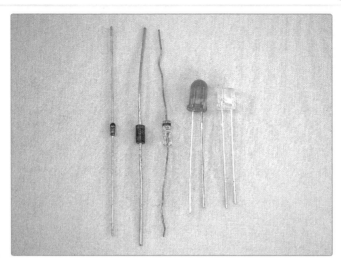
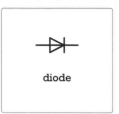

diode

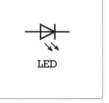

LED

FIGURE A-7: *Diodes and LEDs (left), and their symbols (right)*

Thus far, we've discussed only *passive components*, which obstruct or modulate the flow of electricity, decreasing or slowing a signal. *Transistors* and *integrated circuits* are *active components*—they can be used to construct circuits that produce a *power gain* and increase the level of a signal. They can do this because they're specially configured *semiconductors*. Simply put, *conductors*, like a piece of wire or a resistor, allow current to pass through them; *insulators*, such as the plastic coating on a 24-gauge wire, stop all flow of current; and *semiconductors* pass current in special ways.

Transistors are diodes' fancy big brothers. Like diodes, they let current flow only if it exceeds the transistor's forward voltage and only in one direction. The three-legged transistor works like a valve or switch: a low-current signal at the *base* controls a higher current that flows from the *collector* to the *emitter*. There are many kinds of transistors, but because the projects in this book use only common *negative-positive-negative (NPN)* transistors, that's shown in Figure A-8.

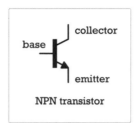

NPN transistor

FIGURE A-8: *An NPN transistor and its symbol*

All semiconductors are sensitive to heat and static electricity. If you're concerned about damaging them, use a *heat sink* when soldering them (see "Helpful Additions to the Standard Soldering Kit" on page 343).

Integrated Circuits

Integrated circuits (ICs) are compact packages of semiconductors and passive components, prewired to perform a specific task (see Figure A-9). Anything an IC can do, a fistful of transistors, diodes, resistors, and caps—plus a few yards of bus wire and some circuit boards—can also accomplish. Because ICs rely on semiconductors, they're also sensitive to heat and static, so handle them carefully. Avoid directly soldering to them and store them in the antistatic bags or tubes they're shipped in.

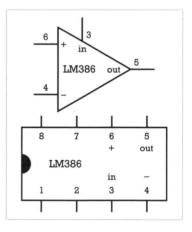

FIGURE A-9: *Two DIP-style ICs (left) and symbols for the lower chip (right), which is the LM386 op-amp used in the Droid Voicebox (Project 6)*

An IC in a circuit diagram is often depicted as a triangle or as a series of triangles for each of its logic sections. The numbers labeling the connections to the triangle correspond to the numbers of the IC pins. For clarity, I've used little drawings of the ICs in circuit diagrams throughout this book (a convention popular among hobbyists).

When ordering ICs from a catalog or website, always be sure to get the *dual in-line package (DIP)* version, like those in Figure A-9, as opposed to the *surface-mount* version. DIPs look like little rectangular insects standing on 8, 14, 16, or more straight little legs. They are small but still relatively easy to work with. Surface-mount chips are tiny little flat midges with short legs. They're usually mounted on circuit boards by machines—you'll find them on your computer's motherboard—and you'll go insane trying to learn to solder on them.

Wire

Electrical wire is described in terms of *gauge*, *insulation*, and *core*. The gauge of a wire is its thickness. Thinner wire has a higher gauge than thicker wire: 42-gauge winding wire used in commercial guitar pickups is as thin as a human hair, while 18-gauge copper wire is nearly as thick as a plastic cocktail straw. Conducting wire is labeled using the American Wire Gauge (AWG) Standard—for example, guitar pickup winding wire is labeled *42 AWG*. Roughly speaking, when a wire's diameter is doubled, the AWG decreases by 6, so 12 AWG wire is actually twice as thick as 18 AWG wire. This is why you can't wind a very good pickup from common 20 AWG magnet wire: it's actually four times thicker than real pickup wire and is thus significantly more difficult to induce a current through.

Wire can be either *insulated* or *bare* (uninsulated). Thicker insulated wire has flexible plastic insulation, but finer insulated wire—say around 30AWG and above—is usually *enameled* (coated in a shiny polyurethane varnish).

Finally, a wire is either *solid core* or *stranded*. Solid core wire is composed of a single solid strand, while stranded wire comprises several strands whose total combined cross-sectional area corresponds to the wire's AWG. Solid core insulated wire is often called *bell wire* (especially in older DIY books), and stranded insulated wire is called *hookup wire*.

Speaker wire is a very common variety of 22-gauge stranded wire with thicker vinyl insulation. It's used to connect hi-fi stereo components. You can find big coils of it for cheap in lots of odd places, like garage sales, thrift shops, big-box electronics stores, and so on. Speaker wire can often replace stranded wire in projects like those in this book.

Solid core wire is a little easier to solder into printed circuit boards (PCBs) because it fits nicely through the holes, but it can break easily with repeated bending when, for example, you open and close a case while checking a circuit, changing batteries, or making other adjustments.

Stranded wire, on the other hand, can be fiddly to thread through the holes on some circuit boards, but it holds up better to stress. Use solid core wire for connections that won't move much, such as point-to-point connections on a board or when running between pieces of hardware mounted on the same surface. Use stranded wire if the wires need to flex a lot, like connections to a battery or to hardware running from a circuit mounted on the side of the box to hardware mounted on its top.

* NOTE: *Bare 22- or 24-AWG solid core wire is often called* **bus wire** *and is great for running grounds, repairing broken traces on prefabricated PCBs, making jumpers, replacing lost eyeglass screws, and so on. It's the all-purpose bailing twine of hobby electronics!*

Quarter-Inch Phone Plugs and Jacks

The 1/4-inch jacks you probably know as *guitar jacks* were first developed for manual telephone switchboards and are thus often still called *phone jacks* by hobbyists. (Be careful not to confuse these with RCA plugs and jacks, which are called *phono* jacks because they were the old standard on hi-fi stereo phonograph systems.) Mono 1/4-inch phone jacks are also called *TS connectors* because they have two lugs: one carries the audio signal and makes a circuit with the *tip* of a plug, and another connects the *sleeve* of the plug to the ground (see Figure A-10).

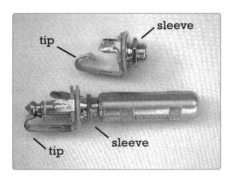 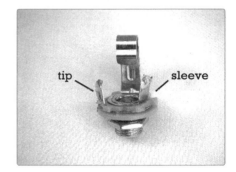

FIGURE A-10: *Left: two 1/4-inch mono jacks—one with a plug inserted. Notice how the jack's bent metal tongue fits into the notch at the plug's tip. Right: a jack with labeled lugs. The right lug is connected to the sleeve, which connects to the circuit's ground, and the other lug connects to the tip.*

Stereo phone jacks and plugs are sometimes called *TRS connectors*. In this configuration, the *sleeve* still connects to ground, but the *tip* and *ring* each connect to an independent audio channel, allowing for stereo sound through a single cable (see Figure A-11). On a conventional cable, the tip carries the left stereo channel, and the ring carries the right.

FIGURE A-11: *A 1/4-inch stereo jack and plug*

Switches

Switches are generally categorized by the number of circuits they can complete (how many *poles* they have), the number of positions they can switch between (their *throws*), and the mechanism for this switching (toggle, pushbutton, rotary, pull-chain, and so on). Your bedroom light switch, for example, is almost certainly a *single-pole single-throw (SPST)* toggle switch, where a single circuit is completed by a single throw of the switch. Other common toggle-switch configurations include the following:

▶ **Single-pole dual-throw (SPDT)**   Throw the switch one way to complete (or *close*) one circuit, and throw it the other way to turn off (or *open*) the first circuit and close a second circuit.

▶ **Dual-pole single-throw (DPST)**   Throw the switch to close two separate circuits simultaneously.

▶ **Dual-pole dual-throw (DPDT)**   Throw the switch one way to turn on two circuits and the other way to turn off the first two and turn on two more circuits.

Figure A-12 shows all four common varieties of toggle switch.

FIGURE A-12: *From left to right: an SPST, SPDT, DPST, and DPDT switch. On the second and fourth switches, the common terminals—those shared by the circuits hooked to each pole—are in the middle.*

Dual-throw switches come in two styles: some flip directly between two on states—one circuit is always on, but never both—and others allow you to park the switch in the middle, with both circuits off. Switches of the first type are often called *on-on* or *changeover* switches; those of the second type are *center-off* or *on-off-on* switches.

As for pushbutton switches, anything beyond a single throw is fairly rare. The big distinction here is between switches that are *normally off* so nothing happens until you press the button, like a doorbell, and those that are *normally on*, letting electricity through until you depress the button, like the switch that controls your refrigerator light. Switches that are off until pushed are also called *normally open*, which is abbreviated as *NO*. Switches that are on until pushed are also called *normally closed*, abbreviated as *NC*.

Batteries, Clips, and Holders

Figure A-13 shows a variety of battery clips and holders. Make a practice of buying a bunch of these when you find them cheap so that you have plenty on hand when you need them.

 ✳ **WARNING:** *Never solder directly to a battery! Heating a battery can cause it to burst, releasing a mélange of chemicals that'll burn exposed skin and that, depending on the type of battery, can catch fire when they contact air or moisture!*

Invest in some rechargeable batteries. Newer nickel-metal hydride (NiMH) cells might seem pricey, but they're worth it. They're less toxic than the old nickel-cadmium rechargeable batteries, can be recharged hundreds of times, and actually outperform regular alkaline batteries in high-drain applications, such as ultra-bright flashlights and digital cameras.

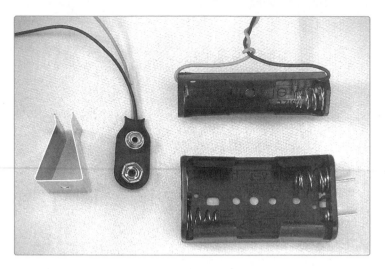

FIGURE A-13: *Assorted battery holders and clips. The piece of metal at the far left is a 9-volt battery clip holder that keeps the battery from rattling around the enclosure.*

Tools

The only tools you'll absolutely need to build the electronics projects in this book are those included in the standard soldering kit, and even some of those can be faked.

The Standard Soldering Kit

Get a 25-watt soldering iron with a chisel or pencil tip, as shown in Figure A-14. You can work on electrical projects with irons of 15 to 40 watts, but a 25-watt iron is probably the handiest. Cooler irons are great for soldering little components but will be a royal pain for soldering hardware, and hotter irons are great for quickly soldering plugs and jacks but can ruin ICs, LEDs, and diodes in a flash.

In terms of value, I recommend a Weller soldering iron. This brand is often sold at hardware stores, and a $25 Weller—the entry-level model—will vastly outperform any $10 generic counterpart. Weller irons typically last longer and heat more regularly, and the replacement tips are cheap and easy to find. You should also acquire a soldering iron stand; otherwise, you *will* burn yourself. In a pinch, you can lay the hot iron across a china plate or tile, but keep that up and you *will* eventually absentmindedly pick up the 700°F barrel instead of the insulated handle. (I vividly recall the hot July night that I seared my index finger and thumbprints into the chrome of my first soldering iron. Fortunately, my hands were so sweaty that I dropped the iron before all the moisture had seared off my fingers. I bought a stand the next day, but my prints were etched into that iron a decade later when I finally invested in a decent Weller.)

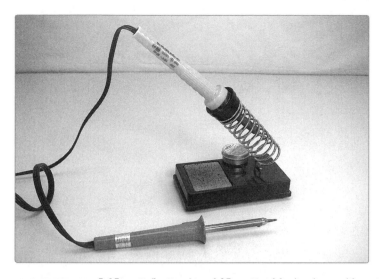

FIGURE A-14: *A 15-watt (bottom) and 25-watt soldering iron with a stand (top). The damp sponge in the stand is used for cleaning the tip, although many tinkerers prefer to use a copper scrubbing pad or old bolt instead.*

Use thin *rosin-core solder*.[3] Anything that's specified for electrical work will fit the bill. It's actually difficult to get the wrong kind of solder nowadays, as long as you avoid the acid-core solder sold at hardware stores. That stuff is used for working with copper plumbing, and it'll ultimately corrode your electronics projects. Lead-free solder is a good idea, but even old leaded solder doesn't pose much of a health threat, as long as you aren't chewing on it. Soldering irons don't get hot enough to vaporize lead. As a matter of good habit, always wash your hands (front and back, with soap and water) after soldering and before eating. (In general, you should always wash your hands—front and back, with soap and water—prior to eating. C'mon, people!)

The diagonal cutters, wire strippers, needle-nose pliers, and clamps in Figure A-15 are also vital when soldering.

Diagonal cutters are fine snips used to cut the legs off of components after you've soldered them. Don't confuse diagonal cutters with the heavy-duty wire snippers sold in hardware stores—these are a different sort of diagonal cutter, often called *dikes*. Dikes are too beefy to get in close to a circuit board. On the other hand, fine diagonal cutters are too delicate to be used on thicker-gauge wires, such as guitar strings, which will notch and ruin them. In a pinch, you can often use a pair of nail clippers in place of diagonal cutters.

Wire strippers are notched snips used for cutting through the plastic insulation of electrical wires while leaving the core intact. You can substitute a small pair of scissors, provided you're very careful not to nick the wire's core. Wire strippers are a very good investment.

----

3. Some Catholic hobbyists have noted that the smell of melting solder always reminds them of Midnight Mass. The rosin used in solder is a pine product that's similar to the resins used in making the incense used during Catholic and Greek Orthodox services. In solder, the rosin serves to clean the contacts while you work, thus helping to ensure a good joint.

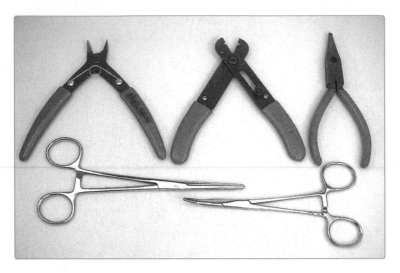

FIGURE A-15: *Clockwise from top left: diagonal cutters, wire strippers, needle-nose pliers, and two types of clamps*

Needle-nose pliers are great for bending wires and component legs, seating components into hard-to-reach places, and loosening stubborn nuts, plugs, jumpers, and the like. Any set of small pliers will work well, and any cheap pair of needle-nose pliers will probably fit the bill.

The tools at the bottom of Figure A-15 are *locking clamps* of the same sort used by doctors, who call them *forceps*, *hemos*, or *hemostatic clamps*. They come curved or straight in a variety of sizes and are really handy for holding components while soldering. You can find them in many hardware stores, as well as lots of hobby and sports shops (folks also use them when building models or tying fishing flies).

Although a *desoldering tool* is not mandatory, if you expect to be working on more than a project or two, you should invest in one. Desoldering tools come in a variety of forms, including *desoldering braids*—also called *solder wicks*—and comical squeezy bulbs, but the spring-loaded *desoldering vacuum* is by far my favorite (see Figure A-16). You'll use your desoldering tool infrequently, but the moment you realize you've soldered a transistor backward, you'll be very glad you bought one. (For more, see "Desoldering" on page 349.)

Finally, buy a roll of *electrical tape*. This black, rubberized tape can be used to secure wires, cover their soldered joints, and insulate bare solder points from each other. Insulation becomes especially important as you cram your projects into enclosures. In general, any enclosure that seems to be the perfect size when you're planning a project will turn out to be a hair too small.

A tidy but trickier alternative to electrical tape is *heat-shrink tubing*. This is nylon or polyolefin tubing that you cut to length and slide over a wire prior to soldering. Once the joint is complete, slide the heat shrink up over the bare connection. Then heat the tubing by gently rubbing the barrel of your soldering iron along its length until it snugly hugs the connection.

FIGURE A-16: *A desoldering vacuum, or "solder sucker"*

Heat shrink comes in a variety of diameters and can be used for more than just insulating connections: it's handy for bundling runs of wire, tidily packaging small projects, and sealing individual components together, as I did when building the optocoupler for the Bleepbox 8-Step Analog Sequencer (Project 16), which you can check out in Figure 16-23 on page 311. While electrical tape often gets gummy or falls off as it ages, especially if it's exposed to heat or wildly fluctuating humidity, heat shrink makes for a rugged covering that basically lasts forever. You can see heat-shrink tubing in action in the Slinkiphone (Project 1); check out Figure 1-8 on page 10.

Helpful Additions to the Standard Soldering Kit

A *heat sink*, as shown in Figure A-17 (left), is clipped into the leads of a heat-sensitive component between the solder point and component body to prevent the component from getting overheated during the soldering process. (An alligator clip or a spare set of clamps will do as stand-ins.) You probably won't need a heat sink very often in the course of normal soldering, but it comes in handy during desoldering, which tends to go slower and take more heat than soldering, or when you're working with an expensive component you're worried about frying.

Although you can hold all your soldering with clamps, clothespins, your elbows, and stacks of books, most hobbyists eventually buy a *helping hands* jig, as shown in Figure A-17 (right). These are also used by model builders and fly fishers, so they might pop up for cheap at garage sales. They come with and without the integrated magnifying glass. I've never had to use the built-in magnifier, so I don't consider it mandatory. But, as I get older I do increasingly find myself reaching for a small 10x magnifier of the sort used by field biologists (google *doublet 10x pocket loupe* for examples; you can get a great one for less than $15). They're very handy when trying to read the labels on ceramic caps, diodes, small ICs, and so on.

FIGURE A-17: *A heat sink (left) and helping hands (right)*

Figure A-18 shows several common hand tools that make frequent appearances on the electronics tinkerer's workbench. *Jeweler's screwdrivers*[4] are great for tiny screws or for prying ICs out of their sockets. The round side of a *tapered half-round file* is great for clearing burrs from holes drilled in metal cases, and the flat side takes the rough edges off of trimmed pot shafts. Finally, you can use a small *hacksaw* to cut down those pot shafts.

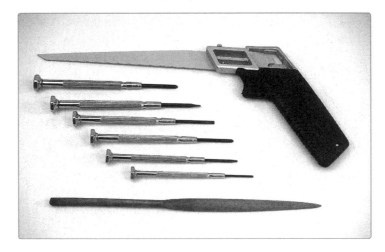

FIGURE A-18: *A hacksaw (top), jeweler's screwdrivers (middle), and a tapered half-round file (bottom)*

---

4. If electronics tinkering is really turning you on, invest in a multibit Torx driver, like the Husky 8-in-1 Precision Screwdriver. Torx-headed screws are sometimes called *star-driver screws*, *star keys*, or *those stupid things!* and are increasingly popular on small consumer electronics that the manufacturer would prefer you just throw away and replace, rather than attempt to fix or monkey with yourself.

Once you've decided you like creating electronic gadgets, go out and buy a *digital multimeter*. Figure A-19 shows one with an LCD display. Do not get an analog meter with the old-timey needle-and-dial gauge. They're far more confusing to use and no longer cheaper than digital meters, which start at around $20.

Even the cheapest multimeters will accurately and precisely measure DC voltage, resistance, and current. Slightly nicer meters have special settings for testing continuity, which is handy, as well as forward voltage on diodes and LEDs, which only comes up occasionally but is a big time-saver. These nicer meters also can be used on AC circuits, which is useful if you need to fix something around the house or install a light fixture. For the projects in this book, a cheap multimeter will do fine.

Most multimeters come with a normal set of *probes* that look like thin knitting needles with bulky insulated handles and serve most purposes as you poke around testing batteries and troubleshooting circuits (see Figure A-20). Occasionally, you'll need a pair of *clip leads*, which will free up your hands as you run tests—again, you won't need them often, but you'll be glad to have them when you do.

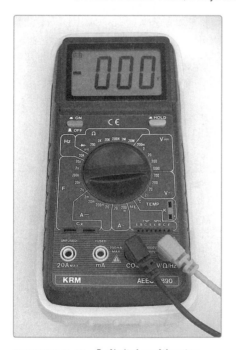

FIGURE A-19: *A digital multimeter*

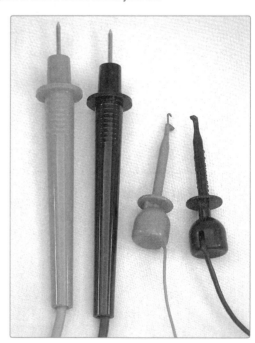

FIGURE A-20: *A set of probes (left) and mini-clip leads (right). The latter are retractable hook leads, also called* mini-clip jumpers, *and are easy to clip to the leg of a single PCB-mounted component without creating a short circuit.*

Also handy for troubleshooting or mocking up circuits are several sets of *insulated test leads*. These are also called *jumpers*, *clip jumpers*, *alligator jumpers*, and *clip leads*, and they look a lot like miniature automotive jumper cables. You can buy these ready-made—for example, Mouser part #290-1942-ND is a set of 10 color-coded insulated test leads at a great price. Or you can buy a bunch of individual alligator clips—such as Digi-Key part #290-1942-ND or Mouser part #835-501784—and get some practice soldering by making your own.

Skills

The keystone skills for an electronics tinkerer are soldering and its sinister twin, desoldering. This section will also discuss using a multimeter, building circuits, and troubleshooting projects.

Soldering

If your soldering iron has a brand-new tip, *tin* the tip before using it: plug in the iron, and when the tip is nice and hot, melt some fresh solder onto it to coat the tip. This will prevent corrosion, which will otherwise eat away the tip pretty quickly.[5] Then, you're ready to start! Figure A-21 shows well-soldered and badly soldered joints.

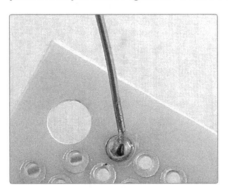 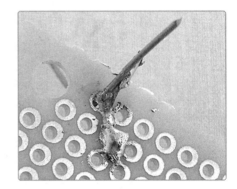

FIGURE A-21: *Good (left) and absurdly bad (right) solder joints*

A good solder joint is smooth and shiny, with enough solder to cover the connection, but not so much that the joint is encased in a round bead or droplet of solder. If you are soldering components to a circuit board, the joint will end up looking like a little volcano. A bad solder joint is dull, lumpy, or clumpy. Good soldering requires four steps, illustrated in Figures A-22 and A-23. Before you start, preheat your iron. Then, do the following:

Step 1 **Make a solid mechanical connection** between the components to be soldered.

Step 2 **Wipe the iron's tip clean** of excess solder using a damp sponge or some scrap metal.

---

5. Re-tinning the tip at the end of each soldering session will prolong its life, but it'll still wear out eventually. Once a tip gets rough, pitted, or lumpy, it should be replaced.

Step 3   **Heat the joint** for a few seconds and then touch some solder to the joint. The heated joint is what melts the solder, not the iron. Solder will flow naturally over and into a properly heated joint, like quicksilver. It's very pretty. Don't apply too much solder—once the joint is coated, you don't need to add more—and don't touch the solder directly to the hot tip of the soldering iron. Forcing hot solder onto cold joints will always make a bad solder joint.

Step 4   **Let the joint cool** and then snip off the excess wire. Leave the joint alone until it has cooled: don't wiggle it to see whether it's ready, and don't blow on it to speed up the process.

Those are the basics! Now let's look at the two most common soldering scenarios: point-to-point and circuit board.

Point-to-Point Soldering

If you're doing point-to-point construction,[6] as shown in Figure A-22, Step 1 means twisting the legs of the components together.

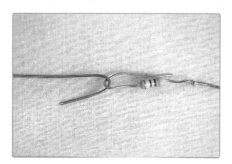
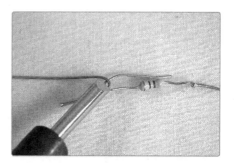
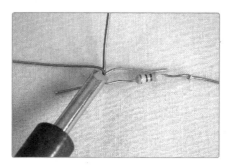
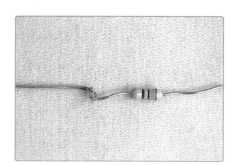

FIGURE A-22: *Point-to-point soldering in four steps*

Circuit-Board Soldering

If you're using a circuit board, as shown in Figure A-23, Step 1 means you slide the component's lead through the appropriate hole in the circuit board. Then, you bend

....

6. Point-to-point construction is sometimes called *dead-bug* construction because when you are directly soldering to an IC socket, you lay it on your workbench with its legs sticking up like a dead bug.

the leg to one side so that it's making contact with the board's copper pad without touching other components or pads (molten solder will tend to slide up the hot wire and stick to whatever other metal it finds).

 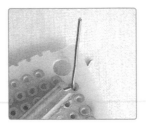 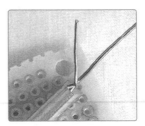 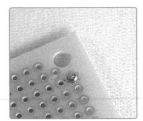

FIGURE A-23: *Circuit-board soldering in four steps*

Tinning Wires

Anytime you use stranded wire, start by *tinning* the wires: strip off the insulation, twist together the strands, heat the wire, and melt solder into it (see Figure A-24). This will make the wire easy to solder to other components or a PCB. The same should be done for the lugs and terminals on switches, jacks, and pots. Get each one nice and hot, and then coat it in a thin layer of solder. This will make it *much* easier to reliably connect wires or components to the lugs later. Small components, whose thin legs have very little mass, heat and solder very quickly. The lugs on hardware—especially switches and 1/4-inch jacks—are much beefier and take a good long while to heat up enough to melt the solder. By tinning lugs in advance, you can assure a good solder joint without abusing the smaller component.

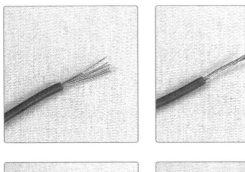 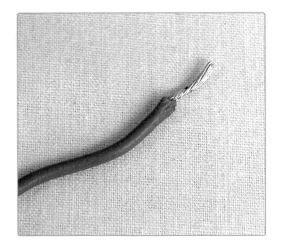

 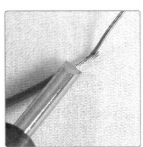

FIGURE A-24: *Tinning a wire in four steps: strip, twist, heat, and apply solder. Voilà!*

If you've invested in a chisel-tip soldering iron, you'll find that the sharp edge of the tip is best for working with small or heat-sensitive components, while the broad, flat face, which has more surface area to conduct heat, will make soldering hardware much faster.

Desoldering

Desoldering can be a real pain, but you'll do it much less frequently, and being good at it will save you time, misery, and money:

Step 1   Heat up your soldering iron, clean off the tip, and then apply heat to the soldered joint.

Step 2   When the solder flows, use your desoldering vacuum to suck up as much of the liquid solder as you can. (The tips of these vacuums are usually made of heat-resistant plastic.)

Step 3   Repeat once or twice.

Step 4   Reheat and wiggle the component free with a pair of needle-nose pliers.

You can desolder hardier components without the aid of a solder vacuum: just put on goggles, heat the solder joint, and tug at the component with your pliers, being careful not to flick hot beads of molten metal anywhere sensitive. The soldering vacuum can also be used to remove a *solder bridge*, which is anyplace that hot solder has inadvertently flowed between two components, causing a short circuit. Just heat the offending solder and suck it up. Killing a solder bridge without the aid of a desoldering tool, however, can be very tricky.

If you're desoldering a heat-sensitive component, such as a transistor, play it safe and clip a heat sink to the component's leg between the solder joint and the body of the component.

Using a Multimeter

For the most part, you'll use a multimeter for three tasks: testing batteries; identifying resistors, pots, and speakers; and testing continuity.

Testing Voltage

Set your meter for the proper DC voltage range. On most meters, you'll need to use the 20-volt range for 9-volt batteries because they'll overwhelm the lower 2-volt range and blow your fuse. Connect the black lead to the negative battery terminal and the red to the positive (see Figure A-25). Check whether the voltage is close to the value it ought to be. For most applications, a 9-volt battery is "dead" once it dips below 7.5 volts, and AAs or AAAs, which come out of the pack at 1.38 to 1.5 volts, are "dead" when they go below 1 volt.

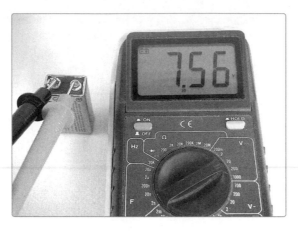

FIGURE A-25: *Testing a dying 9-volt battery*

Slightly advanced maneuver: if a circuit is misbehaving, power it up and set the multimeter for the voltage range of the circuit's power supply. For example, use the 20-volt range for a project powered by a 9-volt battery. Then, connect the multimeter's black lead to the circuit's ground and start poking around with the red lead to see whether the voltage levels throughout the circuit make sense. If you find an area that's being starved of voltage, a short circuit or broken wire is likely the culprit.

Identifying Resistance

To find the maximum resistance of an unlabeled variable resistor, first set the meter to an ohm range corresponding to your best guess about the resistor. Don't worry if you're wrong; there's no way to damage the meter or pot doing this. Connect the leads to the edge lugs, ignoring the center lug (see Figure A-26). It doesn't matter which lead goes to which edge lug. The multimeter will read the maximum resistance for the pot.

FIGURE A-26: *Identifying a potentiometer's resistance*

Advanced Trick:
Using Resistance to Identify Mystery Transformers

Small *audio transformers* are used to step up or step down voltage and current in order to, for example, drive a speaker. This is a shade more advanced, but if you get into building your own amps or experimenting with inductor-based wah pedals, you'll probably end up mucking around with small audio transformers. You can use the ohms setting on your multimeter to figure out the *turns ratio* of an unlabeled mystery transformer. The ratio of turns between the transformer's primary and secondary coils—which is also the ratio of the input-to-output voltages—is roughly equivalent to the ratio of the square roots of the resistances of each coil (round to two significant digits or so). To generalize this, where *P* stands for *primary* and *S* for *secondary*:

$$\frac{\text{Turns}_P}{\text{Turns}_S} = \frac{\text{Volts}_{in}}{\text{Volts}_{out}} = \sqrt{\frac{\Omega_P}{\Omega_S}}$$

So if you need to figure out the turns ratio on a transformer, measure the resistance on each side, take the square root of each, and then reduce the fraction so that the smaller term is a 1—in other words, divide the big number by the small one. For the classic RadioShack 8–1000 ohm audio output transformer—that is, good ole part #273-1380, beloved by hobbyists for decades—this calculation looks like so:

$$\sqrt{\frac{8\Omega_P}{1000\Omega_S}} = \frac{2.83}{31.62} = \frac{1}{11.17}$$

Is it exactly right? Clearly not, but it's a very useful approximation and spot on after you round to the nearest turn.

You can use this same method to identify unlabeled speakers. Most projects like the ones in this book call for 8 ohm speakers, so you should be looking for speakers that measure 8 ohms between their terminals.

Testing Continuity

If two points in a circuit have *continuity*, then there's an unimpeded flow of electricity between them. To test continuity, choose an ohms setting[7] and connect one lead to each end of something that should conduct electricity, such as the tips at the ends of a guitar cable or two ends of a dodgy length of wire (see Figure A-27). If the meter reads anything other than 0, there's a problem. This is also a good way

---

7. No mutimeter handy, but still need to check continuity? I've got you covered: there's a simple LED-based continuity tester in Appendix B: see Figure B-1 on page 358. Enjoy!

to figure out which lugs are which on a complicated switch, such as the many-pole, many-throw selector knobs that can be salvaged from old medical and computer equipment.

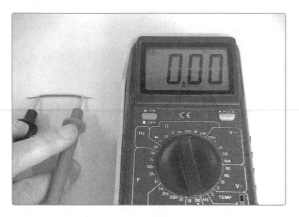

FIGURE A-27: *Testing continuity*

If your multimeter has a built-in continuity tester, it's even easier. Set the meter to test continuity—a mode that often doubles as an LED/diode tester and is represented on the dial as either the diode symbol, a little "circle making a noise" glyph akin to a Wi-Fi symbol, or both. Then, touch your leads to the two points that you're inspecting. If the multimeter makes a continuous tone, you have continuity. If it remains silent, you don't.

＊ **NOTE:** *If you are checking continuity on a circuit, pull the batteries first, lest you damage your meter or mislead yourself with nonsense readings.*

Building a Circuit

A *circuit diagram* is a map of the functional relationship between electrical components. It's drawn using symbols, such as those shown in Figure A-28. A circuit diagram doesn't tell you the actual physical layout of the circuit. In most simple projects, if the components are connected correctly, it doesn't really matter what is where. In some designs, especially for audio circuits, the physical layout of some parts of the circuit are sensitive; if that's the case, it'll be noted in the build notes, or the designer will provide a separate *layout diagram* showing exactly which physical component should be placed where on the circuit board.

Most circuit diagrams flow either from top to bottom or left to right, with positive voltage or signal sources at the top or left and ground connections and signal outputs at the bottom or right. To minimize ambiguity, designers do their best to avoid having lines cross or touch unnecessarily in their circuit diagrams, but it's sometimes unavoidable. If you're looking at a schematic and there's a dot at the intersection where two lines meet, you can assume this denotes an actual electrical connection between those components. If the lines simply cross with no dot, this probably doesn't represent a connection. For an example of this distinction, flip back to the circuit diagram for the Droid Voicebox, Figure 6-4 on page 77.

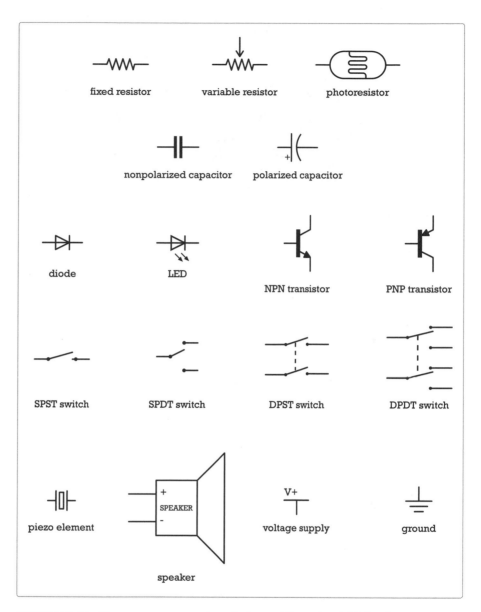

FIGURE A-28: *Common component symbols used throughout this book*

Before getting out your soldering iron, it's a good idea to test circuits on a *solderless breadboard* (see Figure A-29—you won't find this breadboard on your kitchen counter). Breadboards are cheap and come in many sizes. Most come packaged with a selection of precut insulated wire jumpers to connect everything together as you prototype your project—handy! But if yours doesn't, no problem: you can just use little snips of wire. A nice feature of the breadboard is the central divide, which is spaced to hold an IC, giving easy access to each IC pin. This makes it a lot easier for you to hook things up correctly when you start tinkering with ICs.

Once your circuit is working as expected, you'll want to make it permanent by soldering all of the connections. If you use a permanent breadboard–style circuit board, like the one shown at the top of Figure A-30 or one of the Adafruit Perma-Proto PCBs, then you can directly transfer your prototype breadboard layout to the PCB with no muss or fuss.

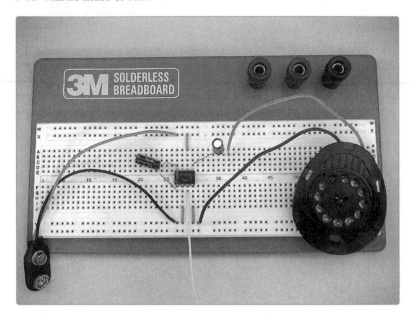

FIGURE A-29: *A breadboard with a sample circuit. Specifically, this is the "Dirt-Cheap Amp" on page 363, with the optional gain-maximizing capacitor running from pin 1 to 8. The white wire running from pin 3 out of frame at the bottom carries the input signal. Notice that the bottom and top rails of the breadboard are being used as a common ground and power rail, respectively.*

To build a circuit, begin by gathering and sorting all the components you need for the project. You don't want to inadvertently grab the wrong resistor or IC in the heat of the moment. A cardboard egg carton is great for sorting components for a small project, but avoid Styrofoam cartons if you're using ICs. Styrofoam can hold a static charge, and many ICs are sensitive to static electricity. After you've sorted your components—and labeled them with scraps of paper at the bottom of each egg compartment—start building the circuit. It's usually easiest to place the ICs or transistors first and then build out from there.

After you have the circuit working on the breadboard and you understand why it works, you can build it for real, either dead-bug style—soldering the components directly to each other, which is very compact but often ugly and difficult to repair or troubleshoot—or on a *printed circuit board (PCB)*, a sturdy, nonconductive board with copper pads and traces etched onto one side (see Figure A-30).

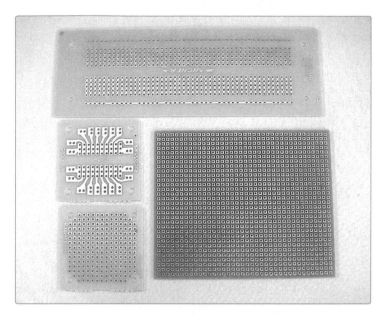

FIGURE A-30: *A selection of PCBs. A permanent breadboard–style circuit board is shown at the top of the figure.*

General Troubleshooting

If a circuit works fine on the breadboard but malfunctions when it's built, check for the following:

▶ **ICs improperly seated in their sockets**: Carefully pull any offending chips out of their sockets, orient correctly, and reseat them so that they're flush to the socket.

▶ **Short circuits**: Look for scraps of stray wire resting on the circuit board or uninsulated components, especially hardware, touching each other inappropriately. Remove or separate.

▶ **Solder bridges**: Reheat and remove excess solder.

▶ **Bad solder points**: A good solder point is smooth and shiny; bad ones are rough, brittle, or chunky (see the right photo in Figure A-21). Redo bad joints.

▶ **Reversed polarized components** (electrolytic caps, diodes, LEDs, transistors, and ICs): Remove and replace. If you powered the circuit with the component oriented incorrectly, it may have been damaged.

▶ **Missing ground connections or incomplete ground**: Complete now.

▶ **Reversed potentiometer connections**: Redo them now.

▶ **Bad hardware**: Disconnect any suspect hardware and check it individually. Sometimes you'll discover a faulty switch, damaged pot, or bent jack. Replace malfunctioning parts.

- **Shorts to ground**: This is especially annoying in tight enclosures. The circuit may function normally on your workbench, but when you close the box, components or wires get squeezed together, creating a temporary short. Prevent shorts by running lengths of shrink tube over bare wires or by using a judicious application of nail polish or Plasti Dip, which will act as an insulator when it dries. You can also cover the bottom of the PCB or conductive portions of the enclosure with strips of duct tape.

- **Ground-tip continuity**: With the batteries removed and nothing plugged in, check whether you have continuity between the ground and tip on the input or output jack. If so, look for short circuits, especially around the jacks, switches, pots, and so on. To remedy this, tighten or reorient the hardware, or wrap a bit of electrical tape around obvious shorts. Also check for bent jacks causing a short circuit. Replace damaged hardware.[8]

- **Voices from beyond**: If an audio project is noisy or picks up AM radio, baby monitors, intercoms, and so on, check that the audio signal lines are either shielded, as discussed in the footnote on page 215, or less than 8 inches long. Err on the side of caution and keep all wires carrying audio signals under 6 inches. The exception is lines running from an amplifier circuit to its speaker, as this amplified signal is strong enough to drown out any interference that might creep in at that point.

Finally, *always check your batteries*! Folks tend to assume that a circuit with good batteries works, one with dead batteries doesn't work, and one with dying batteries works intermittently. This is rarely the case. Dying batteries (especially those hovering right around the threshold I mentioned in "Testing Voltage" on page 349) can cause circuits to perform erratically or exhibit really odd behavior. This is especially the case with any circuit that includes an IC. So if your circuit is being just straight up bonkers, check the battery voltage.

---

8. N.B.: Many continuity testers will beep—indicating continuity—when hooked to the tip and ground of a properly functioning Playing-Card Pickup (Project 9). This is perfectly normal, because that pickup offers less than 20 ohms of resistance, and most meters consider any circuit with resistance that low "continuous." Many meters will give a hint that the circuit isn't perfectly continuous by showing something other than all zeros in the digital read-out. Others offer no such clues, but if you switch to your lowest ohm range, you will indeed find that the pickup is putting up the appropriate amount of resistance (8 ohms or more).

 **B**

EXTRA CIRCUITS

This appendix provides a handful of schematics for circuits that might come in handy as you continue to tinker and explore, including generic versions of all of the oscillators we've built in this book, a basic amplifier, a stripped-down preamp, and some useful filters and effects. Enjoy!

Super-Basic Continuity Tester

Have a switch you aren't sure is working right? Or a multi-pole switch whose terminals you need to sort out and label? What about a pile of instrument cables that look fine but might have gotten slammed in a car door too many times? A *continuity tester*, which tells you whether two points in a circuit are electrically connected, is your best friend in these situations.

Most digital multimeters have a built-in continuity tester, but maybe yours doesn't, or maybe your meter isn't handy. No worries! In about a minute you can build the simple continuity tester shown Figure B-1.

Use a red LED in this circuit. If you power this with a 9-volt battery, then use a 470 ohm resistor for R. If you use a 3-volt supply—that is, a pair of 1.5-volt AA or AAA batteries, or a single 3-volt watch battery, such as a CR2032 or CR2016—then use a 100 ohm resistor.

The probes are just insulated wire in any old gauge, so cut them as long as you like. You could even use a pair of alligator jumpers clipped directly to the circuit. Either way, attach one probe to the red lead from your battery pack and the other to the stray leg of resistor R. Now touch the probes together; the LED lights up. Separate them, and the LED goes dark. To test for continuity between two points (such as either end of a suspect guitar cable), touch one probe to each. If the LED lights up, you have continuity. If not, then something is broken between points A and B.

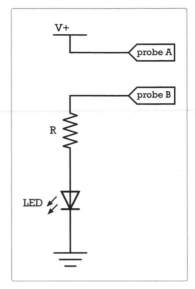

FIGURE B-1: *A simple continuity tester circuit*

Stereo Jack-Power Switch

Jack power is an easy way to eliminate power switches from effects and electronic instruments. In order to accomplish this, swap out the existing mono jack for a stereo jack wired to function both as an input/output and as an automatic power switch (see Figure B-2). Then, just plug in the instrument or effect to turn it on and unplug to turn it off.

To install jack power, follow these steps:

Step 1   Omit the power switch and run the positive battery lead directly to the circuit.

Step 2   Instead of connecting the battery's negative lead to the circuit's ground, connect it to the ring lug of the stereo jack.

Step 3   Wire the circuit's input or output to the tip of the stereo jack, just as you would with the original mono jack.

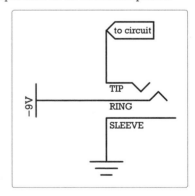

FIGURE B-2: *To add jack power, wire a stereo jack to function as both a power switch and an input/ output.*

Now when you plug the instrument into your standard guitar cable, which is a mono cable, it will complete the ground, activating the circuit.

Oscillators and a Metronome

Meditating upon simplified versions of the oscillators used in this book's synth projects can be instructive. They might come in handy when you start building your own sound sculptures, custom synths, cookie-jar burglar alarms, or other noise projects. With any of these, you can swap the pitch or rate pot for some other resistor. You might try a different potentiometer value; use a single fixed resistor to hardwire a specific tone; or swap in a photoresistor (thus transforming your synth into a photo-theremin), a set of body contacts, a force-sensitive resistor, a flexible resistor, or some other variable-resistance scheme. Whatever suits your project will suit these oscillators just fine.

LM386 Oscillator Circuit

The LM386 is conventionally used to build small power amps, like the Dirt-Cheap Amp in Figure B-7 on page 363. But it can also be used to build a variety of oscillators; the simplest of these is shown in Figure B-3. (Find an LM386 datasheet online for other examples of pretty cool basic audio oscillators you can build around this IC.)

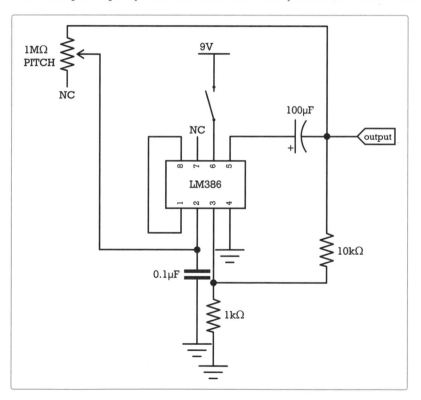

FIGURE B-3: *A bare-bones version of the LM386-based square-wave oscillator used in the Droid Voicebox (Project 6).*

This circuit has a higher part count than similar square-wave oscillators built around other chips, but it can also directly drive a wider variety of speakers (no amplifier needed) and can be a good deal louder. Toy with any of the component values to see whether the results please you.

CD4093 Oscillator Circuits

Both the Single-Chip Space Invader Synth (Project 15) and the Bleepbox 8-Step Analog Sequencer (Project 16) are built around the CD4093, a standard logic IC that we mildly abuse in order to make it chirp and warble for our pleasure. The CD4093 has four identical logic sections; thus far, we've always used two sections per oscillator to guarantee a nice, crisp square wave, even at very high rates. But you can get a perfectly usable oscillator from just one section of the chip.

Single Square-Wave Oscillator

Figure B-4 shows the simplest single square-wave oscillator you can build on a CD4093. Despite its very low part count, this single square-wave oscillator is very reliable and perfectly suitable for many projects. A single-section oscillator like this results in a square wave that gets a little distorted at higher rates, leading to a fuzzier voice, but otherwise works fine.

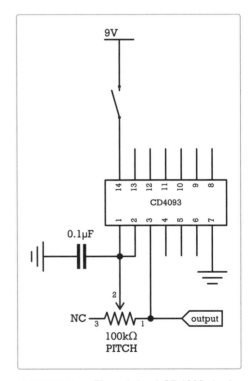

FIGURE B-4: *The minimal CD4093 single square-wave oscillator*

Four Square-Wave Oscillator

Figure B-5 shows the circuit that gives you the most bang for the buck from a CD4093: four independently controlled oscillators on a single chip.

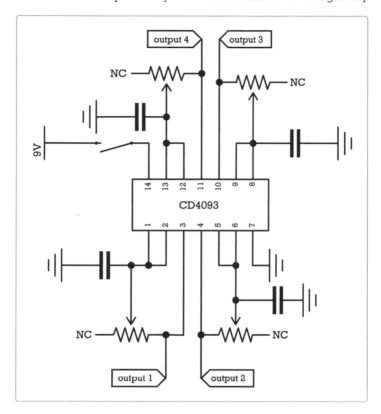

FIGURE B-5: *The maximal CD4093 four square-wave oscillator*

Each oscillator in this circuit is a lo-fi single-section oscillator, but having more oscillators gives you a pretty wide sound palette. You can run each output to its own jack/amp/whatever—resulting in four completely independent synthesizers on a single chip. Or you can mix the outputs via diodes, as we do in the Single-Chip Space Invader Synth (Project 15) and the Bleepbox 8-Step Analog Sequencer (Project 16), to sculpt richly textured cascades of chiptune glitchtastrophe. Or you can mix them using resistors (start with 10k ohm, either fixed or variable) for a thick, roaring wall of sound. As always, experiment with the potentiometer and capacitor values to find combinations you like.

555 Timer Oscillator and Metronome Circuits

Because the 555 timer chip is so cheap and common, it's where many folks begin with their oscillator experiments. The much-loved Atari Punk Console synth is built on a dual-555 chip—that is, a single chip with two independent 555 timers built in. The result is something like a junior version of our CD4093-based dual-oscillator synth (the one we used in the Space Invader Synth and the Bleepbox). If you're looking to build a cheap and effective tone-generating audio oscillator using a 555, there's a circuit for just that in Figure B-6 (left). Figure B-6 (right) provides a simple 555-based metronome circuit, very similar to the clock we built for the Universal LFO (Project 13).

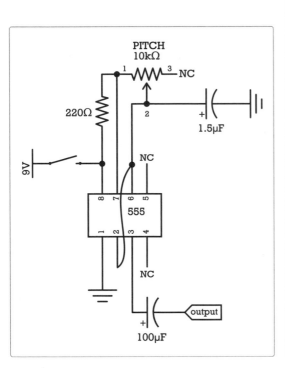

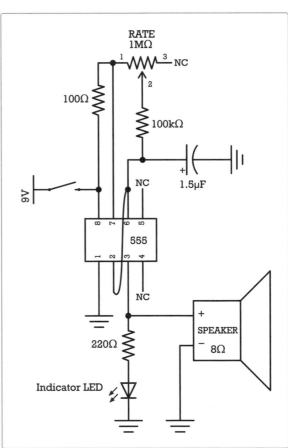

FIGURE B-6: *A 555 tone-generating square-wave oscillator (left) and a basic 555-based metronome (right). Note that the tone generator can output either to an amplifier or directly to an 8 ohm speaker.*

Amps and Preamps

Most of the projects in this book are more fun when you can make them louder. This section provides circuits for simple, crafty amps and preamps.

Dirt-Cheap Amp

Figure B-7 is my go-to, all-inclusive battery-powered amplifier, also called a *power amp*. This is what you're thinking of when you think of a guitar amp: you plug an instrument in, and it amplifies the signal and pumps it out of a built-in speaker (which in this case is an 8 ohm speaker, although values between about 3 ohms and 16 ohms often work fine). Full build instructions for this amplifier are included in my first book, *Snip, Burn, Solder, Shred*. One quick mod: want to make this amp louder? Instead of running a plain piece of wire from pin 1 to pin 8, use a 10 μF capacitor, with the negative cap leg connected to pin 8 and the positive to pin 1.

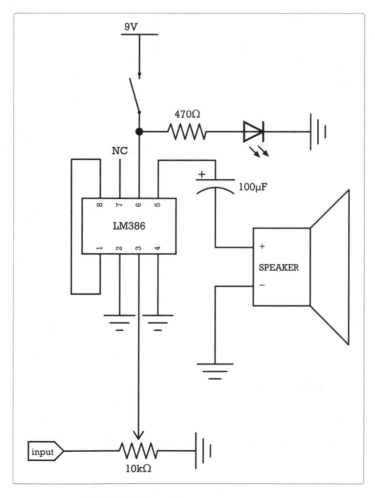

FIGURE B-7: *The Dirt-Cheap Amp*

Basic Transistor-Based Preamp

The Mud-n-Sizzle Preamp (Project 12) combines a simple transistor-based preamp and basic low-pass filter. Figure B-8 shows the preamp section as a standalone circuit, which can be easily built into another project or used as a basic preamp on its own.

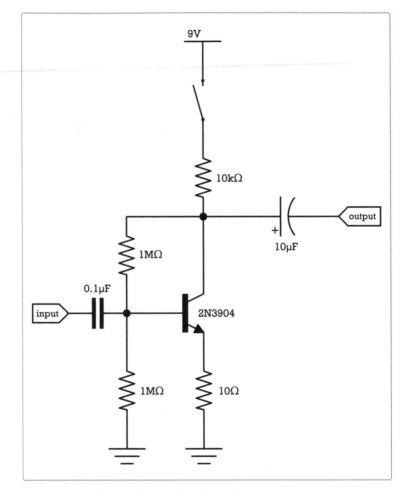

FIGURE B-8: *A generic transistor-based preamp*

Two-Transistor Fuzztone

Figure B-9 shows the Two-Transistor Fuzztone, a souped-up hot-rod version of that basic transistor preamp shown in Figure B-8, optimized for gain, fuzz, and stadium-rocking bravado. As with the Dirt-Cheap Amp, full build instructions for the Two-Transistor Fuzztone—as well as some other electronics effects and instruments—are included in my first book, *Snip, Burn, Solder, Shred.*

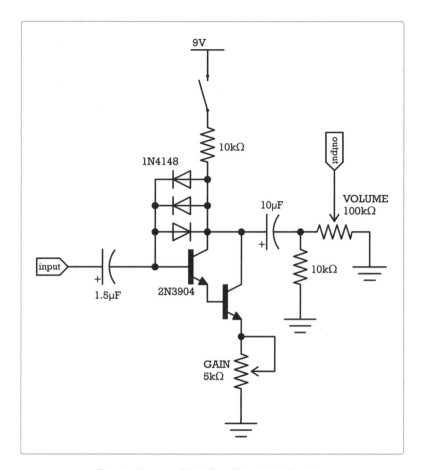

FIGURE B-9: *The stadium-rocking Two-Transistor Fuzztone*

Filters, Mixers, Panners, and Splitters

These filters and mixers are super simple, but they can really transform your sound. The filters allow you to alter the ratio of high- and low-frequency elements in your signal, modifying an instrument's voice or shifting the prominence of different instruments in a mix. The mixers allow you to mix several sound sources together to a single output—very handy if you have three instruments but only one PA amp or input on your recorder. The panners/splitters are specialized, single-knob mixers, making it possible to run two sound sources into a single input or to divide a single source between several destinations.

Low-Pass Filter

When you "crank the bass" on your car stereo, you're really cranking a low-pass filter: it cuts the highs and lets the lows through, leaving you with a window-rattling rumble. Figure B-10 (left) shows a robust little variable low-pass filter—the same one that's integrated into the Mud-n-Sizzle Preamp (Project 12).

In addition to boosting the bass in a mix or giving your guitar a smooth, jazzy tone, low-pass filters are enormously handy in synth projects. Simple square-wave synths like the ones we've built in this book tend to be very buzzy or reedy, which can get annoying in certain frequency ranges or at high volume. A low-pass filter will smooth out a square wave, imparting a smoother, rounder tone more characteristic of a sine-wave synth or an old-school 1950s theremin.

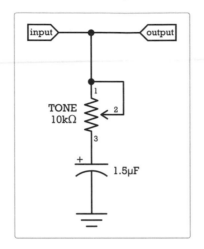 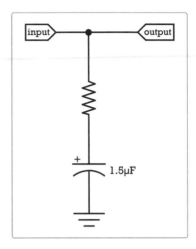

FIGURE B-10: *A generic low-pass filter, adjustable (left) and fixed (right)*

You can package a low-pass filter as a stand-alone passive effect (no batteries necessary!) or build it into projects. It can be adjustable, or you can use a fixed resistor of your choosing to lock in a specific sound, as shown in Figure B-10 (right). Fixed low-pass filters are generally built into projects, while adjustable low-pass filters can either be built in or serve as fun and useful stand-alone effects. Most commercial amps include a built-in adjustable low-pass filter.

Extra-Presence Volume Control

A standard volume control is a rudimentary filter all by itself: it doesn't reduce all frequencies equally. You tend to lose the highs first, which muddies the tone and leads to a loss of *presence*— that is, the guitar doesn't just sound quieter but also less important. One solution long embraced by guitarists— and included in many commercial electric guitar designs—is to short a cap across the volume control (as shown in Figure B-11).

This makes the volume control into a sort of specialized high-pass filter: it leaks extra highs into the output as you turn down the volume, maintaining the instrument's

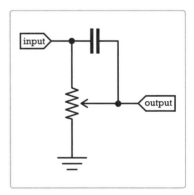

FIGURE B-11: *The classic extra-presence volume control mod*

presence even as you decrease the volume. The cap can be any of a variety of values, depending on the instrument and your tastes. Start with something in the 0.001 µF to 0.002 µF range and experiment.

Crossfade-Style Mixers

A single pot can be used as a simple *crossfade*-style mixer. Crossfade mixers work in one of two ways: you can mix two inputs to a single output, as we did with the two pickups in the Twang & Roar Kalimba (Project 11), or you can divide a single input between two outputs. Imagine the latter as a splitter, instead of a mixer, dividing your signal between two effects.

Figure B-12 shows both options in their most generic forms.

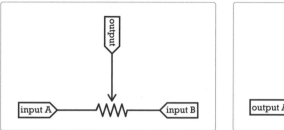 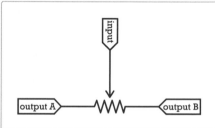

FIGURE B-12: *A two-in/one-out pan-pot mixer like the one in the Twang & Roar Kalimba (left) and a one-in/two-out pan-pot splitter (right)*

One cool use of the one-in/two-out pan-pot splitter is as a *panner,* dividing the output from a single battery-powered power amp—like the Dirt-Cheap Amp in Figure B-7—between two speakers (see Figure B-13).

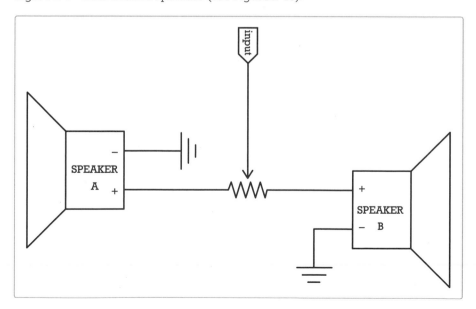

FIGURE B-13: *Controlling two speakers with a single pan-pot*

For example, build the Dirt-Cheap Amp, but instead of directly connecting a speaker, run that output—the negative leg of the 100 μF cap—to the input of the pan-pot circuit in Figure B-13. Then, use especially long runs of speaker wire to connect the pan-pot's two speakers. Separate the two speakers by a good distance—for example, put them on opposite sides of a room—and pan between them for a swooshing rush or fast ping-pong effect. If you put the speakers back to back, panning across them repeatedly—perhaps with a modified Universal LFO (Project 13)—you can create a rudimentary "Leslie amp"–style organ swelling effect.

Simplest Passive Multi-Channel Mixer

If you have several lo-fi instruments or effects you want to pump into the same piece of gear—perhaps in order to record them together or so they can share an amp or mesh into a greater sound system or electro-musical scheme—you need a *multi-channel mixer* to combine their signals. The simple passive (i.e., unpowered) mixer shown in Figure B-14 will mix several signals with a minimum of fuss. Just plug the sound sources into your inputs and then the destination (such as the amp or recording gear) into the output. Added bonus: you probably already have all the necessary parts to build this kicking around the bottom of your components bin.

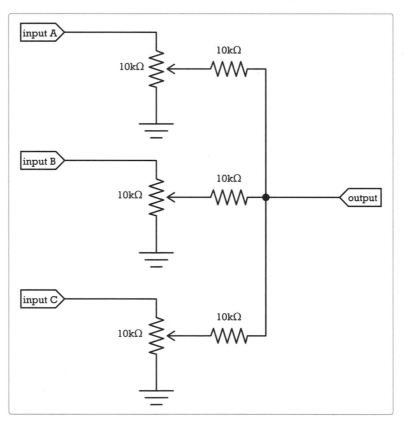

FIGURE B-14: *A simple passive multi-channel mixer*

MUSIC THEORY CRASH COURSE

At its core, a piece of music is a structure in time built out of units of sound and silence. There are four predominant elements that make up any given sound:

▶ **Pitch** is how high or low the sound is compared to the preceding or following sound.

▶ **Duration** is how long the sound lasts.

▶ **Intensity** is how loud or quiet the sound is and how that changes during its duration.

▶ **Timbre** is the "texture" of the sound. Timbre allows you to differentiate sounds with the same pitch, duration, and intensity. For example, you can immediately tell the difference between a trumpet and piano playing the same note for the same amount of time with the same intensity because the two instruments have very different timbres.

Likewise, silence isn't just the thing that happens between sounds. Our silences also have duration and even a sense of intensity and timbre: how stark is the silence, and what does the silence make the listener feel?

When structures of sound and silence fall into recognizable patterns—usually because of some combination of repetition and regularity—folks tend to be willing to say, "Yeah, okay. That's music. I might not like it, but I agree that it's some kind of musical something."

In other words, "making music" means *creating an intelligible structure of sounds and silences, usually by offering the listener some sort of recognizable regularity.* This is absolutely the most generic, culturally neutral way to talk about making music.

That said, if you're a native English speaker reading this book, you're likely to think of music in terms of *Western popular music*, predominantly structures of sound and silence rooted in the artistic output of the peoples of the Western Hemisphere, with a special emphasis on the results of pale-skinned Europeans crashing into and generally abusing dark-skinned non-Europeans. The orderly sounds we enjoy every day wouldn't exist if not for people like Johann Sebastian Bach, Ludwig van Beethoven, Igor Stravinsky, Ada Lovelace, Charles Babbage, Thomas Edison, Alan Turing, Leon Fender, George Beauchamp, Adolph Ricken-backer, Leon Theremin, Clara Rockwell, Robert Moog, Wendy Carlos, Phil Spec-ter, Raymond Scott, George and Ira Gershwin, Howling Wolf, and Alan Lomax. But more importantly, our music could not exist without the contributions of an untold number of serfs, slaves, indigenous persons, displaced persons, nomadic peoples, traders, scoundrels, thieves, agents, publishers, flim-flam artists, fans, and fanatics. I'll fall back on my betters and quote Stephen Jay Gould here: "I am, somehow, less interested in the weight and convolutions of Einstein's brain than in the near certainty that people of equal talent have lived and died in cotton fields and sweatshops."

All that said—and being sort of heavy politics for a book about homebrew synths and garden-hose trumpets—here are some concepts that might help you in your musical explorations.

Beats, Bars, and Time Signatures

If you grew up with Western music in your ear, you're accustomed to music having a regular, equally divided beat. Even when you hear a totally new song, I'll bet you almost immediately sense when to tap your toes or clap your hands. This feels automatic for the same reasons speaking feels automatic: beginning before you were born and extending through your formative years, these were the sounds and patterns that you heard in your environment.

Beat is essential to musicality. There's a whole slew of terms we use when approaching the idea of what the beat is:

▶   **Tempo** is the speed, and often mood, of a piece of music. It might carry a specific name—usually Italian, like *largo* or *allegro*—or the composer might note how a piece is to be played: *solemn, lively, steady rock, laid back, heavy,* and so on.

▸ **Meter** is usually given as a *time signature,* often rendered as a fraction, like 4/4 or 3/4. There are also specific named *rhythms,* like the blues shuffle, the Bo Diddley beat, or the clave rhythm I mentioned in Junkshop Percussion (Project 8, on page 121).

▸ **BPM** is the exact number of *beats per minute* at which a song plays. This is especially common now because electronic music and computerized composing, recording, and listening are such a huge part of our lives.

The basic unit of any Western piece of music is the *bar* (also called a *measure*). A bar is a group of sounds and silences of specific durations that lasts a specific number of beats. Western music, from classical concertos to death metal ballads, is very often composed of four-beat bars, meaning each musical sentence in that composition consists of four beats.[1] This is written as *4/4* because there are four beats per bar, as represented by the upper four in the fraction, and each beat corresponds to one *quarter note,* as represented by the lower four. (I'll explain what a quarter note is in the next section). For example, sing "Twinkle, Twinkle Little Star" while clapping. Each line of the song occupies one four-beat measure: the words *star, are, high,* and *sky* land on the fourth and final beat of each bar. On sheet music, this division into bars is represented with a vertical line, as shown in Figure C-1.

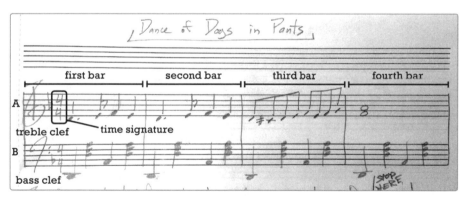

FIGURE C-1: *An annotated sample of piano sheet music showing the first four bars of "The Dance of the Dogs in Pants" (an original composition). The upper staff, marked A and headed with a* treble clef, *shows the melody, or the higher-pitched bit played by the right hand; the lower staff, marked B and headed with a* bass clef, *shows the lower-pitched bass part, which incorporates both single notes and chords.*

Not everything, however, fits into a 4/4 pattern. A waltz, for example, is three beats per bar, not four. If you try to count off a waltz as 1-2-3-4, you'll find yourself stumbling as you run over into the next bar. A waltz is in 3/4 time, which means

1. In fact, this four-beat-per-bar format is so common that it's also called *common time.* It can be represented with a *C* at the beginning of a piece of sheet music.

there are three beats per bar and each beat lasts for one quarter note. You count a waltz as 1-2-3, 1-2-3, 1-2-3, with the *1* stressed.

Now, listen to a few polkas and marches on YouTube. Counting these off as 1-2-3-4 probably feels funny, like you're running two musical sentences together. That's because you are. Polkas and marches are often in 2/4 time, meaning they have two beats per measure and each beat is represented by a trusty quarter note. You count music in 2/4 time as 1-2, 1-2, 1-2.

Ready for a challenge? Search YouTube for the Allman Brothers Band's "Whipping Post." Give it a listen and try to figure out the time signature. If you jump to the middle of the song, you'll hear a bluesy three beats per bar in 3/4 time. If you jump back to that incredible intro, it at first seems like 3/4 time, too. But then, at what your brain initially takes to be a fourth bar, something happens that brings the bass line around like a whipcrack. It certainly isn't 3/4 time. Whatever happens as the riff comes back around, it's vital to what makes the song work. As it turns out, this intro pattern isn't three bars of 3/4 plus something else; it's best understood as a single bar of 11/4. Count it out, and you'll see.

Even though the pattern is most rationally understood as 11/4, that's not how the writer, Gregg Allman, understood it. In fact, he had no clue what 11/4 time was until his brother and bandmate, Duane, listened and said, "That's good, man; I didn't know you understood 11/4."[2] Greg had no idea what Duane was talking about, and Duane had to sketch it out on scrap paper to get the concept across. When Greg wrote the intro to "Whipping Post," he'd counted it in his head as 1-2-3, 1-2-3, 1-2-3, 1-2—that is, three bars of 3/4 and a bar of 2/4, a pretty damn unconventional meter.

The point is that there are tons of possible time signatures, and the most important thing is to understand them to be descriptive, not prescriptive: you can play in whatever crazy divisions you want. Notation is a system for you to describe to others what you've done, not to limit what you're doing.

Notes, Scales, Chords, and Intervals

Thinking of the beat as the smallest unit of music is really convenient, because doing so allows you to map out every moment in a song and ask yourself, "What am I doing at this particular beat? Am I making a sound? If so, how loud is it, what pitch is it, and so on?" We mark out these sounds as *notes* or sets of notes and the not-sounds as *rests*. A note shows the duration and pitch of a sound you make, and a rest just gives the duration of the silence. You can play pitches individually for a fixed duration (that is, as individual *notes*), or you can play pitches simultaneously in groups, which we call *chords*. You can play only one rest at a time. (Think about it—and if you have a counterclaim, email me!)

Keeping Time

Durations are described in terms of the portion of the bar they take up. A *whole* note sounds for an entire bar, and a whole rest is silent for an entire bar. Therefore, in

2. As recounted in Randy Poe's *Skydog: The Duane Allman Story*

common time (4/4), a whole note or rest lasts for four beats. A *half* note or rest lasts for half a bar, which is two beats when playing in 4/4. A *quarter* note or rest lasts for one-quarter of a bar (one beat), an *eighth* for one-eighth of a bar (one half beat), a *sixteenth* for one-sixteenth of a bar (I'm sure you can do the math yourself), and so on.

We think of everything in terms of quarter notes when we play. I hold a whole note for a full four-count, a half note for a two-count, and so on. When I'm playing *eight to the bar* in common time—that is, rocking out, with everything in quick eighth notes—I count 1-and-2-and-3-and-4-and, which gives me eight things to say out loud to keep time to. If you're playing sixteenth notes, count 1-ee-and-a-2-ee-and-a-3-ee-and-a-4-ee-and-a. Playing thirty-second notes? That's too fast! Cut it out!

If I'm playing in waltz time (3/4), I have only three quarter notes per bar. In this context, a half note doesn't last for one-and-a-half beats but instead is worth two quarter notes, taking up two-thirds of the bar.

Pitches and Octaves

Pitches are assigned letters *A* through *G*, and they repeat in this pattern as the pitch gets higher or lower. In Western music, we tend to use *A440* as the baseline to calibrate our other notes against. This is a 440 Hz pitch that we call *A*. Incidentally, the A string on your uke plays at 440 Hz. That said, we actually *arrange* our music around middle C, which is also called C_4. This is the white key in the middle of a standard 88-key piano keyboard.

The letters we use to map notes repeat themselves in *octaves*, which are groups of the seven letters A through G. These groups correspond to groups of seven white keys on the piano. The repetition of these notes is based on their frequency: a given note has twice the frequency of the same note one octave lower. For example, A_4 has a frequency of 440 Hz. The next higher A—called A_5—is eight white keys to the right on a standard piano, and it has a frequency of 880 Hz.

Scales and Chords

A *scale* is a set of pitches of specific *intervals*. An *interval* is the difference between the frequencies of two pitches—and thus, intervals map to distances between piano keys, ukulele frets, and so on.

Just as with time signatures, there are tons of different scales, and you don't need to learn a specific scale or limit yourself to a given scale in order to play. But when you do chose to stick to a given scale, it's typically easier to make patterns of sound and silence that are easily intelligible (and thus more easily recognized as music). Because lots of Western folk tunes, like "When the Saints Go Marching In" and "Camptown Races," are confined to a single *major diatonic scale*,[3] most of the eight notes in a diatonic scale can be made to sound pretty good next to each other.

A *pentatonic* scale is a scale of five notes that, for reasons involving brains and physics, basically always sound good together—even cross-culturally! In Western music, we tend to think of the pentatonic scale as a subset of the diatonic scale (see Figure C-2), but that's just one perspective. I'm sure plenty of Laplanders, West

---

3. Remember that "do-re-me-fa-sol-la-ti-do" thing from grade school music class? That's a diatonic scale.

Africans, Native Americans, auditory neurologists, and hip-hop professors would argue that with you.

Scales and chords are named for their *root*, which is the note they're constructed on top of. The root is usually the lowest note in the scale or chord and the first played in the scale. Play all the white keys between middle C and C_5, and you've played a C major scale, which is a *diatonic scale* and therefore a staple of Western music. Play the middle C, the D above it, and the G above that simultaneously, and you've played a C major chord.

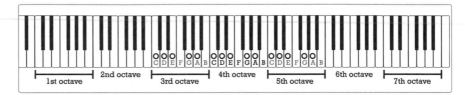

FIGURE C-2: *The circles indicate three octaves of the C major pentatonic scale. Want three octaves of the C major diatonic scale? Just play all of the white keys. (Keyboard image by Artur Jan Fijałkowski)*

Sharps and Flats

Pianos also have five black keys per octave. These keys are named for the white keys to either side. So, for example, the little guy between C and D can be called either *C-sharp* (C♯), meaning it's a touch higher-pitched than C, or *D-flat* (D♭), meaning it's a touch lower-pitched than D.

If you play a full octave from lowest to highest and hit each white key, you've played a diatonic scale. Play a full octave from lowest to highest hitting every key, black and white, and you've played a *chromatic scale*. It's tricky to play a fully chromatic composition that sounds good,[4] but you'll hear the term as you dabble in music, so I wanted to make sure we covered it. In music, *chromatic* just means we use every possible note in a given octave.

Because I love you, I've summarized the concepts covered in this section in Figure C-3.

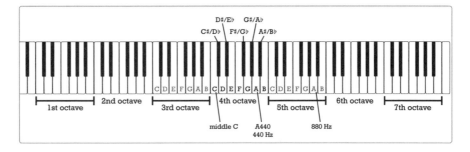

FIGURE C-3: *An annotated 88-key piano keyboard*

---

4. That said, if you want your mind totally blown, search YouTube for *Vi Hart 12 tone*. Twinkle, twinkle, laser bat!

The Circle of Fifths

The *circle of fifths*, shown in Figure C-4, is a geometric way of representing some neat relationships among notes in the full 12-note chromatic scale. Remember that each note is a specific frequency. For example, A_4 is 440 Hz, and its octave is 880 Hz. Play any two frequencies simultaneously, and you'll notice that some sound sweet together, while others are annoying. Pitches that sound good together are said to be *consonant*, with some intervals being more consonant than others. The more consonant two notes are, the safer it is to play them next to or on top of each other. Try mashing a couple of piano keys, plucking a pair of uke strings, or setting off two different brands of smoke detectors, and you'll quickly develop a sense of consonance.

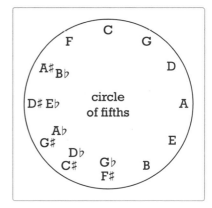

FIGURE C-4: *The circle of fifths. Pick any pitch. Move one click clockwise, and you have its fifth. Move one click counterclockwise, and you have its fourth. Most blues, folk, punk, and rock songs are composed of a root, fourth, and fifth.*

There are varying degrees of consonance because of the way the two tones' sound waves mesh or crash together against your eardrum. If the frequencies don't divide evenly into each other, they sound bad because they're hard for your brain to decode. A note and its octave always sound fine together because their frequencies divide so easily: the higher note always has twice the frequency of the lower note—a ratio of 2:1. Meanwhile, playing any piano key together with the key directly next to it always sounds pretty bad because that interval is too narrow: there's no way for your brain to resolve one pitch into the other. Brains *hate* that.

The *fifth* is the next most consonant interval after the octave. The frequency ratio of a fifth to its root is pretty close to 3:2. For example, G and C are a fifth apart; middle C is 261 Hz, and the G above that is 391 Hz. So playing a fifth—either both tones at once or moving from one to the other—will always sound fine. It will *resolve*.

The circle of fifths encodes tons of cool relationships—for example, the versatile and popular *power chord* is just a root and its fifth. But let's start with the most basic relationship illustrated by the circle of fifths, one that's really handy for building your own compositions. Pick a root note. We'll say C (if you have an actual or virtual keyboard handy, now is a good time to fire it up). Flip to Figure C-4 and find C on the circle of fifths. Now move one pitch clockwise, and you have G, the fifth that meshes with that root. You now know C and G will always sound good together. Try hitting them on your keyboard—simultaneously and then one after the other—and you'll hear how compatible they are.

Flip back to the circle of fifths diagram and move one click counterclockwise from your root to find its *fourth*. With C as your root, that puts you at F. Play them together. The fourth won't sound quite right—not terrible, but certainly less consonant than the octave or fifth. Your ear has more trouble fitting it with the C, which creates *tension*. Moving from the root to the fourth creates the sense that something should come next because the fourth and root just don't quite resolve. Go from the fourth to the fifth or back to the root, and that tension releases.

Pick a root, fourth, and fifth and play around with them. You'll find that different amounts of tension release when you move between notes, with a partial release when you move from a fourth to the fifth and a full release when you return to the root. Try making a pattern using C as your root, G (its fifth), F (its fourth), and C# (the sourest interval you can get). If you pin your structure of sound and silence to the C and G, sprinkling in a few C#s to create lots of tension and then using the F to smoothly transition back toward G, you'll find lots of workable musical patterns— which is to say, you'll be composing a song very much in the Western tradition. Congratulations!

✳ **NOTE:** *When folks talk about the fundamental patterns that underlie a melody and chord structure, these intervals are usually represented by roman numerals. The root is* I, *the fourth is* IV, *the fifth is* V, *and so on.*

Applied Theory: The 12-Bar Blues

Let's put theory into practice. If you have an instrument you're comfy with, swell! Otherwise, find a piano or music keyboard and use Figure C-3 as a crib sheet. In a pinch, download a digital keyboard app, whether on your phone, tablet, or computer. I like the "Virtual Keyboard" at *http://www.bgfl.org/virtualkeyboard/*; it has a limited range, but it assigns the keys rationally to the computer keyboard, making it possible to play a few chords.

Musically speaking, the blues is a sonic equation that your brain is very used to solving. If something bizarre has happened and you've somehow never heard the blues, please proceed directly to YouTube and listen to Howlin' Wolf's "Evil," Bessie Smith's "Empty Bed Blues," Big Bill Broonzy's "C.C. Rider," Elvis Presley's "Hound Dog," Big Mama Thornton's "Hound Dog," Johnny Cash's "Folsom Prison Blues," and the Georgia Satellites' "Keep Your Hands to Yourself." That's 60 years of the blues being used to diverse effect, and damn are you in for a treat!

The Gory Details:
The Music History of Two Hound Dogs

Presley's "Hound Dog" and Thornton's "Hound Dog" are the same song. It was originally written for Thornton—a Southern-born musician with gospel roots who made her R&B career in Harlem—by the songwriting team of Leiber and Stoller, a pair of teenaged New York Jews. Leiber and Stoller worked closely with Thornton to craft a signature comeback song that drew on Thornton's sexy, take-no-crap stage persona. Leiber/Stoller/Thornton's "Hound Dog" was an enormous hit, spawning dozens of covers, spoofs, and "answer songs." Presley's version is actually a sanitized parody of the song, with modified lyrics made famous by Freddie Bell and the Bellboys, a Vegas lounge act composed of Italian Americans transplanted from Philadelphia. When Presley licensed "Hound Dog" from Leiber and Stoller and made it a number one hit (again), they knew him only as "some white kid" from Tennessee. There's something about this story—its diverse cast of characters, their fundamentally incompatible lived experiences drawn tightly around a fairly vapid pop confection, the fact that there's no reasonable way to apportion authorship to this iconic tune—that, for me, quite aptly summarizes all of American music.

In songs that follow a 12-bar blues format, you play 12 musical sentences, which comprise a verse and chorus, and then repeat—likely with new lyrics—*ad libitum*. Take a look at the blues *chord progression* illustrated in Figure C-5 (use the roman numerals as your guide).

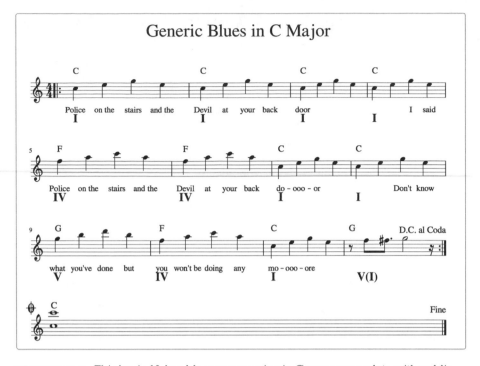

FIGURE C-5: *This basic 12-bar blues progression in C comes complete with public domain lyrics (use them as you like!) and a snappy turnaround at the 12th bar, which creates a transition to loop back to the 1st bar. The letters above each staff indicate which chord should be played when. The roman numerals under the lyrics highlight the generic chord structure of the 12-bar blues. Once you've repeated the first 12 bars as many times as you want, wrap up the song by continuing into the 13th bar, which is the tune's coda.*

For this exercise, your root is C. Let's start by treating those C, F, and G chords as just single notes (that is, play just a C note in place of the C chord, just an F for the F chord, and so on). It's conventional to anchor melodies in the fourth octave, and chords or bass parts in the third. So fire up your piano keyboard, check out Figure C-6 for help navigating, and then find the fourth octave with its C, F, and G notes.

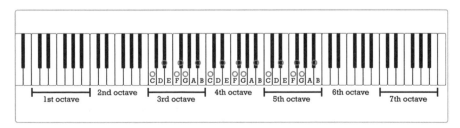

FIGURE C-6: *Three octaves of the blues scale in C on a piano keyboard*

Let's begin! Count out some four-beat 1-2-3-4 bars. For the first four bars, hit C_4—the C at the beginning of the 4th octave—on the *1*. For the 5th and 6th bars, play F_4 on the *1* instead. Then, hit C_4 on *1* for the 7th and 8th bars. Hit G_4 on *1* for the 9th, F_4 on *1* for the 10th, C_4 on *1* for the 11th, and, finally, play your 12th-bar turnaround. I went with G here because it's snappy, but following the generic blue progression (rendered in roman numerals), you could just as easily use a C instead. Repeat as long as you like. Even the first time through, and despite being stripped to its bare bones, you'll certainly hear that this is indeed the blues. It's just that easy!

Now let's flesh out those bare bones, starting with chords instead of just the root notes of each. Once again, the circle of fifths proves handy: a power chord is composed of a root and its fifth. Flick back to Figure C-4 to see that the C's fifth is G. Play C_3 and G_3 together for a C power chord—perfectly suitable to serve as the root, or I, in a blues progression. Check the figure again, and you'll see that F's fifth is C, so playing an F_3 and C_4 for an F power chord serves as our IV. Finally, G's fifth is D, so play G_3 and D_4 together for a G power chord, which is our V.

Go through the "Generic Blues in C Major" again, this time playing your power chords instead of just single notes. You'll immediately hear the difference: where before you just had bones, now you've got a little muscle, too. Play it several times.

Once you get a feel for the song, try playing power chords several times per bar. Hit your chord on every quarter note—that is, at each number on the count. It's a nice, driving beat. Then, try hitting the chord just on the 1 and 3; it'll feel laid back. After that, go through one more time, now hitting your chords on the 2 and 4; it might feel slow but aggressive, like a tiger pacing, ready to pounce. Now, try something different. Make it swing, make it grind, or nail it down eight-to-the-bar.

When you tire of pounding out those power chords, try doing the same thing with the three *major chords* illustrated in Figure C-7.

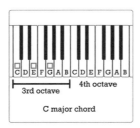
C major chord

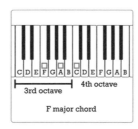
F major chord

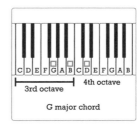
G major chord

FIGURE C-7: *The C major, F major, and G major chords*

Finally, you'll want to add a melody. I'm not going to walk you through the simple one I've scored in Figure C-5—reading sheet music is beyond the scope of this appendix, and there are plenty of online resources if you're interested. However, I'll give you a clue: each measure of the melody (save for the 12th) is just the three notes for that measure's chord played evenly for the four-count. For example, in the 1st bar, the four notes in the melody are C, E, G, E.

If you want to graduate up to full-blown blues, learn the blues scale illustrated in Figure C-6. Practice playing the C blues scale backward and forward. Compared to the C major diatonic scale and the major pentatonic scale (shown way back in

Figure C-2), the blues scale is sort of crazy, right? But deliciously so—feel how wonderfully it builds and releases tension compared to those humdrum major diatonic and pentatonic scales. Also, feel how this tension changes when played against different chords: the first two notes of the blues scale, played one after the other, clash hard against each other when played over the C major chord, but they almost-sorta-kinda work over F major. How do they feel when matched with G major? You can build a whole song with just those two notes and three chords.

Armed with the "Generic Blues in C Major" chord progression, your fistful of chords, the lyrics in Figure C-5, and your blues scale, make up your own melody for this sad tale of being caught between a rock and a hard place. Then, make up a new verse. Then, make up whatever you want.

Go, cat, go: make a good noise now!

Junkyard Jam Band is set in Rockwell. The book was printed and bound by Lake Book Manufacturing in Melrose Park, Illinois. The paper is 60# Husky Opaque Offset, which is certified by the Sustainable Forestry Initiative (SFI).

The book uses a layflat binding, in which the pages are bound together with a cold-set, flexible glue, and the first and last pages of the resulting book block are attached to the cover. The cover is not actually glued to the book's spine, and when open, the book lies flat and the spine doesn't crack.

RESOURCES

Visit *https://www.nostarch.com/jamband/* for resources, errata, and more information.

More no-nonsense books from **NO STARCH PRESS**

SNIP, BURN, SOLDER, SHRED
Seriously Geeky Stuff to Make with Your Kids
by DAVID ERIK NELSON
NOVEMBER 2010, 360 PP., $24.95
ISBN 978-1-59327-259-3

THE SPARKFUN GUIDE TO PROCESSING
Create Interactive Art with Code
by DEREK RUNBERG
AUGUST 2015, 312 PP., $29.95
ISBN 978-1-59327-612-6
full color

THE SPARKFUN GUIDE TO ARDUINO
by DEREK RUNBERG *and* BRIAN HUANG
WINTER 2016, 200 PP., $29.95
ISBN 978-1-59327-652-2
full color

THE MAKER'S GUIDE TO THE ZOMBIE APOCALYPSE
Defend Your Base with Simple Circuits, Arduino, and Raspberry Pi
by SIMON MONK
FALL 2015, 300 PP., $24.95
ISBN 978-1-59327-667-6

ARDUINO WORKSHOP
A Hands-On Introduction with 65 Projects
by JOHN BOXALL
MAY 2013, 392 PP., $29.95
ISBN 978-1-59327-448-1

LEARN TO PROGRAM WITH MINECRAFT
by CRAIG RICHARDSON
FALL 2015, 304 PP., $29.95
ISBN 978-1-59327-670-6
full color

PHONE:
800.420.7240 OR
415.863.9900

EMAIL:
SALES@NOSTARCH.COM

WEB:
WWW.NOSTARCH.COM